Communications in Computer and Information Science

2605

Series Editors

Gang Li ⓘ, *School of Information Technology, Deakin University, Burwood, VIC, Australia*

Joaquim Filipe ⓘ, *Polytechnic Institute of Setúbal, Setúbal, Portugal*

Zhiwei Xu, *Chinese Academy of Sciences, Beijing, China*

Rationale
The CCIS series is devoted to the publication of proceedings of computer science conferences. Its aim is to efficiently disseminate original research results in informatics in printed and electronic form. While the focus is on publication of peer-reviewed full papers presenting mature work, inclusion of reviewed short papers reporting on work in progress is welcome, too. Besides globally relevant meetings with internationally representative program committees guaranteeing a strict peer-reviewing and paper selection process, conferences run by societies or of high regional or national relevance are also considered for publication.

Topics
The topical scope of CCIS spans the entire spectrum of informatics ranging from foundational topics in the theory of computing to information and communications science and technology and a broad variety of interdisciplinary application fields.

Information for Volume Editors and Authors
Publication in CCIS is free of charge. No royalties are paid, however, we offer registered conference participants temporary free access to the online version of the conference proceedings on SpringerLink (http://link.springer.com) by means of an http referrer from the conference website and/or a number of complimentary printed copies, as specified in the official acceptance email of the event.

CCIS proceedings can be published in time for distribution at conferences or as post-proceedings, and delivered in the form of printed books and/or electronically as USBs and/or e-content licenses for accessing proceedings at SpringerLink. Furthermore, CCIS proceedings are included in the CCIS electronic book series hosted in the SpringerLink digital library at http://link.springer.com/bookseries/7899. Conferences publishing in CCIS are allowed to use Online Conference Service (OCS) for managing the whole proceedings lifecycle (from submission and reviewing to preparing for publication) free of charge.

Publication process
The language of publication is exclusively English. Authors publishing in CCIS have to sign the Springer CCIS copyright transfer form, however, they are free to use their material published in CCIS for substantially changed, more elaborate subsequent publications elsewhere. For the preparation of the camera-ready papers/files, authors have to strictly adhere to the Springer CCIS Authors' Instructions and are strongly encouraged to use the CCIS LaTeX style files or templates.

Abstracting/Indexing
CCIS is abstracted/indexed in DBLP, Google Scholar, EI-Compendex, Mathematical Reviews, SCImago, Scopus. CCIS volumes are also submitted for the inclusion in ISI Proceedings.

How to start
To start the evaluation of your proposal for inclusion in the CCIS series, please send an e-mail to ccis@springer.com.

Askhat Diveev · Vasily Fomichev ·
Aleksander Ilin · Ivan Zelinka · Elena Sofronova
Editors

Intelligent Systems

16th International Conference on Intelligent Systems, INTELS 2024
Moscow, Russia, December 2–4, 2024
Proceedings, Part III

Editors
Askhat Diveev
Federal Research Center Computer Science
and Control of the Russian Academy
of Sciences
Moscow, Russia

Aleksander Ilin
Lomonosov Moscow State University
Moscow, Russia

Elena Sofronova
Federal Research Center Computer Science
and Control of the Russian Academy
of Sciences
Moscow, Russia

RUDN University
Moscow, Russia

Vasily Fomichev
Lomonosov Moscow State University
Moscow, Russia

Ivan Zelinka
FEI VSB Ostrava
Ostrava, Czech Republic

ISSN 1865-0929　　　　　　　ISSN 1865-0937　(electronic)
Communications in Computer and Information Science
ISBN 978-3-032-04763-2　　　ISBN 978-3-032-04764-9　(eBook)
https://doi.org/10.1007/978-3-032-04764-9

© The Editor(s) (if applicable) and The Author(s), under exclusive license
to Springer Nature Switzerland AG 2026

This work is subject to copyright. All rights are solely and exclusively licensed by the Publisher, whether the whole or part of the material is concerned, specifically the rights of translation, reprinting, reuse of illustrations, recitation, broadcasting, reproduction on microfilms or in any other physical way, and transmission or information storage and retrieval, electronic adaptation, computer software, or by similar or dissimilar methodology now known or hereafter developed.
The use of general descriptive names, registered names, trademarks, service marks, etc. in this publication does not imply, even in the absence of a specific statement, that such names are exempt from the relevant protective laws and regulations and therefore free for general use.
The publisher, the authors and the editors are safe to assume that the advice and information in this book are believed to be true and accurate at the date of publication. Neither the publisher nor the authors or the editors give a warranty, expressed or implied, with respect to the material contained herein or for any errors or omissions that may have been made. The publisher remains neutral with regard to jurisdictional claims in published maps and institutional affiliations.

This Springer imprint is published by the registered company Springer Nature Switzerland AG
The registered company address is: Gewerbestrasse 11, 6330 Cham, Switzerland

If disposing of this product, please recycle the paper.

Preface

The 16th International Conference on Intelligent Systems (INTELS 2024) took place at the Lomonosov Moscow State University, Faculty of Computational Mathematics and Cybernetics, on 2–4 December 2024. The conference has been held for 32 years and is conducted biennially. It was established by a well-known control theory specialist, Konstantin Pupkov, who led several chairs dedicated to control theory at the Bauman Moscow State Technical University, MIREA – Russian Technical University, and Peoples' Friendship University of Russia. The name of the conference – "Intelligent Systems" – has not changed ever since. The first chairman chose a prospective subject for the conference, as he had a clear understanding that this topic would be dominant for many years.

INTELS 2024 was organized by several institutions: the Federal Research Center "Computer Science and Control" of the Russian Academy of Sciences, the Lomonosov Moscow State University, the Ivannikov Institute for System Programming, the St. Petersburg State Electrotechnical Institute (LETI), and the Bauman Moscow State Technical University.

The majority of conference participants were control specialists who research complex systems of various natures. The goal of the conference was to enable people to share their experience in the investigation and development of intelligent systems, as this term is understood nowadays.

The conference included four plenary presentations. Genaro Martinez presented research on microrobots and their ability to solve complex tasks when combined into a structure, very similarly to how cellular automata behave. Nikolay Kuznetsov demonstrated a phenomenon of hidden oscillations, occurring in systems of higher dimension and sometimes being extremely difficult to spot or model numerically. Sergey Bezrodnykh elaborated on the use of hypergeometric functions and their application to optimization problems. Askhat Diveev proposed a number of crucial problems that require finding a function in the form of an explicit mathematical expression and proposed a method of symbolic regression that may be applicable to these problems.

A significant part of the materials presented at the conference was dedicated to the use of artificial neural networks in various spheres: object control, identification and parametric optimization, text and image analysis, object recognition, etc. Among these works, a new type of network should be emphasized: the development of a binary perceptron-based neural network. Traditionally, works have been presented concerning the development of intelligent systems based on symbolic regression.

Among the spheres of application, significant attention was paid by the participants to medical topics in the context of artificial intelligence and big data analysis, which is an essential part of intelligent systems.

A new but promising direction of research is the problem of trust in decisions made by or with the help of artificial intelligence. It has been noted that, along with the rise of the impact of AI systems on everyday life, the problems of their security cannot be neglected.

The educational mission of the INTELS Conference must also be mentioned. Throughout its more than 30-year history, it has attracted students, postgraduates, and young scientists whose scientific and professional future is tied to their knowledge in the sphere of control theory, optimization, and artificial intelligence. In 2024, the organizers of the conference conducted a workshop on software and applied solutions for artificial intelligence, where young scientists had a chance to collaborate with their more experienced colleagues.

Since 2016, the proceedings of the INTELS Conference have been published by leading international publishers and are indexed in global scientific databases – Scopus and Web of Science. In 2024, the conference gathered over 250 participants from Belarus, China, Czech Republic, India, Japan, Kazakhstan, Mexico, Russia and Vietnam.

December 2024　　　　　　　　　　　　　　　　　　　　　　　　　　Askhat Diveev

Organization

General Chair

Diveev A. I. — Federal Research Center "Computer science and control" of RAS, Russia

Vice Chairs

Avetisyan A. I. — Federal Research Center "Computer science and control" of RAS, Russia
Fomichev V. V. — Lomonosov Moscow State University, Russia
Ilin A. V. — Corresponding Member of RAS, Lomonosov Moscow State University, Russia
Sofronova E. A. — Federal Research Center "Computer science and control" of RAS, Russia

Program Committee

Zelinka Ivan — FEI VSB Ostrava, Czech Republic
Kusiak Andrew — University of Iowa, USA
Das Swagatam — India Statistical Institute, Kolkata, India
Chadli Mohammed — University Paris-Saclay, IBISC Lab-UEVE, France
Silva Geraldo Nunes — Universidade Estadual Paulista, Brazil
Sokolov I. A. — RAS Member, Federal Research Center "Computer Science and Control" of Russian Academy of Sciences, Lomonosov Moscow State University, Russia
Vassilyev S. N. — RAS member, Lomonosov Moscow State University, Russia
Shananin A. A. — RAS member, Lomonosov Moscow State University, Russia
Grigorenko N. L. — Lomonosov Moscow State University, Russia
Kalyaev I. A. — RAS Member, Scientific Research Institute of Multiprocessor Computing Systems, Russia
Zhabko A. P. — St. Petersburg State University, Russia

Kupriyanov M. S.	St. Petersburg Electrotechnical University "LETI", Russia
Demidova L. A.	Ryazan State Radioengineering University, Russia
Khranilov V. P.	Nizhny Novgorod State Technical University, Russia
Shvetsov A. N.	Vologda State Technical University, Russia
Stepanov M. F.	Saratov State Technical University, Russia
Fomichev A. V.	Bauman Moscow State Technical University, Russia
Proletarskiy A. V.	Bauman Moscow State Technical University, Russia
Novikov D. A.	RAS Member, V. A. Trapeznikov Institute of Control Sciences of Russian Academy of Sciences, Russia
Gubko M. V.	V. A. Trapeznikov Institute of Control Sciences of Russian Academy of Sciences, Russia
Galyaev A. A.	Corresponding member of RAS, V. A. Trapeznikov Institute of Control Sciences of Russian Academy of Sciences, Russia
Arutyunov A. V.	V. A. Trapeznikov Institute of Control Sciences of Russian Academy of Sciences, Russia
Pashchenko F. F.	V. A. Trapeznikov Institute of Control Sciences of Russian Academy of Sciences, Russia
Chernousko F. L.	RAS Member, A. Ishlinsky Institute for Problems in Mechanics of Russian Academy of Sciences, Russia
Bolotnik N. N.	Corresponding member of RAS, Ishlinsky Institute for Problems in Mechanics of Russian Academy of Sciences, Russia
Zatsarinnyy A. A.	Federal Research Center "Computer science and control" of Russian Academy of Sciences, Russia
Daryina A. N.	Federal Research Center "Computer science and control" of Russian Academy of Sciences, Russia
Nikulchev E. V.	MIREA – Russian Technological Institute, Russia
Serov V. A.	MIREA – Russian Technological Institute, Russia
Massel L. V.	Melentiev Energy Systems Institute, Siberian Branch of Russian Academy of Sciences, Russia
Semenkin E. S.	Siberian Institute of Applied System Analysis named after A.N. Antamoshkin, Russia
Sopov E. A.	Siberian Institute of Applied System Analysis named after A.N. Antamoshkin, Russia

Kobzev A. A.	Vladimir State University named after A.G. and N.G. Stoletovs, Russia
Turdakov D. Yu.	Ivannikov Institute for System Programming of the RAS, Russia
Arkhipenko K. V.	Ivannikov Institute for System Programming of the RAS, Russia

Organizing Committee

Diveev A. I.	Federal Research Center "Computer Science and Control" of Russian Academy of Sciences, Russia
Fomichev V. V.	Lomonosov Moscow State University, Russia
Atamas E. I.	Lomonosov Moscow State University, Russia
Maltseva A. V.	Lomonosov Moscow State University, Russia
Tochilin P. A.	Lomonosov Moscow State University, Russia
Vostrikov I. V.	Lomonosov Moscow State University, Russia
Artemyeva L. A.	Lomonosov Moscow State University, Russia
Dryazhenkov A. A.	Lomonosov Moscow State University, Russia
Sofronova E. A.	Federal Research Center "Computer Science and Control" of Russian Academy of Sciences, Russia
Daryina A. N.	Federal Research Center "Computer Science and Control" of Russian Academy of Sciences, Russia
Kazaryan D. E.	Federal Research Center "Computer Science and Control" of Russian Academy of Sciences, Russia
Konstantinov S. V.	RUDN University, Russia
Belotelov V. N.	Federal Research Center "Computer Science and Control" of Russian Academy of Sciences, Russia
Voronin E. A.	Federal Research Center "Computer Science and Control" of Russian Academy of Sciences, Russia
Gromov I. A.	Federal Research Center "Computer Science and Control" of Russian Academy of Sciences, Russia
Prokopiev I. V.	Federal Research Center "Computer Science and Control" of Russian Academy of Sciences, Russia

Contents – Part III

The Use of Neural Networks Based on the YOLO Architecture
to Automatically Determine Indicators of Crop Growth Using Photographs
Taken from UAVs 1
  Dmitriy Poleshchenko, Ilia Mikhailov, and Vladislav Petrov

Residues Control for Perishables in Retail Through Dynamic Pricing 15
  Anna Kitaeva, Alexandra Zhukovskaya, and Yu Cao

Control Algorithm for Reducing the Influence of External Load
on the Steering Drive System 29
  Pham Van Tuan, N. D. Khanh, and N. N. Hung

Software Framework for EEG Signals Analysis Using Machine Learning
Methods 41
  N. Shanarova, M. Lipkovich, M. Pronina, V. Knyazeva, A. Sagatdinov,
  A. Aleksandrov, J. Kropotov, and V. Ponomarev

A Review on Facial Expression Recognition Approaches, Datasets
and Technologies 53
  M. S. Lavanya, Vanishri Arun, Mayura Tapkire, and Yulia Shichkina

An Open Source Laboratory Information Management System for Biobank
Optimization in a Robotic Environment 69
  Anna Nozdracheva, Alexey Nozdrachev, Larisa Rybak,
  Vladislav Cherkasov, and Dmitry Malyshev

Methods of Objects Recognition System Enhancement in Computer Vision 83
  Andrey N. Kokoulin and Rostislav A. Kokoulin

Reconstruction Derivatives from Values of Functions Belonging
to Nikolskii-Besov Classes of Mixed Smoothness in Domains of a Certain
Kind 95
  S. N. Kudryavtsev

A Neural Network Approach to Longitudinal Vehicle Acceleration Control 108
  Ivan Gromov

Algorithms of ECG Time Series Processing in EDF-Format 118
  Sinan V. Kurbanov, Denis A. Andrikov, and Aleksandr E. Khramov

On Homogeneous Observers for Linear Systems 131
    V. Zhdanov, D. Galkina, D. Konovalov, A. Kremlev, and K. Zimenko

Intelligent Systems for Urban Planning: Well-Being and Residents'
Perception 143
    Nailia Gabdrakhmanova and Maria Pilgun

The Principle of Trajectories Separation for the Optimal Control Problem
in a Feedback System 159
    A. N. Daryina

Fault Detection and Isolation for USV with INS Using Genetic Algorithm 167
    Dmitry Bazylev, Alexey Margun, and Radda Iureva

The Necessity of Using Technical Analogues of Thinking
and Consciousness for the Creation of AGI 182
    Alexey Podoprosvetov, Vladimir Smolin, and Georgy Malinetsky

Solving Boundary Value Problems Based on a Fractional Differential
Equation of Diffusion Type with Arbitrary Values of the Orders
of the Derivatives 196
    Dmitry O. Zhukov and Konstantin K. Otradnov

Automation Systems for Scientific Calculations for Multiscale Modeling
of Composite Materials 216
    K.K. Abgaryan and E.S. Gavrilov

Automated Generation of Educational Video Material 224
    B. S. Ksemidov and K. K. Abgaryan

Application of Stochastic Petri Nets in a Network-Centric Approach
to the Organization of the Emergency Response Process 235
    Viktor A. Drogovoz

A Self-organization Intelligent Model in Network-Centric Systems Based
on Coalition-Structural Synthesis 247
    V. A. Serov, D. A. Kozlov, O. V. Trubienko, S. A. Skaev, and S. A. Tepsikoev

The Mathematical Model Identification Problem and Its Solving
by Symbolic Regression 255
    Askhat Diveev, Sergey Kozlov, and Igor Prokopiev

Hierarchal Compromise in Decision Making 266
   Felix Ereshko

Author Index 279

The Use of Neural Networks Based on the YOLO Architecture to Automatically Determine Indicators of Crop Growth Using Photographs Taken from UAVs

Dmitriy Poleshchenko, Ilia Mikhailov(✉), and Vladislav Petrov

Automated and Information Control Systems Department Stary Oskol Technological Institute n.a. A.A. Ugarov (branch) National University of Science and Technology MISIS, Stary Oskol, Russia
mikhaylov.is@yandex.ru

Abstract. Determining the indicators of crop development is an important task in the agricultural sector. This study proposes a method for determining the planting density and the leaf area index of sunflower plants using RGB images and deep learning techniques obtained using unmanned aerial vehicles (UAVs). For image analysis, the instance segmentation task was solved to count the detected objects and highlight their boundaries. Modern architectures such as YOLOv8 and YOLOv9 were considered. YOLOv8x demonstrated better accuracy on the test data. The analysis of the impact of image preprocessing methods such as Histogram Equalization and Contrast Accumulated Histogram Equalization on model performance was carried out. The results of the experiment indicate that the use of these methods can enhance the generalization ability of models.

Keywords: Planting density · Leaf Area Index · YOLOv8 · YOLOv9 · Instance segmentation

1 Introduction

In the last decade, computer vision has become an increasingly relevant tool in agriculture and has been actively developed by a large number of researchers worldwide [1].

In [2], a camera system based on UAVs and an image analysis technique using a fully convolutional neural network (FCN) was introduced to automate the counting of plants in fields of sugar beets, corn, and strawberries. The results showed that when counting sugar beets using UAVs, the error rate was less than 4.6%. The paper emphasized the importance of automatic crop counting to reduce the manual labor burden on farmers.

The study [3], a camera system based on UAVs and an image analysis technique using the FCN was introduced to automate the counting of plants in fields of sugar beets, corn, and strawberries. The result presents a method for identifying and quantifying maize foliage by means of advanced neural networks and color imagery acquired via UAVs. Mask R-CNN is utilized to distinguish corn sprouts from heterogeneous backgrounds and mitigate the influence of weed presence on the accuracy of leaf enumeration. A new loss smoother function is proposed to enhance the segmentation process. YOLOv5 is subsequently utilized to identify and quantify individual leaf structures of maize seedlings upon completion of the segmentation process. The results show that Mask R-CNN with ResNet50 and the proposed loss smoother function outperforms LI loss in terms of segmentation efficiency. An average precision of 96.9% was obtained for bounding boxes and 95.2% for masks. YOLOv5 outperforms both Faster R-CNN and SSD in the task of leaf detection. The largest YOLOv5 variant shows the best detection rates, with an accuracy of 92.0% for fully developed leaves and 68.8% for newly emerged leaflets.

This paper [4] focuses on the study of deep convolutional neural network (DNN) models for the detection of defective rice seedlings in aerial photographs. The EfficientNet, ResNet50, and MobileNet V2 models are considered in this research. The findings of the experiments indicate that the methods developed on the foundation of pre-trained efficient data networks (D1 and EfficientNet) are able to detect defective seedlings with F1 indices of 0.83 and 0.77, respectively.

The paper [5] presents the results of a study on automatic counting of rape seed inflorescences using the YOLOv5 model with a convolutional block attention module (CBAM), using RGB images. Experiments on the collected dataset showed that R2 and mAP values were greater than 0.96 and 92%, correspondingly. CenterNet, TasselNetV2+, Faster R-CNN, and YOLOv4 were also tested, but the proposed approach demonstrated better accuracy and efficiency.

The authors of the study [6] suggest using YOLOv5 in conjunction with various simplified strategies to detect lettuce plants. A lightweight YOLOv5s-ShuffleNetv2 model was developed to count lettuce plants, and it achieved a high level of accuracy. The model's mAP was 99.42%, recall was 99.13%, precision was 98.24% and F1 score was 0.99.

In another study [7], a system for determining crop development indicators was considered using sunflower as an example. Images from UAVs were used to train a neural network, and the YOLOv5 architecture was adopted owing to its proficiency in counting number of plants. The study also showed that the segmented pixel area within the image can serve as a basis for calculating the leaf surface index using the Feature Pyramid Network (FPN).

The literary analysis revealed a significant interest in addressing the challenge of counting different plants in fields using deep learning techniques. Two main tasks can be identified: detection and segmentation. Recently, studies have investigated the use of the YOLO detection algorithm in a single step, while separate models have been developed to address the segmentation issue.

Deep neural networks (DNNs) necessitate substantial datasets to ensure convergence throughout the training process. However, when solving specific tasks with DNNs, there is often a problem with a lack of training data due to the limitations of data collection. To address this issue, transfer learning can be used, which makes use of models pre-trained on large-scale data collections. Transfer learning allows the knowledge gained from solving one task to be transferred to another task, thus reducing the need for additional training data. For high-quality object recognition and generalization ability, it is important to train the model on images taken under various shooting conditions. This can be achieved through artificial data modification, such as image preprocessing. By using these techniques, the model can better generalize to new data and improve its performance on unseen examples. In this study, we propose to investigate the latest YOLO architectures, v8 and v9, which combine solutions for detection and segmentation, simplifying the image processing algorithm. Our objectives are to evaluate the effectiveness of sunflower plant recognition in the field using a YOLO approach and images in RGB format from UAVs. We will also consider image preprocessing techniques to enhance the models' accuracy and robustness to new data.

2 Materials and Methods

2.1 General Scheme of Work

To automatically determine the indicators of crop development using RGB images from UAVs, a new approach is proposed. This approach is shown in Fig. 1. (1) Using UAVs, we collect images from a sunflower field at an early stage. These images are then combined to create orthophotographs of the field. (2) We cut these orthophotopmaps to the required size and use them as input for the neural network. (3) To train the network, we annotate the images using the Roboflow [8] service and combine them into a dataset. (4) After that, we divide the dataset into training and testing sets to train YOLOv8 and YOLOv9 networks. (5) Based on the output of these algorithms, we calculate indicators such as planting density and the leaf area index. The planting density is calculated by counting the number of plants in a given area. The leaf area index is determined by measuring the ratio of green pixels to the total area of the image. (6) The obtained indicators can be presented to the user in a format that is convenient for them (reports, markers on the map).

2.2 Dataset

The experimental sunflower field was located in the Starooskolsky district of the Belgorod region, Russia. The soil type was chernozem, and the photographs were taken in June 2021 on a cloudy day. At the time of photography, the sunflowers were in the 2–4 growth stages, with two to four leaves.

A DJI Phantom 4 Pro V2.0 UAV was employed to acquire the images, which were subsequently merged into orthomosaics. These orthomosaics were then split

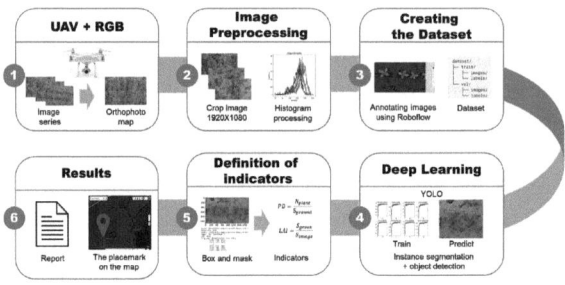

Fig. 1. The scheme of work.

into tiles measuring 1920 × 1080 pixels. The resulting tiles were annotated using the Roboflow service for instance segmentation, resulting in 150 unique tiles with 618 annotations. Using augmentation techniques, the dataset was expanded to 370 tiles, providing a more comprehensive dataset for training and testing.

2.3 Overview of Neural Network Architectures

The YOLO (You Only Look Once) object detection and image segmentation model was developed by Joseph Redmond and Ali Farhadi at the University of Washing-ton. Since its introduction in 2015, YOLO has become popular due to its speed and accuracy [9].

YOLOv8. YOLOv8 is an upgraded YOLO by Ultralytics. YOLOv8 is an advanced, modern (state-of-the-art (SOTA). The model based on the success of previous versions, with updates and refinements to increase productivity, flexibility and efficiency [10].

The YOLOv8 architecture consists of three primary components:

- **Backbone.** This is a convolutional neural network (CNN) that is used to identify informative features within the input image. YOLOv8, as the backbone, uses CSPDarknet53. CSPDarknet53 uses Cross-Stage Partial (CSP) in order to improve the flow of information between different levels of the network and increase accuracy.
- **Neck.** The neck is used to highlight features by combining function maps from different stages of the backbone, thereby capturing information at multiple scales. In YOLOv8, the traditional Feature Pyramid Network has been replaced by an innovative C2f module, which merges semantic features of higher levels with spatial details of lower levels. This integration leads to greater detection accuracy, particularly when identifying small objects.
- **Head.** The head is responsible for generating predictions. In YOLOv8, multiple detection modules are utilized to estimate bounding boxes, objectness scores, and class probabilities for each grid cell within the object map. These outputs are subsequently aggregated to yield the final detection outcomes.

Compared to YOLOv5, YOLOv8 uses a 3×3 convolution instead of a 6×6 one. Module C3 has been replaced with C2f. By integrating high-level semantic data with low-level spatial cues, this module improves the precision of small object detection [11]. A diagram of the new C2f is shown in Fig. 2.

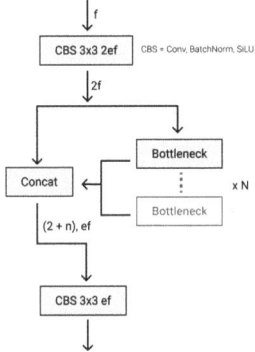

Fig. 2. C2f is the YOLOv8 module. "f" is the number of functions, "e" is the expansion rate, and CBS is a block consisting of a Conv, BatchNorm and SiLU of later versions.

The YOLOv8 architecture, as visualized by OpenMMLab [12], is shown in Fig. 3.

YOLOv9. A new framework is outlined in [13], consisting of the following fundamental parts:

– Information Bottleneck Principle;
– Reversible Functions;
– Programmable Gradient Information (PGI);
– Generalized Efficient Layer Aggregation Network (GELAN).

Information Bottleneck Principle. According to this principle, data X may lead to loss of information during conversion, as demonstrated by Eq. 1 [14]:

$$I(X,X) \geq I(X, f_\theta(X)) \geq I(X, g_\phi(f_\theta(X))), \qquad (1)$$

where I refers to mutual information, with f and g being transformation functions, and θ and ϕ representing their respective parameters.

As X passes through the layers of the DNN, represented by f_θ and g_ϕ, some essential information required for accurate prediction may be lost. The use of this loss may cause gradient instability and suboptimal convergence during model training. A common solution to this problem is to increase the size of the network in order to improve its ability to process data and thereby preserve more information. However, this does not completely solve the problem of gradient instability in DNNs.

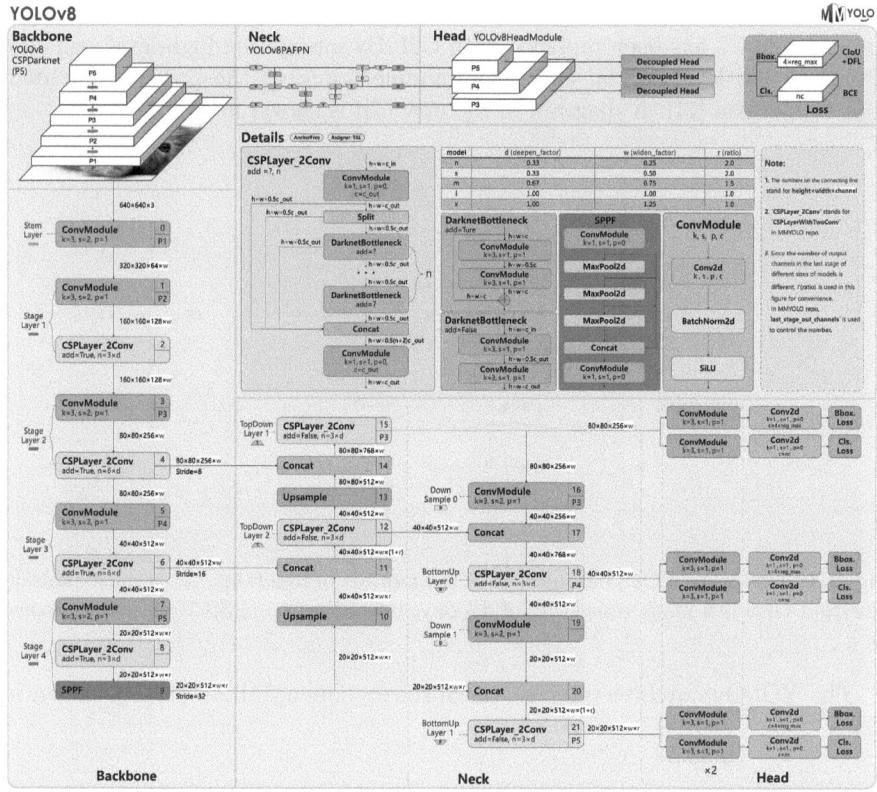

Fig. 3. YOLOv8-P5 model structure and visualization are presented by OpenMMLab.

Reversible Functions. The reversible function can help solve the bottleneck issue:

$$X = v_\zeta(r_\psi(X)), \qquad (2)$$

where ψ and ζ represent the respective parameters of r and v.

The information contained in data X is fully retained when they are transformed using an invertible function (see Eq. 3):

$$I(X, X) = I(X, (r_\psi(X)) = I(X, v_\zeta(r_\psi(X))) \qquad (3)$$

Storing input data at all levels helps to ensure more accurate gradient calculations for updating the model, which leads to more reliable results.

Programmable Gradient Information (PGI) is an approach that improves the reliability of gradients by using auxiliary reversible branches. This method works well for models of all sizes, from small to large. The PGI architecture is illustrated in Fig. 4.

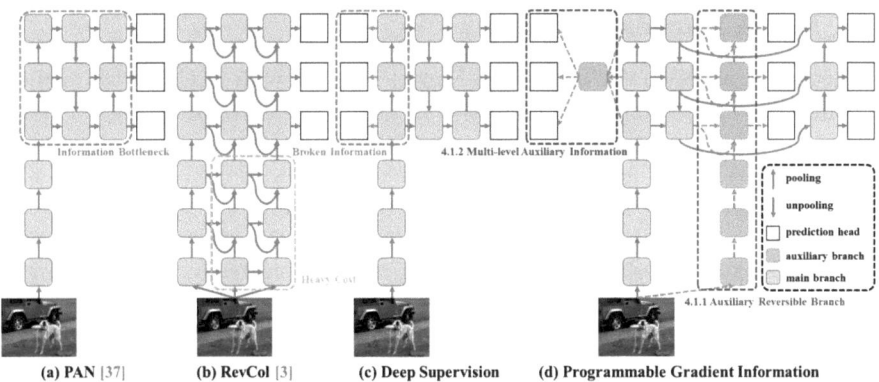

Fig. 4. Overview of architectures: (a) Path Aggregation Network (PAN), (b) Reversible Columns (RevCol), (c) standard deep supervision, and (d) the proposed Programmable Gradient Information (PGI) approach.

Generalized Efficient Layer Aggregation Network (GELAN) is an original architectural design developed on the basis of the PGI framework, which enhances the model's ability to efficiently process and learn from data. Because PGI tackles the issue of retaining critical information in deep neural networks, GELAN incorporates this methodology and offers a versatile, efficient architecture compatible with diverse computational modules. The GELAN architecture is illustrated in Fig. 5.

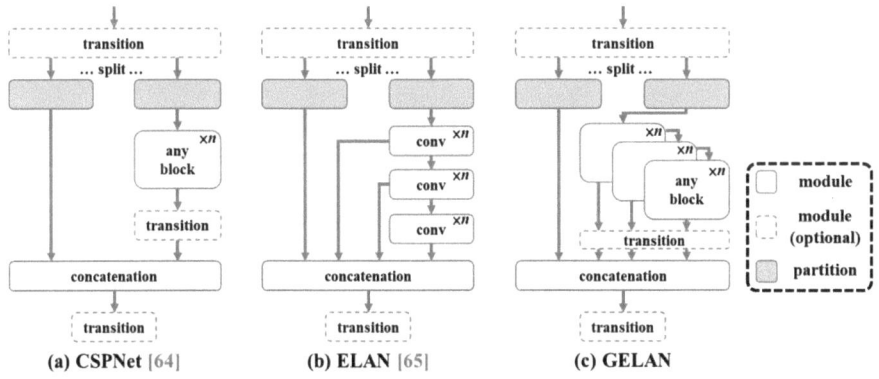

Fig. 5. The GELAN architecture, as illustrated, comprises: (a) CSPNet, (b) ELAN, and (c) the proposed GELAN model.

2.4 Model Training

In this study, an NVIDIA TITAN RTX graphics card with 24 GB video memory and an Intel Core i9-10920X processor @ 3.5 GHz were used as training hardware. The YOLOv8 and YOLOv9 models were trained using transfer learning, which is the initialization of a model using feature extractors trained beforehand with the COCO data. This significantly reduced training duration. Trained networks were saved in .pt format. The smallest model, yolov8n-seg, is 6.8 MB in size, while the heaviest, yolov9e-seg, weighs 347.7 MB. The training of the largest model took up to 2.5 h.

Table 1 below presents the parameter configurations employed during model training.

Table 1. Hyperparameter configuration employed in this study.

Model	Epochs	Batch size	Optimizer	Learning rate	Momentum factor
yolov8n-seg	43	12	AdamW	0.001	0.9
yolov8m-seg	43	4	AdamW	0.002	0.9
yolov8x-seg	48	2	AdamW	0.002	0.9
yolov9c-seg	46	2	AdamW	0.002	0.9
yolov9e-seg	48	1	AdamW	0.002	0.9

2.5 Evaluation Metrics

The performance of the models is assessed using the following metrics.

Precision – determines the proportion of true positive results among all positive forecasts. This metric evaluates the ability of the model to avoid false positives (Eq. 4):

$$Precision = \frac{TP}{TP + FP}, \tag{4}$$

where TP is True Positive, FP is False positive.

Recall – calculates the proportion of correctly identified positive instances among all positive instances in the dataset, by measuring the model's ability to detect all instances of a given class (Eq. 5):

$$Recall = \frac{TP}{TP + FN}, \tag{5}$$

where FN is False negative.

Average Precision (AP) - metric represents the area under the accuracy–completeness curve, serving as an aggregate indicator of the model performance in terms of both accuracy and completeness (Eq. 6):

$$AP = \int_0^1 Precision(Recall)dRecall \qquad (6)$$

Mean Average Precision (mAP) is an extension of average precision (AP), calculated as the mean of AP values across multiple classes. It is useful in multi-label object detection scenarios as it provides a more comprehensive assessment of a model's performance (Eq. 7):

$$mAP = \frac{\sum_{i=1}^{C} AP(i)}{C}, \qquad (7)$$

where C is the number of classes.

The Intersection over Union (IoU) metric evaluates how much the predicted bounding box overlaps with the actual bounding box (Eq. 8):

$$IoU = \frac{Intersection\ area}{Union\ area} \qquad (8)$$

In object detection applications, mAP^{50} and mAP^{50-95} metrics are frequently utilized to assess model performance. mAP^{50} quantifies the average precision of detections where the Intersection over Union exceeds 0.50, taking into account simpler object detections. This metric provides a good representation of the overall accuracy of the model. mAP^{50-95} extends this concept by calculating the average accuracy over a broader range of IoU values, from 0.50 to 0.95. This facilitates a more detailed assessment of the effectiveness of the model and provides a better understanding of its detection capabilities.

2.6 Image Preprocessing

Preprocessing of input data can help to improve the generalization ability of a classification system by reducing the differences between the training and testing data. The main goal of preprocessing is to minimize the variability of input data based on available data.

Various types of normalization and histogram equalization-based methods are used for image preprocessing [15]. A histogram is a graphical representation of the distribution of pixel brightness in an image, which allows us to accurately evaluate the color rendition and exposure of an image. One extended version of histogram equalization is Contrast-Accumulated Histogram Equalization, or CACHE [16].

Histogram Equalization (HE) is a technique for adjusting the intensity of an im-age in order to increase contrast.

Contrast-Accumulated Histogram Equalization (CACHE) is a method that differs from histogram equalization in that it calculates several histograms for different areas of the image and uses them to redistribute the brightness values. This improves the local contrast and sharpness of borders in each area.

To analyze the impact of preprocessing on model performance, we apply CACHE to images. The results are shown in Fig. 6, where the images with histogram adjustments are presented.

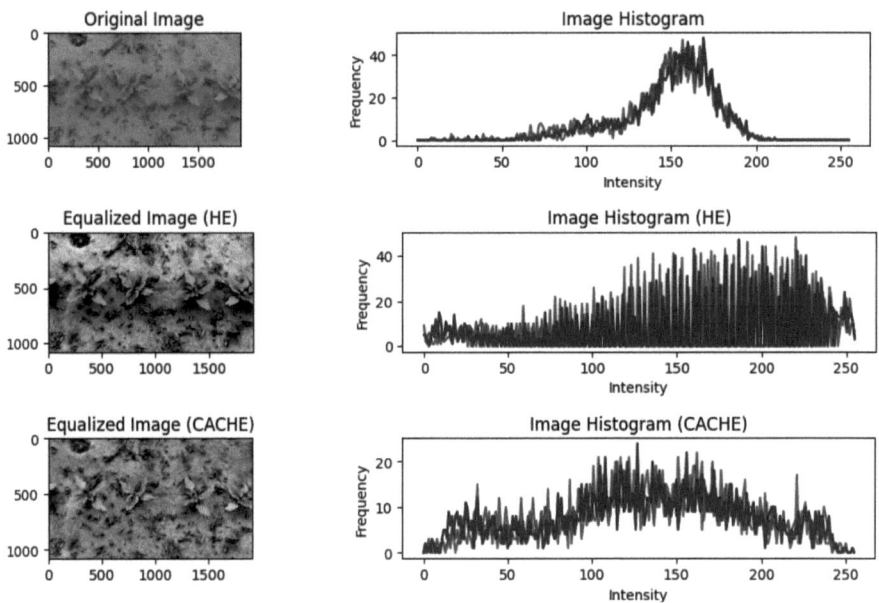

Fig. 6. Examples of the original image and images after histogram alignment.

3 Results

3.1 Model Performance Results

In this paper, we considered various models based on the YOLOv8 and YOLOv9 architectures and trained them to detect sunflower plants. To avoid overfitting, we introduced a limit on the number of training iterations without improvement in the verification metrics before early stopping. The loss curves for the trained models are presented in Fig. 7.

The findings from the neural network testing are detailed in Table 2.

Figure 8 illustrates the performance outcomes of the trained models.

During the testing of the trained models, we found that five of the models had the segmentation accuracy of mAP^{50} greater than 0.95 and the mAP^{50-95} greater than 0.85.

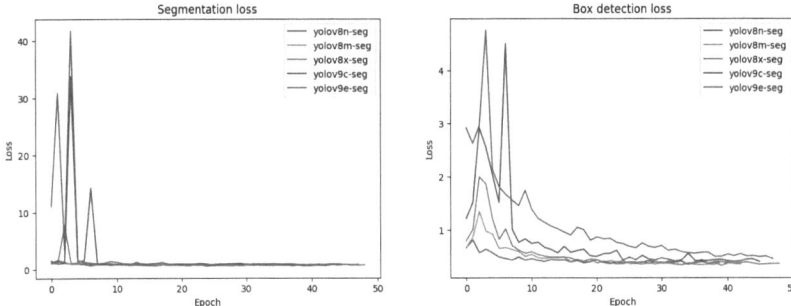

Fig. 7. Loss graphs on the validation dataset.

Table 2. Comparative analysis of models.

Models	Box				Mask				Speed, (ms)
	P	R	mAP^{50}	mAP^{50-95}	P	R	mAP^{50}	mAP^{50-95}	
yolov8n-seg	0.987	0.932	0.962	0.896	0.987	0.932	0.962	0.872	0.49
yolov8m-seg	1.00	0.926	0.972	0.912	1.00	0.926	0.972	0.855	2.69
yolov8x-seg	1.00	0.934	0.954	0.895	1.00	0.934	0.954	0.856	8.30
yolov9c-seg	0.961	0.938	0.956	0.882	0.948	0.925	0.948	0.839	4.60
yolov9e-seg	0.956	0.956	0.961	0.876	0.956	0.956	0.958	0.855	9.13

3.2 Model Performance Results

To investigate the effect of image preprocessing on model performance, the original dataset was processed using two different methods: CACHE and HE. Three data sets were created: the original dataset, the one processed with HE, and the one processed with CACHE.

For the test images, we artificially altered the shooting conditions to create warm and cold shade images (see Fig. 9).

The yolov8x-seg model was further trained on data from HE and CACHE, resulting in three models: yolov8x-seg, yolov8x-seg(HE), which was trained on data from HE, and yolov8x-seg(CACHE), which was pre-trained on data from CACHE. The performance of these models was then evaluated on new test data. No preprocessing was done before uploading the data to the yolov8x-seg network. HE and CACHE were also used for yolov8x-seg(HE) and yolov8x-seg(CACHE), respectively. The experimental results can be found in Table 3.

The analysis showed that preprocessing images by aligning the histogram can help increase the accuracy and generalization ability of a model. A model that has not been preprocessed loses up to 1.3% in accuracy on average when submitted with photos that have been modified. The accuracy of pre-trained models on images with preprocessing decreases: HE – up to 0.5%, CACHE – up to 0.8%. This way, image processing can strengthen the model's generalization performance.

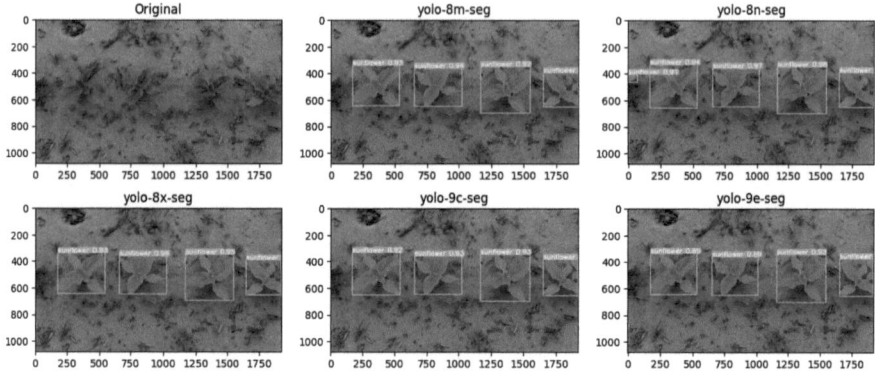

Fig. 8. Results produced by the models.

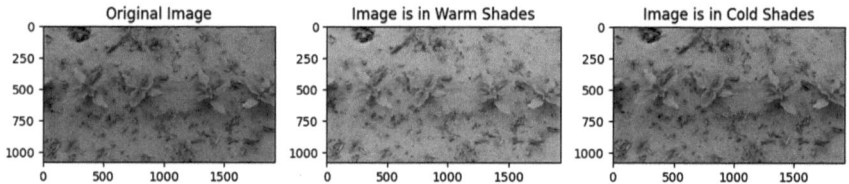

Fig. 9. Examples of images with artificially modified shooting conditions.

Table 3. Comparative analysis

Preprocessing	Image	Box mAP50-95	Mask mAP50-95
No preprocessing	Original	0.895	0.856
	Warm Shaders	0.881	0.829
	Cold Shaders	0.900	0.842
HE	Original	0.909	0.863
	Warm Shaders	0.907	0.863
	Cold Shaders	0.896	0.857
CACHE	Original	0.909	0.856
	Warm Shaders	0.908	0.852
	Cold Shaders	0.895	0.843

4 Discussion

The practical experiments carried out in the context of this study confirmed the effectiveness of using neural networks to assess the growth indicators of agricultural crops. It was demonstrated that the YOLOv8 and YOLOv9 architectures achieve high accuracy in evaluating vegetation features, with an mAP^{50-95} score exceeding 0.85. This indicates a high precision in detecting and segmenting plant elements.

To achieve maximum accuracy, larger models such as yolov8x-seg and yolov9e-seg can be used. However, it should be noted that training YOLOv9 requires significantly more computational resources and takes longer compared to YOLOv8 models of similar size. In the experiments, models were trained for up to 50 epochs; training yolov9e-seg on a dataset of fewer than 300 training images took less than 2.5 h, while yolov8x-seg took less than 2 h. The smaller yolov8n-seg model achieved the required accuracy in approximately 30 min. Therefore, the choice of model depends on the available training resources and performance requirements at the inference stage.

In the future, we plan to study various growth stages of sunflower, which may improve the model's accuracy and reliability in assessing growth indicators at different stages of the plant's life cycle. This enhancement will be particularly useful for precise biomass estimation and yield prediction. In addition, we intend to extend the developed pipeline to other crop types, including potatoes and sugar beets, to evaluate its versatility and adaptability to different types of agricultural crops.

5 Conclusions

This study presents the approach for estimating the density of seedlings and leaf surface area in sunflower plants using images in RGB format acquired from UAVs and DL techniques. The instance segmentation task is used for image analysis, which enables the counting of detected objects and highlighting of their boundaries. Modern YOLOv8 and YOLOv9 neural network architectures are considered for this purpose.

The effect of image preprocessing techniques, such as histogram equalization and contrast-accumulated histogram equalization, is analyzed in the context of model performance.

The findings indicate that this approach can effectively determine the desired indicators based on image data. Future work will focus on extending the application of this approach to other crop species.

References

1. Yang, X., Sun, M.: A survey on deep learning in crop planting. In: IOP Conference Series: Materials Science and Engineering, vol. 490. IOP Publishing (2019)
2. Baretto, A., et al.: Automatic UAV-based counting of seedlings in sugar-beet field and extension to maize and strawberry. Comput. Electron. Agric. **191** (2021)
3. Xu, X., et al.: Detection and counting of maize leaves based on two-stage deep learning with UAV-based RGB image. Remote Sens. **14**(21) (2022)
4. Anuar, M.M., Halin, A.A., Perumal, T., Kalantar, B.: Aerial imagery paddy seedlings inspection using deep learning. Remote Sens. **14**(274) (2022)
5. Li, J., et al.: Automatic counting of rapeseed inflorescences using deep learning method and UAV RGB imagery. Front. Plant Sci. **14** (2023)

6. Zhang, P., Li, D.: Automatic counting of lettuce using an improved YOLOv5s with multiple lightweight strategies. Expert Syst. Appl. **226** (2023)
7. Poleshchenko, D., Petrov, V., Mikhailov, I.: Development of a system for automated control of planting density, leaf area index and crop development phases by UAV photos. In: 2023 7th International Conference on Information, Control, and Communication Technologies (ICCT), pp. 1–6. IEEE (2023)
8. RoboFlow: Computer vision tools for developers and enterprises. https://roboflow.com/. Accessed 04 Apr 2024
9. Joseph, R., Santosh, D., Ross, G., Ali, F.: You only look once: unified, real-time object detection. In: Proceedings of the IEEE Conference on Computer Vision and Pattern Recognition (CVPR), pp. 779–788 (2016)
10. Ultralytics YOLOv8 Docs. https://docs.ultralytics.com/. Accessed 04 Apr 2024
11. Talaat, F.M., ZainEldin, H.: An improved fire detection approach based on YOLO-v8 for smart cities. Neural Comput. Appl. **35**(28) (2023)
12. OpenMMLab. YOLOv8. https://github.com/open-mmlab/mmyolo/tree/main/configs/yolov8. Accessed 04 Apr 2024
13. Wang, C.-Y., Yeh, I.-H., Liao, H.-Y.M.: YOLOv9: learning what you want to learn using programmable gradient information. arXiv preprint arXiv:2402.13616 (2024)
14. YOLOv9: Advancing the YOLO Legacy. https://learnopencv.com/yolov9-advancing-the-yolo-legacy/. Accessed 4 Apr 2024
15. Sada, A., Kinoshita, Y., Shiota, S., Kiya, H.: Histogram-based image pre-processing for ma-chine learning. In: 2018 IEEE 7th Global Conference on Consumer Electronics (GCCE), pp. 272–275. IEEE (2018)
16. Wu, X., Liu X., Hiramatsu, K., Kashino K.: Contrast-accumulated histogram equalization for image enhancement, pp. 3190–3194. IEEE (2017)

Residues Control for Perishables in Retail Through Dynamic Pricing

Anna Kitaeva[1(✉)], Alexandra Zhukovskaya[2], and Yu Cao[1]

[1] Tomsk State University, Tomsk, Russia
kit1157@yandex.ru
[2] Tomsk State Pedagogical University, Tomsk, Russia

Abstract. Modern Information and Communication Technologies allow to adjust retail prices in real time maximizing profitability and inventory turnover. For perishable goods, lowering prices as the product approaches its expiration date helps increase sales and reduce waste. Four models of a perishable product's demand control in retail through the dynamic prices during one fixed sales cycle are described. Demand follows a compound Poisson process, and no replenishment is allowed during the sales cycle. Within the diffusion approximation framework for the inventory level dynamics, the problems of maximizing expected revenue and profit are addressed. The proposed three models allow to sell the product almost surely during its lifetime, that is, realize the zero-ending inventory strategy. The fourth model is more flexible and makes it possible to optimize the size of residues depending on the cost of disposal.

Keywords: Dynamic Pricing · Perishables · Zero-ending Inventory · Price Sensitive Demand · Compound Poisson Demand · Diffusion Approximation · Residues · Digitalization

1 Introduction and Problem Statement

Today, governments around the world place a high priority on sustainability. In recent decades, intelligent technologies have played a significant role in sustainable social development.

Digitalization leads to economic sustainability by making it possible for socio-environmental technologies to be adopted. Nowadays, a lot of literature is devoted to sustainable supply chain management and green technologies; see, e.g., [1,2].

Achieving environmental sustainability largely depends on effective waste management, with particular emphasis on the waste hierarchy: from waste prevention at the top to recycling at the bottom, creating a sustainable and waste-free environment [3].

It is well known that the retail sector has a significant negative impact on the environment. In an era of growing consumer demand and environmental restrictions, retailers face a challenge: maximizing revenue and reducing waste.

The proper solution can be achieved by implementing of a dynamic pricing strategy that prevents an accumulation of unsold perishable goods resulting in financial losses and environmental footprint. Prices can be changed in real time using modern technologies. A reduction in price near the expiration date of perishable goods increases sales and reduces waste [4]. Dynamic pricing can provide a significant increase in retail profits; see, e.g., key paper by Gallego and van Ryzin [5], and [6,7].

These models prove particularly applicable to digital commerce platforms, where menu costs become negligible. "Implementation of an effective pricing strategy is among the most important aspects of running an online business. In today's environment, dynamic pricing is one of the strategies that offers enough flexibility to adjust to any changes." [8]. Here, also the real-life examples of dynamic pricing implementation are given, such as Amazon, Walmart, Target.

Let us briefly consider the expiration date-based retail price control models in stochastic environment introduced and studied in [9–11]. These papers deal with product that must be sold within a limited time frame due to its perishable nature. Demand is expected to exhibit high price elasticity, such that sufficiently low pricing will ensure the product's sale. This scenario is typical for fast-moving consumer goods.

In [9], we proposed a basic stochastic dynamic price control model, which allow us, triggering purchases, to sell all the perishable product at hand during its lifetime almost surely, that is, to solve the problem of residues. In [10], we introduced a multiplicative coefficient into the basic model. In [11], we considered more complicated model introducing a power-low coefficient. The presence of adjustable coefficients allows us to employ a first-order estimate to determine the strength of the price-customer flow correlation and optimize the zero-ending inventory retailing process. Moreover, the coefficients can help us to solve the problem of fitting the demand rate in a real-life situation. This problem is the most challenging one in dynamic pricing; see, e.g., [12].

We begin by establishing the foundational assumptions and mathematical notations. Consider a vendor who purchases an initial inventory quantity Q_0 at the commencement of sales period T, with no possibility of replenishment. The demand process follows a compound Poisson distribution characterized by intensity $\lambda(c)$, where $c = c(t)$ denotes the unit retail price that changes over time. Customer orders are modeled as independent and identically distributed random variables, possessing first moment a_1 and second moment a_2. For analytical tractability, we employ a diffusion approximation approach to represent the inventory level process $Q(\cdot)$, which can be effectively captured through the following stochastic differential equation:

$$dQ(t) = -a_1\lambda(c(t))dt - \sqrt{a_2\lambda(c(t))}dw(t),$$

where $w(\cdot)$ is the Wiener process. Diffusion approximation is often used in inventory management models and gives adequate results for large lots; see, e.g., [13–15].

2 Basic Retail Price Control Model

Let us consider the following model of the retail price control:

$$a_1\lambda(c(t)) = \frac{Q(t)}{T-t}. \tag{1}$$

This formulation establishes the fundamental equilibrium condition that the instantaneous sales rate at time t (the left-hand term) must precisely match the time-integrated average sales rate over the entire selling period $[0, T-t]$ (the right-hand term).

The cumulative distribution function characterizing the clearance duration of lot Q_0, denoted as τ, is defined by equation:

$$F_\tau(t) = P(\tau \leq t) = \exp\left(-\beta\frac{T-t}{t}Q_0\right),$$

where parameter $\beta = 2a_1/a_2$. Statistical moments of the inventory level process $Q(t)$ are equal:

$$\overline{Q}(t) = Q_0(1 - t/T),$$

$$V_Q(t) = \frac{a_2}{a_1}Q_0\left(1 - \frac{t}{T}\right)\frac{t}{T}.$$

Thus, Q_0 will be depleted with probability one before the terminal time of the sales period. For small deviations of the price from the basic price c_0 defined by equation $a_1\lambda(c_0) = Q_0/T$, the expected cycle revenue:

$$\overline{S} \approx a_1 c_0 \lambda(c_0) T \left(1 + \frac{\lambda(c_0)}{c_0\lambda'(c_0)} \cdot \frac{1}{\beta Q_0}\right). \tag{2}$$

Note that the second term in (2) is negative but for $\beta Q_0 \gg 1$ its impact on revenue is insignificant. It is the payment for the zero-ending inventory.

This model is explored in detail in [9].

Figure 1 presents the simulation results of the inventory level dynamics $Q(t)$ in the basic model. A thinning algorithm is used to generate a heterogeneous compound Poisson process [16]. Ten iterations were completed for $T = 10$ and $Q_0 = 2000$. Uniformly distributed over the intervals $[0, 15]$ and $[0, 10]$ purchases were considered. Solid black lines represent the simulation outcomes, while solid and dashed red lines indicate the theoretical expectations and the intervals within two standard deviations respectively, $\sigma_Q(t) = 2\sqrt{V_Q(t)}$. According to the simulation results the diffusion approximation works well for large lot sizes.

As to exact analytical results, that is, without using the diffusion approximation, the Laplace transforms of the conditional probability density function of the session time and the conditional mean of this time for an exponential batch size's distribution managed to get [17].

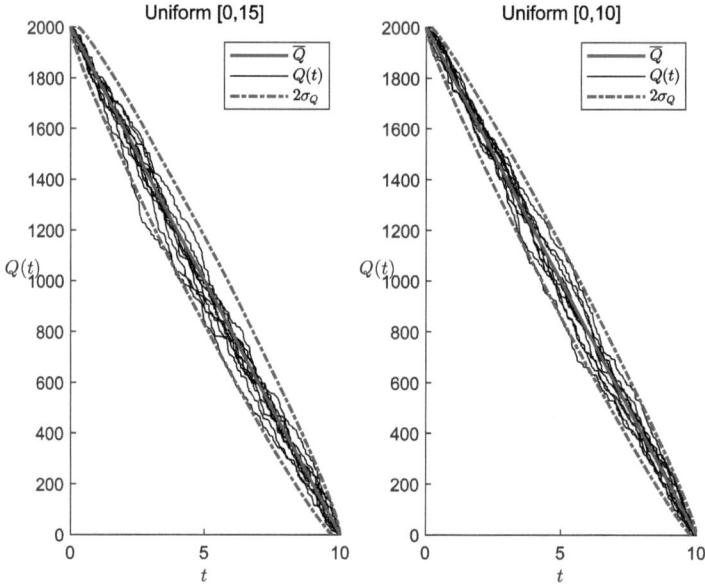

Fig. 1. Inventory level simulation $Q(t)$.

3 Retail Price Control Models with Adjustable Coefficients

3.1 Adjustable Proportional Coefficient

Let us consider the retail price control model following the equation:

$$a_1 \lambda(c(t)) = \kappa \frac{Q(t)}{T-t}, \tag{3}$$

that is, the rates are proportional to each other with coefficient $\kappa > 0$.

In this case, the cumulative distribution function of τ is given by

$$F_\tau(t) = \exp\left(-\beta Q_0 \frac{(1-t/T)^\kappa}{1-(1-t/T)^\kappa}\right).$$

The first and second central moments of the inventory level process are equal respectively:

$$\overline{Q}(t) = Q_0(1-t/T)^\kappa,$$

$$V_Q(t) = \frac{a_2}{a_1} Q_0 \left(1 - \frac{t}{T}\right)^\kappa \left(1 - \left(1 - \frac{t}{T}\right)^\kappa\right).$$

It turns out that the presence of the coefficient has a significant impact on the characteristics of the process $Q(\cdot)$ and makes it possible to use a linear approximation of the intensity on price dependence for $\kappa > 1$:

$$\lambda(c) = \lambda_0 - \lambda_1 \frac{c(t) - c_0}{c_0}, \tag{4}$$

where c_0 represents the initial price associated with baseline intensity λ_0, while parameter $\lambda_1 > 0$ quantifies the responsiveness of $\lambda(\cdot)$ to proportional price deviations from the initial price c_0. The linear approximation is widely used in the literature; see, e.g., [18–20].

The expected revenue over the cycle for $\kappa > 1$

$$\overline{S} = c_0 Q_0 \left(1 + \frac{\lambda_0}{\lambda_1}\right) - \frac{c_0 a_2 Q_0 \kappa^2}{a_1^2 \lambda_1 T}\left(\frac{1}{\kappa - 1} - \frac{1}{2\kappa - 1}\right) - \frac{c_0 Q_0^2 \kappa^2}{a_1 \lambda_1 T(2\kappa - 1)}.$$

If $\kappa \leq 1$, the approximation (4) cannot be used. For a fixed large lot size optimal κ value

$$\kappa \approx 1 + \sqrt[3]{\frac{a_2}{2 a_1 Q_0}}.$$

Due to using the approximation (4) we managed to optimize the expected profit $P = \overline{S} - Q_0 d$ with respect to lot size Q_0; d represents the wholesale unit procurement cost. In this case, the optimal κ value belongs to the interval $1 < \kappa < 2+\sqrt{2}$ and can be find only numerically. In case of large values of $\lambda_1 T$, the optimal κ value tends to 1:

$$\kappa \approx 1 + \sqrt[3]{\frac{a_2}{a_1^2 \lambda_1 T(1 - d/c_0 + \lambda_0/\lambda_1)}}.$$

The optimal lot size

$$Q_0 \approx \frac{a_1 \lambda_1 T}{2}(1 - d/c_0 + \lambda_0/\lambda_1).$$

This model is explored in detail in [10].

3.2 Adjustable Power Coefficient

Let us consider the retail price power-law control model following the equation:

$$a_1 \lambda(c(t)) = \frac{Q(t)}{T(1 - t/T)^\gamma}, \tag{5}$$

where coefficient $\gamma \neq 1$.

The first and second central moments of the inventory level process are equal respectively:

$$\overline{Q}(t) = Q_0 \alpha(t/T),$$

$$V_Q(t) = \frac{a_2 Q_0}{a_1}\alpha(t/T)\left(1 - \alpha(t/T)\right),$$

where $\alpha(t/T) = \exp\left\{\dfrac{(1 - t/T)^{1-\gamma} - 1}{1 - \gamma}\right\}.$

It follows that inventory depletion is almost surely achieved within the sales period for $\gamma > 1$.

Then we get the expected revenue as follow

$$\overline{S} = \int_0^T E\{c(t)\lambda(t)\}\,dt = \frac{c_0 Q_0}{\lambda_1}\left\{(\lambda_0+\lambda_1) - \frac{1}{a_1 T}\left(Q_0 - \frac{a_2}{a_1}\right)\right.$$

$$\left. \times \int_0^1 (\alpha^2(1-z) + q\alpha(1-z))\frac{dz}{z^{2\gamma}}\right\},$$

where linear approximation (4) is adapted and coefficient $q = \dfrac{a_2}{a_1 Q_0 - a_2}$.

The optimal γ values are close to 1 for small values of q, and this is usually the case, because Q_0 is usually large. For example, for $q = 0.001$ optimal $\gamma = 1.002$.

This model is explored in detail in [11], where the results of the numerical optimization are also given.

4 General Retail Price Control Model

Let us consider a generalization of expiration date-based retail price control models in stochastic environment studied in [9–11].

We consider the following general model of the intensity of the customers' flow control through the dynamic price:

$$a_1 \lambda(c(t)) = \frac{Q(t)}{T\varphi(t/T)} \qquad (6)$$

where $\varphi(\cdot)$ represents an unspecified weight function, $t \in [0, T]$.

Accordingly, the inventory level follows the stochastic differential equation:

$$dQ(t) = -\frac{Q(t)}{T\varphi(t/T)}dt + \sqrt{\frac{a_2}{a_1}\frac{Q(t)}{T\varphi(t/T)}}dw(t), \qquad (7)$$

and let us use the approximation (4) to get more advanced analytical results.

From (6) and (4) we get

$$c(t) = c_0\left(1 + \frac{\lambda_0}{\lambda_1} - \frac{Q(t)}{a_1\lambda_1 T\varphi(t/T)}\right).$$

Let us address the expected revenue optimization problem. The expected revenue per time unit during the cycle

$$E\{c(t)a_1\lambda(c)\} = c_0 E\left\{\left(1 + \frac{\lambda_0}{\lambda_1} - \frac{1}{a_1\lambda_1}\frac{Q(t)}{T\varphi(t/T)}\right)\frac{Q(t)}{T\varphi(t/T)}\right\}$$

$$= c_0\left(1 + \frac{\lambda_0}{\lambda_1}\right)\frac{\overline{Q}(t)}{T\varphi(t/T)} - \frac{c_0}{a_1\lambda_1}\frac{\overline{Q^2}(t)}{T^2\varphi(t/T)^2},$$

where $\overline{Q}(t) = E\{Q(t)\}$ and $\overline{Q^2}(t) = E\{Q^2(t)\}$.

Let us find the first and the second initial moments of process $Q(\cdot)$.

From (7) we have
$$d\overline{Q}(t) = -\frac{\overline{Q}(t)}{T\varphi(t/T)}dt \qquad (8)$$
with $\overline{Q}(0) = Q_0$.

We get $\overline{Q}(t) = Q_0 \exp\left\{-\int_0^{t/T} \frac{dz}{\varphi(z)}\right\}$ and $\overline{Q}(T) = Q_0 \exp\left\{-\int_0^1 \frac{dz}{\varphi(z)}\right\}$.

Applying Itô's lemma to Eq. (7) yields
$$d\left(\overline{Q^2}(t)\right) = \left(-\frac{2Q^2(t)}{T\varphi(t/T)} + \frac{a_2 Q(t)}{a_1 T\varphi(t/T)}\right) + 2Q(t)\sqrt{\frac{a_2 Q(t)}{a_1 T\varphi(t/T)}}dw(t)$$

and
$$\frac{d\overline{Q^2}(t)}{dt} = -2\frac{\overline{Q^2}(t)}{T\varphi(t/T)} + \frac{a_2 Q_0 \exp\left\{-\int_0^{t/T}\frac{dz}{\varphi(z)}\right\}}{a_1 T\varphi(t/T)} \qquad (9)$$

subject to $\overline{Q^2}(0) = Q_0^2$. It follows
$$\overline{Q^2}(t) = Q_0^2 \exp\left\{-2\int_0^{t/T}\frac{dz}{\varphi(z)}\right\} + \frac{a_2 Q_0}{a_1}\exp\left\{-\int_0^{t/T}\frac{dz}{\varphi(z)}\right\}$$
$$\times \left(1 - \exp\left\{-\int_0^{t/T}\frac{dz}{\varphi(z)}\right\}\right),$$

and variance
$$Var\{Q(t)\} = \frac{a_2 Q_0}{a_1}\exp\left\{-\int_0^{t/T}\frac{dz}{\varphi(z)}\right\}\left(1 - \exp\left\{-\int_0^{t/T}\frac{dz}{\varphi(z)}\right\}\right).$$

The expected revenue over the cycle
$$\overline{S} = \int_0^T E\{a_1 c(t)\lambda(t)\}\, dt = \frac{c_0 Q_0}{\lambda_1}\left[(\lambda_0 + \lambda_1)\int_0^1 e^{-\Psi(z)}\Psi'(z)dz\right.$$
$$\left. -\frac{1}{a_1 T}\left(Q_0 - \frac{a_2}{a_1}\right)\int_0^1 e^{-2\Psi(z)}\Psi'^2(z)dz - \frac{a_2}{a_1^2 T}\int_0^1 e^{-\Psi(z)}\Psi'^2(z)dz\right] + \eta d\overline{Q}(T),$$

where $\Psi(z) = \int_0^z \frac{dx}{\varphi(x)}$, $\eta d\overline{Q}(T)$ is a salvage value, coefficient $\eta < 1$; see [21]. Here, we assume that η can be less than -1, that is, the penalty for the residues can be very high. Note that $\Psi'(z) = \frac{1}{\varphi(z)}$ and $\Psi(0) = 0$.

Taking into account that $\int_0^1 e^{-\Psi(z)}\Psi'(z)dz = 1 - e^{-\Psi(1)}$, we get
$$\overline{S} = \frac{c_0 Q_0}{\lambda_1}\left[(\lambda_0 + \lambda_1)(1 - e^{-\Psi(1)}) - \frac{1}{a_1 T}\left(Q_0 - \frac{a_2}{a_1}\right)\right.$$

$$\times \int_0^1 e^{-2\Psi(z)} \Psi'^2(z) \mathrm{d}z - \frac{a_2}{a_1^2 T} \int_0^1 e^{-\Psi(z)} \Psi'^2(z) \mathrm{d}z \Big] + \eta dQ_0 e^{-\Psi(1)}.$$

Let us solve the task

$$\widetilde{S} = \int_0^1 \left[\left(\frac{a_1 Q_0}{a_2} - 1 \right) e^{-2\Psi(z)} + e^{-\Psi(z)} \right] \psi'^2(z) \mathrm{d}z \Rightarrow \min_{\Psi(\cdot)}$$

constrained by $\Psi(0) = 0$.

Application of Euler-Lagrange equation reveals that the optimal function $\Psi(\cdot)$ satisfies

$$Ae^{-\Psi}(\Psi'' - \Psi'^2) + \left(\Psi'' - \frac{1}{2}\Psi'^2 \right) = 0, \qquad (10)$$

where $A = \dfrac{a_1 Q_0}{a_2} - 1$.

4.1 Optimization Through the Optimal Weight Function's Approximation

Since Q_0 is normally large enough, we get $A \gg 1$ after neglecting the last term in (10) and $\Psi'' - \Psi'^2 = 0$.

Consequently, the approximation of the solution to (6) takes the linear form $\varphi(z) = C - z$, where C is a constant. The normalized case with $C = 1$ gives $E\{Q(T)\} = Var\{Q(T)\} = 0$, so the lot will be sold during the period almost surely. This case was considered in [9].

If $C > 1$, then the residues remain feasible. The expected terminal inventory level $\overline{Q}(T) = Q_0 \left(\dfrac{C-1}{C} \right)$. Here, we will consider the case $C > 1$.

Denote $1/C = \widetilde{C}, 0 \leq \widetilde{C} < 1$. Then the expected revenue

$$\overline{S}_{C>1} = \frac{c_0 Q_0}{\lambda_1} \left[(\lambda_0 + \lambda_1) \widetilde{C} - \frac{1}{a_1 T} \left(Q_0 - \frac{a_2}{a_1} \right) \widetilde{C}^2 \right. \\ \left. + \frac{a_2}{a_1^2 T} \widetilde{C} \ln(1 - \widetilde{C}) \right] + \eta d Q_0 (1 - \widetilde{C}), \qquad (11)$$

and we derive the following equation in terms of $\widetilde{C}$

$$\frac{\partial \overline{S}_{C>1}}{\partial \widetilde{C}} = \frac{c_0}{\lambda_1} \left\{ \lambda_0 + \lambda_1 - \frac{2}{a_1 T} \left(Q_0 - \frac{a_2}{a_1} \right) \widetilde{C} \right. \\ \left. + \frac{a_2}{a_1^2 T} \left[\ln(1 - \widetilde{C}) - \frac{\widetilde{C}}{1 - \widetilde{C}} \right] \right\} - \eta d = 0. \qquad (12)$$

Equation (12) has a unique solution at $[0, 1)$, which admits exclusively numerical solution.

Let us address the joint optimization of initial lot size and constant $\widetilde{C}$ to maximize the expected profit $P = \overline{S}_{C>1} - Q_0 d$, that is, given the approximate form of the optimal weight function. We need to solve the system of two equations consisting of (12) and the following one:

$$\frac{\partial P}{\partial Q_0} = \frac{c_0}{\lambda_1}\left[(\lambda_0 + \lambda_1)\widetilde{C} - \frac{1}{a_1 T}\left(Q_0 - \frac{a_2}{a_1}\right)\widetilde{C}^2 + \frac{a_2}{a_1^2 T}\widetilde{C}\ln(1-\widetilde{C})\right]$$
$$-d - \frac{c_0 Q_0}{a_1 \lambda_1 T}\widetilde{C}^2 + \eta d(1-\widetilde{C}) = 0. \qquad (13)$$

Taking (12) into account, (13) can be rewritten as

$$\frac{c_0}{\lambda_1}\left[\frac{1}{a_1 T}\left(Q_0 - \frac{a_2}{a_1}\right)\widetilde{C}^2 + \frac{a_2}{a_1^2 T}\frac{\widetilde{C}^2}{1-\widetilde{C}}\right] - (1-\eta)d - \frac{c_0 Q_0}{a_1 \lambda_1 T}\widetilde{C}^2 = 0. \qquad (14)$$

Note, that usually $\lambda_1(T-\eta) \gg 1$. In this case, (14) have the following approximate solution:

$$\widetilde{C}_{opt} \approx 1 - \frac{c_0 a_2}{d a_1^2 \lambda_1 T(1-\eta)}, \quad C_{opt} \approx 1 + \frac{c_0 a_2}{d a_1^2 \lambda_1 T(1-\eta)}. \qquad (15)$$

From (13) we get

$$Q_{0opt} = \frac{a_2}{a_1} + (\lambda_0 + \lambda_1)\frac{a_1 T}{2\widetilde{C}_{opt}} + \frac{a_2}{2a_1 \widetilde{C}_{opt}}\left[\ln(1-\widetilde{C}_{opt}) - \frac{\widetilde{C}_{opt}}{1-\widetilde{C}_{opt}}\right]. \qquad (16)$$

4.2 The Exact Solution for the Optimal Weight Function

Let us return to (10) and try to find the exact solution for the weight function and compare it with the approximate one.

Denote $u(\Psi) = \Psi', \Psi'' = \frac{du}{dz} = \frac{du}{d\Psi}\frac{d\Psi}{dz} = uu'$, and rewrite (10) as following separable equation:

$$(Ae^{-\Psi} + 1)\frac{du}{d\Psi} = \left(Ae^{-\Psi} + \frac{1}{2}\right)u.$$

After integration we get

$$\sqrt{\frac{A+1}{(Ae^{-\Psi}+1)e^{-\Psi}}} = C_1\frac{d\Psi}{dz} \qquad (17)$$

subject to $\Psi(0) = 0$. From (17) it follows

$$\sqrt{\frac{A+1}{(Ae^{-\Psi}+1)e^{-\Psi}}} = C_1\frac{d\Psi}{dz} = \frac{C_1}{\sqrt{A(A+1)}}\left[\sqrt{A(A+1)} - \sqrt{Aw(Aw+1)}\right.$$
$$\left. + \frac{1}{2}\ln\left(2A+1+2\sqrt{A(A+1)}\right) - \frac{1}{2}\ln\left(2Aw+1+2\sqrt{Aw(Aw+1)}\right)\right],$$
$$\qquad (18)$$

where $w = e^{-\Psi}$ and

$$\frac{dz}{d\Psi} = \varphi(z) = C_1 \sqrt{\frac{(Aw+1)w}{A+1}}. \tag{19}$$

Equations (18) and (19) give us a parametric representation of the weight function, parameter $0 \leq w \leq 1$. Tending A to infinity the system asymptotically converges:

$$\begin{cases} z = C_1 - C_1 w, \\ \varphi = C_1 w, \end{cases} \tag{20}$$

that is, the approximate result is the same as in the previous subsection.

Figure 2 shows the plots of weight functions $\varphi(\cdot)$ (dashed lines) for different values of coefficient A and $C_1 = 1$. Straight solid line represents $\varphi(z) = 1 - z$. For $A = 1000$ the exact and approximate solutions practically coincide. The difference between the solutions increases by the end of the cycle. For other values of C_1 the results are analogous.

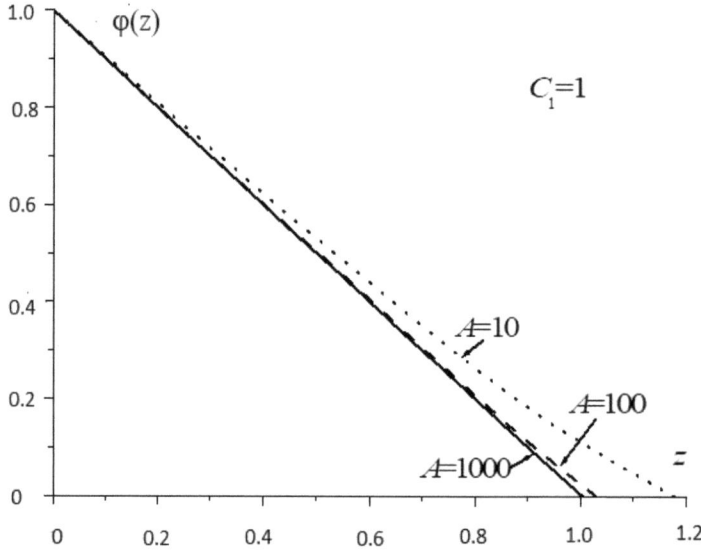

Fig. 2. Weight function $\varphi(\cdot)$ dependence on $t/T = z$ for $A = 10, 100, 1000$ and $C_1 = 1$.

4.3 Numerical Results

In Fig. 3 the results of the simulation of the weighted revenue $\overline{S}_{C>1}/a_1 c_0$ with respect to C are presented. The revenue is simulated for 1000 iterations and we take mean values for each C; $a_2/a_1^2 = 4/3, Q_0/a_1 = 200, \lambda_0 T = 400, \lambda_1 T = 50, \eta d/c_0 = -0.2$, where $Q_0 = Q_{0opt}$ is the optimal inventory value calculated according to (15) and (16). On Fig. 3 black dashed line represents theoretical result (11), red dashed line is the simulation results and black solid line represents theoretical optimal C, where $C_{opt} \approx 1 + \dfrac{c_0 a_2}{d a_1^2 \lambda_1 T(1-\eta)} = 1.0308$.

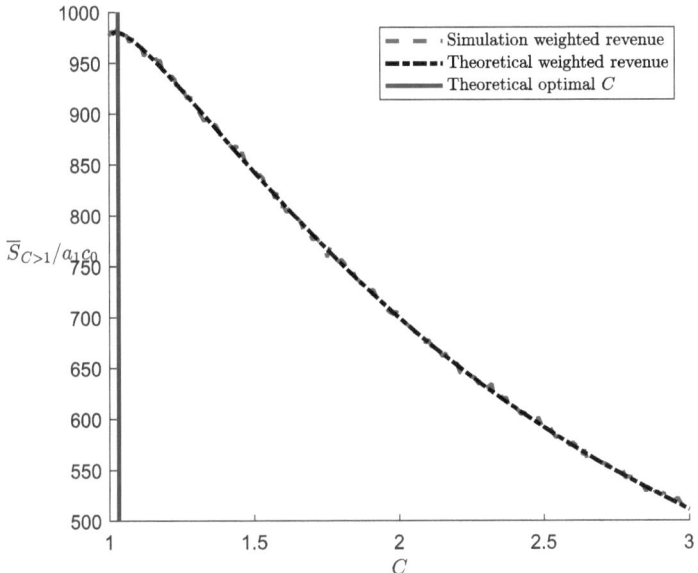

Fig. 3. Simulation of weighted revenue $\overline{S}_{C>1}/a_1 c_0$. (Color figure online)

Figure 4 presents $\overline{S}_{C>1}/c_0 Q_0$ dependence on $C > 1$ for $a2/a_1^2 = 4/3$, $Q_0/a_1 = 400$, $\lambda_0 T = 400$, $\lambda_1 T = 100$, and $\eta d/c_0 = \pm 0.2$. Function is concave with respect to C, and for the considered parameters' values it quickly increases with increasing C and then slowly decreases giving the significant increase in revenue compared to the basic model for small $|\eta|$ values.

In Table 1 numerical results of normalized revenues calculation for the basic model $\overline{S}_{C=1}/c_0 Q_0$ and the general one $\overline{S}_{C>1}/c_0 Q_0$ under different sets of inventory system parameters are presented; $Q_0/a_1 = 1000$, $d/c_0 = 2/3$, $\overline{S}_{C>1}$ is calculated for optimal C value. The revenues in Table 1 are calculated for large utilization's cost, so the differences between the revenues are not so impressive as in Fig 4. For very large utilization's cost (coefficient $\eta = -4$) the revenues are close to each other.

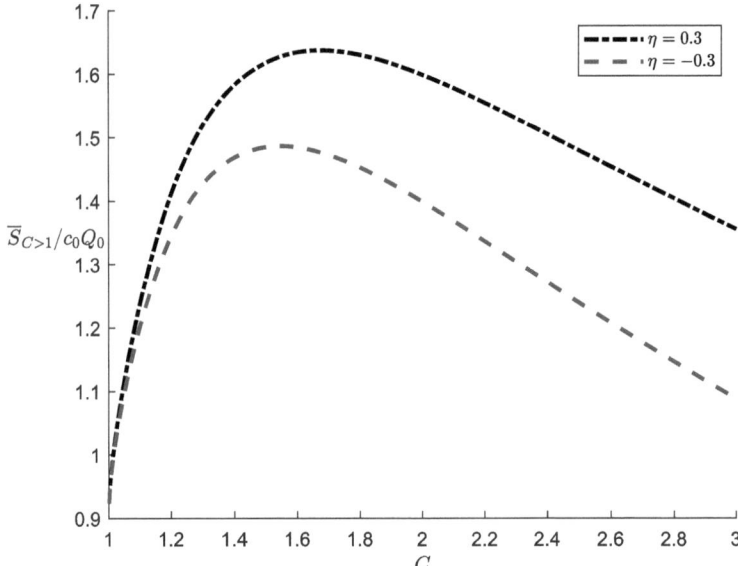

Fig. 4. $\overline{S}_{C>1}/c_0Q_0$ dependence on C value for $\eta = 0.3$ and $\eta = -0.3$.

Table 1. Revenues for the basic and general model.

a_2/a_1^2	7/6	4/3	2	2	2	2
λ_0/λ_1	4	4	4	2	4/3	4/3
$\lambda_1 T$	250	250	250	500	750	750
η	-2	-2	-2	-2	-2	-4
$\overline{S}_{C=1}/c_0Q_0$	0.941	0.933	0.900	0.950	0.967	0.967
$\overline{S}_{C>1}/c_0Q_0$	1.171	1.170	1.169	0.982	0.984	0.981

5 Conclusion

We presented four models of customers' rate control in perishables' retail through dynamic prices adjusted continuously such a way that the rate is proportional to the inventory level and inversely proportional to time to the end of the sales cycle. The real time control with a proper discretization can be realized due to modern digitalization advantages.

The first (basic) model reflects all main ideas involved. First, all results assume sales process behavior follows diffusion approximation, which is suitable for a large lots and number of customers. Secondly, we assume that the demand function is assumed to exhibit high price elasticity. To adjust the demand's sensitivity to different situations or kinds of goods we introduce the adjustable coefficients into the basic model and consider two models with the possibility of adaptation to the real market environment. All three models ensure zero inventory at the end of the cycle with probability one, that is, correspond to the highest waste management requirements. Moreover, the adjustable coefficients permit linear representation of the price-sensitivity in consumer demand intensity.

Finally, the new optimal model is considered taking into account the salvage value, that is, residues are possible. Waste admission has significant positive influence on the revenue, but for high enough cost of disposal, the revenue is close to the basic model's revenue; see Table 1. The difference between revenues decreases with increasing parameter $\lambda_1 T$, which reflects the sensitivity of the demand to the price.

The numerical results in Table 1 are illustrative. Extensive numerical analysis of revenues' sensitivity and accuracy of the approximations as well as the other forms of a price-intensity dependence investigation are planned as a direction for future research.

References

1. Yu, Z., Waqas, M., Tabish, M., et al.: Sustainable supply chain management and green technologies: a bibliometric review of literature. Environ. Sci. Pollut. Res. **29**, 58454–58470 (2022)
2. Chiang, C.-T., Kou, T.-C., Koo, T.-L.: A systematic literature review of the IT-based supply chain management system: towards a sustainable supply chain management model. Sustainability **13**(5), 2547 (2021)
3. Ranjbari, M., Saidani, M., Esfandabadi, Z.S., et al.: Two decades of research on waste management in the circular economy: insights from bibliometric, text mining, and con-tent analyses. J. Clean. Prod. **314**, 128009 (2021)
4. Scholz, M., Kulko, R.-D.: Dynamic pricing of perishable food as a sustainable business model. Br. Food J. **124**(5), 1609–1621 (2022)
5. Gallego, G., van Ryzin, G.: Optimal dynamic pricing of inventories with stochastic demand over finite horizons. Manage. Sci. **40**(8), 999–1020 (1994)
6. Zhao, W., Zheng, Y.-S.: Optimal dynamic pricing for perishable assets with non-homogeneous demand. Manage. Sci. **46**, 375–388 (2000)
7. Sahay, A.: How to reap higher profits with dynamic pricing. MIT Sloan Manag. Rev. **48**(4), 53–60 (2007)
8. ELEKS Homepage. https://eleks.com/blog/technology-enabled-dynamic-pricing-strategy/. Accessed 14 Apr 2024
9. Kitaeva, A.V., Stepanova, N.V., Zhukovskaya, A.O.: Zero ending inventory dynamic pricing model under stochastic demand, fixed lifetime product, and fixed order quantity. IFAC-PapersOnLine **52**(13), 2482–2487 (2019)
10. Kitaeva, A.V., Stepanova, N.V., Zhukovskaya, A.O.: Profit Optimization for zero ending inventories dynamic pricing model under stochastic demand and fixed lifetime production. IFAC-PapersOnLine **53**(2), 10505–10510 (2020)
11. Kitaeva, A.V., Stepanova, N.V., Zhukovskiy, O.I.: Profit optimization with a power-law adjustable coefficient for zero ending inventories dynamic pricing model, sto-chastic demand, and fixed lifetime product. IFAC-PapersOnLine **55**(10), 1793–1797 (2022)
12. Dolgui, A., Proth, J.-M.: Pricing strategies and models. Annu. Rev. Control. **34**(1), 101–110 (2010)

13. Harrison, J.M.: Brownian Models of Performance and Control. Cambridge University Press (2013)
14. Chiamsiri, S., Wee, H.M., Chen, H.C.: Continuous-review inventory models using diffusion approximation for bulk queues. Int. J. Ind. Eng. **19**(10) (2012)
15. Kitaeva, A., Subbotina, V., Zmeev, O.: Diffusion approximation in inventory management with examples of application. In: Dudin, A., Nazarov, A., Yakupov, R., Gortsev, A. (eds.) ITMM 2014. CCIS, vol. 487, pp. 189–196. Springer, Cham (2014). https://doi.org/10.1007/978-3-319-13671-4_23
16. Pasupathy, R.: Generating Homogeneous Poisson processes. Wiley Encyclopedia of Operations Research and Management Science (2010)
17. Kitaeva, A.V., Stepanova, N.V., Zhukovskaya, A.O.: Distribution of selling duration for zero ending inventory dynamic pricing model with exponentially distributed purchases. IFAC-PapersOnLine **54**(1), 999–1004 (2021)
18. Alamri, A.A., Balkhi, Z.T.: The effects of learning and forgetting on the optimal production lot size for deteriorating items with time varying demand and deterioration rates. Int. J. Prod. Econ. **107**(1), 125–138 (2007)
19. Adida, E., Perakis, G.: Dynamic pricing and inventory control: robust vs. stochastic un-certainty models-a computational study. Ann. Oper. Res. **181**(1), 125–157 (2010)
20. Li, S., Zhang, J., Tang, W.: Joint dynamic pricing and inventory control policy for a stochastic inventory system with perishable products. Int. J. Prod. Res. **53**(10), 2937–2950 (2015)
21. Agi, M.A., Soni, H.N.: Joint pricing and inventory decisions for perishable products with age-, stock-, and price-dependent demand rate. J. Oper. Res. Soc. **71**(1), 85–99 (2020)

Control Algorithm for Reducing the Influence of External Load on the Steering Drive System

Pham Van Tuan[1]([✉]) [iD], N. D. Khanh[1] [iD], and N. N. Hung[2] [iD]

[1] VietNam Naval Academy, Nha Trang, Viet Nam
tuanhvhq@gmail.com
[2] Institute of Control Engineering, Le Quy Don Technical University, Ha Noi, Viet Nam

Abstract. This paper investigates steering actuators operating in challenging conditions with significant external forces. A comprehensive mathematical model of the steering actuator system is developed, incorporating nonlinear factors and the impact of unknown external disturbances. A control algorithm is designed to mitigate the effects of external loads on the system using a signal compensation approach. The proposed control strategy is validated through simulations of a civil aircraft steering actuator in the Matlab/Simulink environment.

Keywords: Load compensation structure · external load compensation · hydraulic steering drives

1 Introduction

We are examining control systems for objects operating in complex environments with significant external loads. These systems include aircraft control surface servo drives, hydrofoil control surface drives, coal mine bogie drive systems, and hybrid drives for factory production lines. The challenge lies in effectively compensating for the effects of the load, especially when its value cannot be accurately measured or predicted.

For technical objects, as illustrated above, the Lyapunov function method is commonly employed in developing adaptive control algorithms for dynamic systems. While the Lyapunov function method ensures adaptive stability, it may not fully compensate for external loads acting on the system's actuator.

To implement this concept, we will construct a control structure incorporating a compensation mechanism. This signal compensation structure comprises a virtual model (representing the system without external load), which is equivalent in its description to the actual physical system (considering external load), and a shared controller for both the virtual and real models. The dynamic virtual model's role is to generate and accumulate an error signal representing the discrepancy between the virtual and actual models when subjected to external load and uncertainties. This deviation signal signifies a component of the external load, and the system's uncertainties are addressed by adapting the actual model using the compensated virtual model at the system's input.

2 Construction of a Load Compensation Management Structure

2.1 Description of the General Mathematical Model of the Control Object

The behaviour of a controlled dynamic system, usually called a control object, can in many cases be described as follows [1, 6]:

$$\dot{x} = Ax + Bu + f \tag{1}$$

In Eq. (1) there is a restriction of variables as follows: $x = [x_1, x_2, ..., x_n]^T \in R^n$- is the state vector of the control object; $u(t) \in R^m$- is the control signal vector; $f(t) \in R^n$- vector disturbance function; $A = A(x, t)$, $B = B(x, t)$- respectively, $n \times n$- and $m \times m$-dimensional functional matrices, continuous and bounded together with their partial derivatives for all x, t from a bounded domain.

Let us assume that we have constructed a controller that provides stability conditions:

$$\lim_{t \to \infty} |y(t) - g(t)| = \varepsilon \leq \varepsilon_0 \tag{2}$$

where ε_0 - small positive number, g- desired signal (Fig. 1).

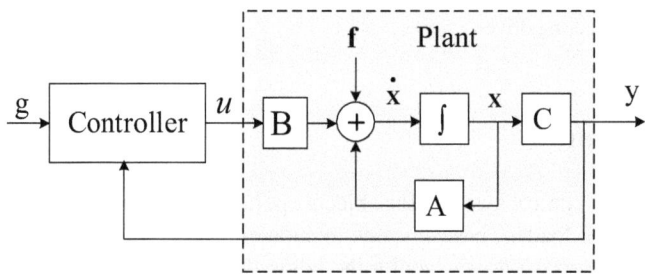

Fig. 1. Structure of the system with a controller

As shown in Fig. 1, the parameter $f(t)$ represents the external load's influence on the system and directly impacts the system's output. If the vector of external disturbances $f(t)$ is significant, such as the aerodynamic load acting on an aircraft's control surface, the deviation in the control surface's inclination angle becomes substantial. Conventional adaptive controllers may not adequately compensate for significant external loads encountered by control objects operating in such environments. This is particularly problematic for systems with low dynamic rigidity that demand precise control at small control signal angles. In these scenarios, substantial static and dynamic errors can manifest in the steering control bodies.

Considering the significant external load, Eq. (2) can be expressed as follows:

$$\lim_{t \to \infty} |y(t) - g(t)| = \varepsilon + \Delta, \tag{3}$$

where Δ is a parameter representing the effect of external loads.

The objective is to develop a control system capable of substantially reducing or entirely eliminating the impact of external loads.

2. Construction of a Load Compensation Management Structure

The external load compensation algorithm is synthesized by constructing a control structure that generates a compensatory effect (Fig. 2).

The proposed structure in this study integrates a virtual model and a real physical system to optimize the control process. The virtual model, described by the state-space equations $\dot{x}_v = Ax_v + Bu_v$; $y_v = Cx_v$, is essentially equivalent to the real system (represented by $\dot{x} = Ax + Bu + f$; $y_v = Cx_v$), but it does not account for the influence of external load f, which is a vector function of disturbances.

The key aspect of this method is the utilization of a common controller for both models. With the presence of external loads (f), this control structure is designed to achieve specific control objectives. This implies that the controller must be capable of maintaining the desired system performance even in the presence of external disturbances, thereby ensuring the stability and accuracy of the system under all operating conditions.

$$\lim_{t \to \infty} (y(t) - g(t)) = \varepsilon, \quad |\varepsilon| \leq \varepsilon_0, \tag{4}$$

where g - desired signal, ε_0 - small positive number.

Indeed, let us assume that an adaptive controller is constructed which, in the absence of an external load ($f = 0$), provides:

$$\lim_{t \to \infty} (y(t) - u(t)) = \varepsilon, \quad |\varepsilon| \leq \varepsilon_0 \tag{5}$$

Where u - is the controller input signal.

In practical applications, when the external load is substantial, conventional controllers may struggle to fully suppress the influence of these significant external load disturbances. This can lead to persistent steady-state errors or degraded system performance. Consequently, the system's output under such conditions, incorporating the uncompensated disturbance effect, can be more accurately represented by the expression (5):

$$\lim_{t \to \infty} (y(t) - u(t)) = \varepsilon + \Delta, \tag{6}$$

where Δ - is a component characterizing the influence of external load. Given the control structure in Fig. 2, the input control signal is defined as $u = g - e$. Then from (6) we obtain the following expression for the error of the real model:

$$\lim_{t \to \infty} [y(t) - (g(t) - e(t))] = \varepsilon + \Delta \tag{7}$$

and for the error in the virtual model:

$$\lim_{t \to \infty} [y(t) - (g(t) - e(t))] = \varepsilon. \tag{8}$$

From (7) and (8) we obtain:

$$\lim_{t\to\infty} [y(t) - (g(t) - e(t))] - \lim_{t\to\infty} [y_v(t) - (g(t) - e(t))] = \varepsilon + \Delta - \varepsilon \quad (9)$$

or

$$\lim_{t\to\infty} (y(t) - y_v(t)) = \Delta \quad (10)$$

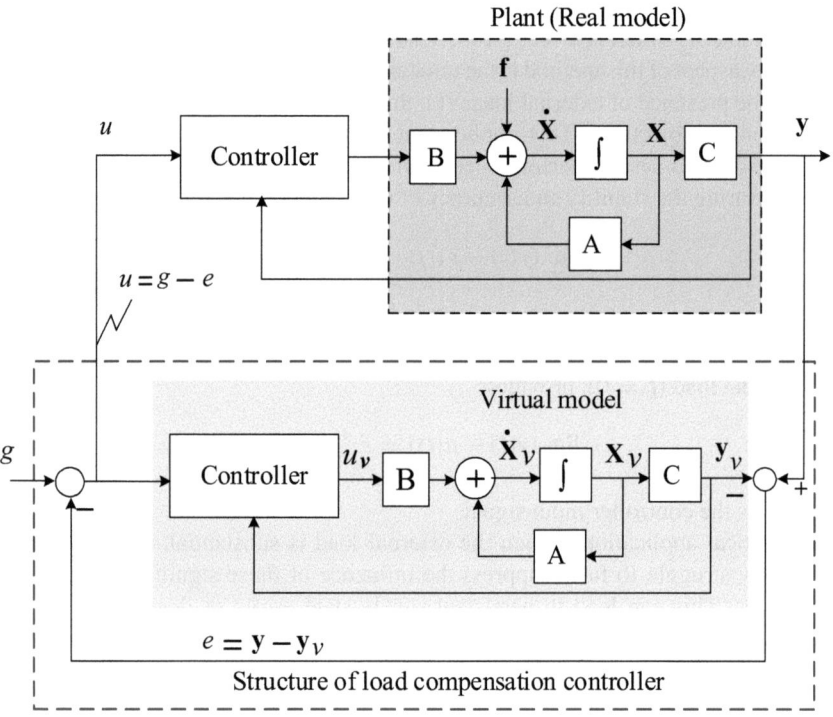

Fig. 2. Structure of the load compensation controller

From Fig. 2, it is obvious that $e = y - y_v$, then taking into account (10), we obtain:

$$\lim_{t\to\infty} e(t) = \Delta. \quad (11)$$

From (7) we have:

$$\lim_{t\to\infty} [y(t) - (g(t) - e(t))] = \lim_{t\to\infty} (y(t) - g(t)) + \lim_{t\to\infty} e(t) = \varepsilon + \Delta. \quad (12)$$

From (11) and (12) we obtain:

$$\lim_{t\to\infty} ((y(t) - g(t)) = \varepsilon, \quad (13)$$

this implies the realization of control objective (4). In other words, the external disturbances are "compensated" by the control signal formed from the error e.

3 Example of Constructing a Regulator for an Aircraft Steering System with a Load Compensation Control Structure

A key trend in contemporary civil aircraft engineering is the escalating demand for reliable flight control actuators. To address this, most commercial aircraft employ a redundant steering system equipped with two electrohydraulic actuators. These actuators, each with two hydraulic cylinders, are connected to the control surface via their shafts and operate in a force summation mode, exerting a combined force on a common output link [1, 3, 7, 9].

To demonstrate the effectiveness of the aforementioned load compensation algorithm, this section considers an example of a redundant aircraft steering system. When operating in this mode, only one channel of the actuator system is active, while the other channel is either inactive or operating in an unloaded mode. This scenario increases the load on the active channel, thus necessitating "load compensation".

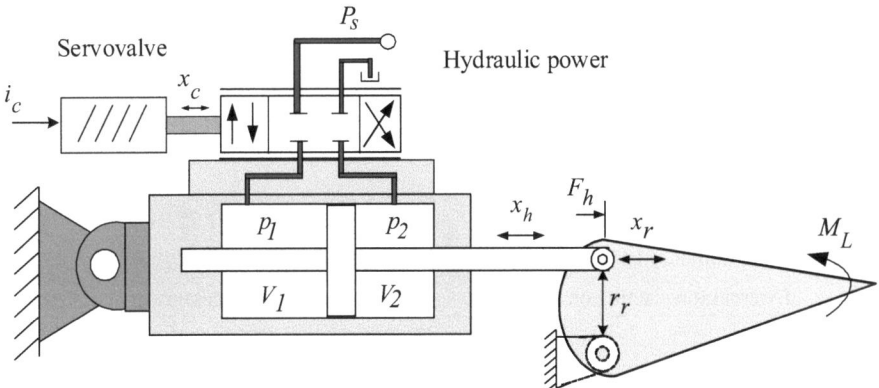

Fig. 3. Schematic diagram of the electrohydraulic steering drive

The electrohydraulic steering drive includes an electrohydraulic power amplifier and a servo drive (Fig. 3). The equations of motion of the electrohydraulic steering drive (EHA) and the equations of motion of the steering surface are described by a system of differential equations as follows [2, 4, 5]:

$$\begin{cases} \tau_c \dot{x}_c = -x_c + k_c i_c; \\ Q_h = C_d b_c x_c \sqrt{\dfrac{1}{\rho}(P_s - \text{sign}(x_c) P_h)} = A_h \dot{x}_h + \dfrac{V}{2E}\dot{P}_h + C_l P_h; \\ A_h P_h = m_h \ddot{x}_h + b_h \dot{x}_h + F; \\ F = K(x_h - x_r); \\ F r_r = J_r \ddot{\theta}_r + b_r \dot{\theta}_r + M_{Load} \end{cases} \quad (14)$$

In the Fig. 2 and Eq. (14), the symbols and parameters are shown in the Table 1 and variables are shown in the Table 2.

Table 1. Parameters for the SHA actuation system [8, 9]

Symbol	Description	Value	Unit
τ_c	Time constant	0.01	s
k_c	Electrohydraulic amplification factor	$3.04\ 10^{-3}$	m/A
C_d	Flow rate coefficient	3.2×10^{-5}	
b_c	A threshold constant	512.5	-
P_s	Supply pressure	27×10^3	kg/cm^2
ρ	Density of the fluid	900	kg/m^3
C_l	Leakage coefficient	1×10^{-11}	m$^3 \cdot$ Pa/s
E	Elastic modulus of pressure fluid	8×10^8	Pa
A_h	Piston square area	2.3×10^{-3}	m^2
m_h	Piston rod weight	25	kg
b_h	Viscous friction of the hydraulic cylinder	10^4	N s/m
J_r	Equivalent moment of inertia	13.5	kg $\cdot$ m^2
r_r	Lever to control surface	0.54	m
K	Stiffness coefficient of EHA coupling with a flight control surface	4×10^8	N/m
b_r	Rudder damping coefficient	51.75	N s/m
F	Force output of EHA	-	-
M_{Load}	External force acting on the rudder	Unknown	-

Table 2. Variables for the SHA actuation system [8, 9]

Variable	Description	Value	Unit
x_c	SHA distribution valve position	-	m
i_c	Torque motor current	-	A
x_h	Displacement of EHA	-	m
x_r	Rudder displacement	-	m
P_h	Rudder displacement	-	kg/cm^2
θ_r	Rudder angular position	-	radian

The synthesis of the load compensation controller for the aforementioned system proceeds in the following stages: First, a model-reference adaptive controller is synthesized to stabilize the system and ensure the desired trajectory of the steering element under conditions of parameter uncertainty and nonlinearity. Subsequently, the load compensation structure depicted in Fig. 2 is employed to 'compensate' for the load's influence on the system.

3.1 Synthesis of the Model-Reference Adaptive Controller

Synthesis of the model-reference adaptive controller is carried out in the following stages [1, 2]:

- Construction of the reference model,
- Construction of the system state observer,
- Develop an adaptive control mechanism for the system.

A. Construction of the reference model.

According to the mathematical description above, the electrohydraulic steering system can be represented in state space by the following non-linear equations:

$$\dot{x} = A(x,t)x + B(x,t)u + f(t), \quad y = Cx \quad (15)$$

where $x = [x_1, x_2, x_3, x_4, x_5, x_6]^T = [\theta_r, \dot{\theta}_r, x_h, \dot{x}_h, P_h, x_c]^T$ - is the state vector of the control object; $u - p$ is dimensional control vector; $A = A(x,t)$, $B = B(x,t)$- respectively, 6×6- and 6×1- dimensional functional matrices, continuous and bounded together with their partial derivatives for all x, t from a bounded domain; $f(t)$ - vector disturbance function. The construction of the reference model in an adaptive system is carried out by linearizing the controlled objects, and then incorporating full-state feedback. As a result of linearizing the second equation of system (14) at the operating point, we obtain [4]:

$$Q_h = K_{Qx} x_c - K_{QP} P_h \quad (16)$$

where K_{Qx} is the flow gain coefficient of the hydraulic control valve, and K_{QP} is the pressure gain coefficient of the hydraulic control valve. At this point, the linearized representation of system (14) at the operating point will take the form:

$$\dot{\tilde{x}} = A_0 \tilde{x} + B_0 u, \quad \tilde{y} = C\tilde{x} \quad (17)$$

where

$$A_0 = \begin{bmatrix} 0 & 1 & 0 & 0 & 0 & 0 \\ \frac{-r_r^2 K}{J_r} & \frac{-b_r}{J_r} & \frac{r_r K}{J_r} & 0 & 0 & 0 \\ 0 & 0 & 0 & 1 & 0 & 0 \\ \frac{Kr_r}{m_h} & 0 & \frac{-K}{m_h} & \frac{-B_h}{m_h} & \frac{A_h}{m_h} & 0 \\ 0 & 0 & 0 & \frac{-2A_h E}{V} & \frac{-2E_h(C_l+K_{QP})}{V} & \frac{2E_h K_{Qx}}{V} \\ 0 & 0 & 0 & 0 & 0 & -\frac{1}{\tau_c} \end{bmatrix}, \quad B_0 = [0\ 0\ 0\ 0\ 0\ k_c/\tau_c\ ]^T,$$

$C = [0\ 0\ 1\ 0\ 0\ 0]$.

The reference dynamics of a plant is given by equation [2]:

$$\dot{x}_m = A_m x_m + B_m u, \quad (18)$$

Here, matrix A_m is in the form $A_m = A_0 - B_0 k$, matrix k is in the form $k = [k_1\ k_2\ k_3\ k_4\ k_5\ k_6]$, and matrix B_m is in the form $B_m = B_0$. These are the state-feedback controller gain matrices and are determined by calculating the criterion to find the controller gains when selecting the desired characteristic equation form.

$H_0(s) = a_0 s^6 + a_1\omega_0 s^5 + a_2\omega_0^2 s^4 + a_3\omega_0^3 s^3 + a_4\omega_0^4 s^2 + a_5\omega_0^5 s + a_6\omega_0^6$, where a_1, a_2, a_3, a_4, a_5, a_6 are coefficients of the chosen Newton polynomial).

B. Construction of the system state observer.
In order to assess the state vector of the control plant $\hat{x}(t)$, full-order state observers are used [2].

$$\dot{\hat{x}} = A_0\hat{x} + B_0 u + L(y - C\hat{x}) \tag{19}$$

where vector L is the observer gain vector, which is determined by the following equation:

$$\det[sI - A_0 + LC] = H_L(s) \tag{20}$$

$H_L(s) = a_0 s^6 + a_1\omega_L s^5 + a_2\omega_L^2 s^4 + a_3\omega_L^3 s^3 + a_4\omega_L^4 s^2 + a_5\omega_L^5 s + a_6\omega_L^6, \omega_L = (3 \div 5)\omega_0$.

C. The adaptive control law for the system is synthesized as follows.
This control law is synthesized with the aim of meeting the control objective.

$$\lim_{t\to\infty} e(t) = \lim_{t\to\infty} (x - x_m) = 0, \tag{21}$$

here, e is defined as: $e = x - x_m$ or $e = \hat{x} - x_m$.

A control law of the following form [1] will be implemented for system (14), where the adaptation algorithm ensures the maximum decrease rate of Lyapunov functions for the system's solutions:

$$u(t) = g + z(t), \tag{22}$$

$$z(t) = -h\,\text{sign}(B_m^T P e), \tag{23}$$

here, g represents the desired signal, $z(t)$ denotes the adaptive control signal, and $h = const > 0$ is the gain coefficient, which characterizes the amplitude of the relay control. The matrix $P = P^T > 0$ is determined from the solution of the equation $A_m^T P + PA_m = -Q$, where $(Q = Q^T > 0)$.

Figure 4 illustrates the schematic diagram of the load compensation controller, integrating a model-reference adaptive controller and a signal compensation structure.

The simulation results are performed in MATLAB/Simulink environment with the parameters in the Table 1.

To assess the effectiveness of the load compensation process when using a signal compensation structure, we examine the simulation results of the transient responses at the EHA (x_h) output. We will consider two cases: first, with the signal compensation structure enabled (switch "k" connected in Fig. 4), and second, with it disabled (switch "k" disconnected). Figure 5 illustrates the impact of external load acting on the aircraft's control surfaces when load compensation is absent. With this result, the static component of the position error of the flight control actuator increases as the external forces intensify.

Figure 6 illustrates the effect of external forces on the flight control surfaces when the load compensation structure is implemented. When there is no static component in

Control Algorithm for Reducing the Influence   37

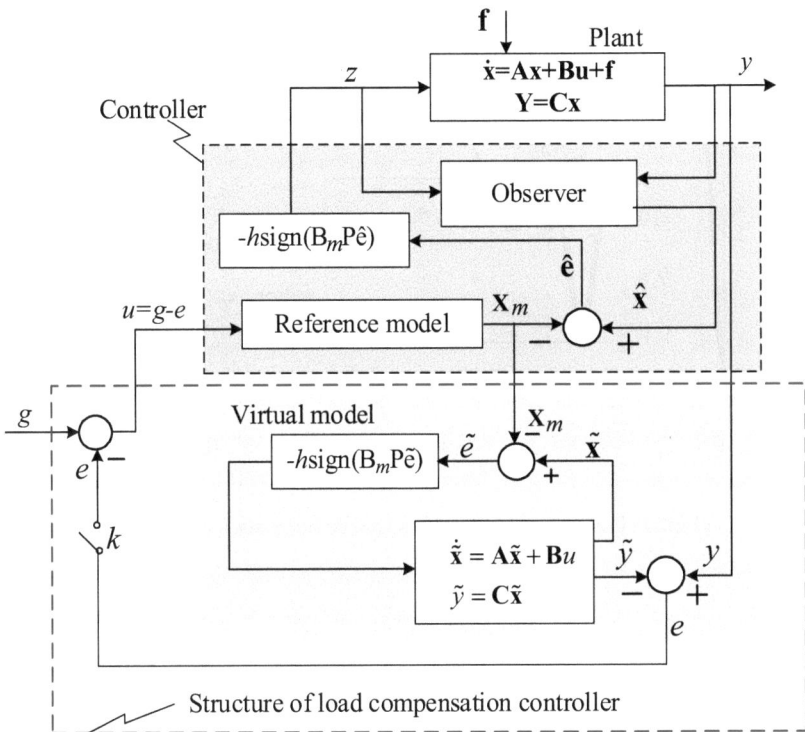

Fig. 4. Schematic diagram of the load compensation controller.

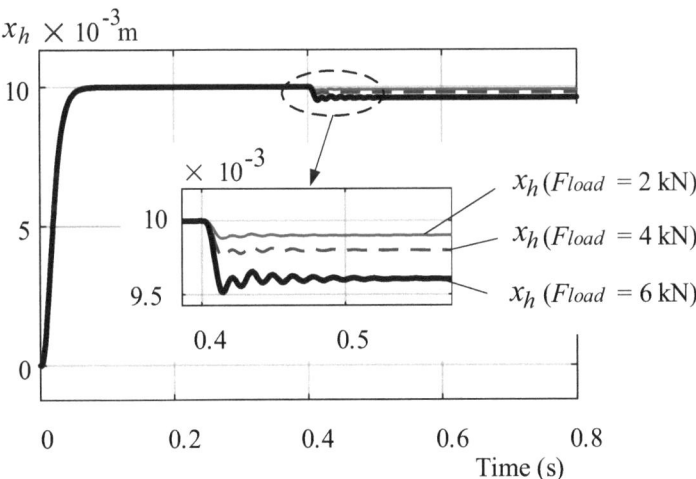

Fig. 5. Position of the x_h without a load compensation controller

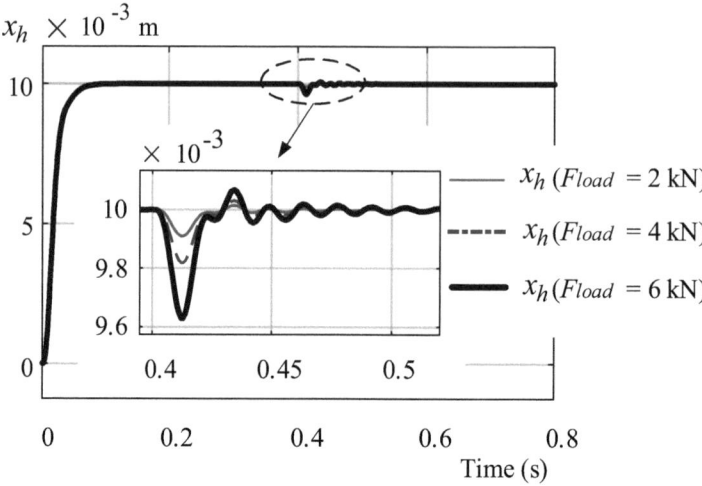

Fig. 6. Position of the x_h with a load compensation controller

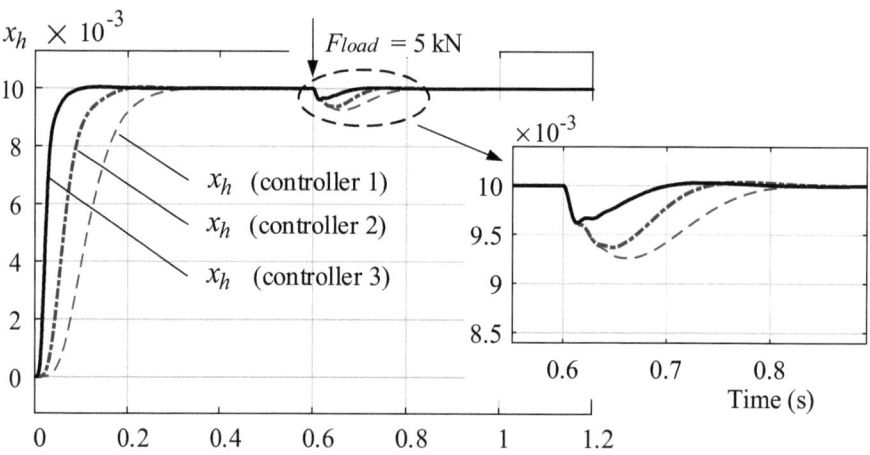

Fig. 7. The rate of the external load compensation of the signal compensation structure

the position error of the flight control actuator, the flight control actuator has either fully or partially compensated for the external load acting on the control surface.

When evaluating the external load compensation ratio of the signal compensation structure, we assess and compare controllers with different parameters. As shown in Fig. 7, the load compensation ratio depends on the maximum descent rate of the Lyapunov functions for the reference dynamics of the plant (or system) represented in Eq. (18). At this point, it, in turn, depends on the gain matrix of the state feedback controller (or state regulator) K in the reference dynamics, when we select the desired characteristic equation.

When subjected to a constant external load, the system's output static error compensation rate aligns with the system's transient process rate. In other words, controllers with shorter transient process times can quickly compensate for static errors.

4 Conclusion

In this paper, we developed a method for compensating the influence of external load on control systems based on an advanced signal compensation structure. Unlike traditional approaches that often necessitate precise models of disturbances or complex observers, our proposed approach eliminates the need to estimate or directly observe external load disturbances. This not only significantly simplifies the design process but also enhances feasibility in practical applications where acquiring disturbance information often presents considerable challenges.

To demonstrate the superior effectiveness of the proposed algorithm, we presented a specific example, clearly comparing its external load compensation capability with and without the signal compensation structure. The results clearly show a significant improvement in control performance, particularly the ability to maintain the steady-state error at a minimum value even when the system is subjected to varying external loads. This unequivocally proves a substantial enhancement in stability and disturbance rejection capability.

A notable special feature of this signal compensation structure is its high flexibility, allowing it to be applied to various types of controllers without requiring fundamental changes to their architecture. This opens up broad integration possibilities within existing control systems, optimizing performance without the need for complete redesign. Importantly, the convergence speed of the steady-state error to its minimum value will directly depend on the convergence speed of the base controller. This implies that with a fast-converging controller, our compensation method will maximize its effectiveness.

Furthermore, another special peculiarity of this method is its ability to increase system control accuracy in the small reference signal region. In many applications, precise control when the reference signal is near zero is crucial, and our signal compensation method has proven capable of meeting this requirement effectively without the need for complex disturbance observers. This affirms the broad application potential of the method in fields demanding high accuracy and robust disturbance rejection.

References

1. Konstantinov, S.V., Kuznetsov, V.E., et al.: Systems of electro-hydraulic flight control actuators of aircraft flight control systems. Publishing House SPbSETU LETI, 518 p. St Petersburg (2019)
2. Khanh, N.D., Kuznetsov, V.E., Vung, C.N.: Synthesizing an adaptive controller to enhance movement quality of a flight control actuator under external load. In: 2021 IV International Conference on Control in Technical Systems (CTS), St. Petersburg, Russia, pp. 146–149 (2021). https://doi.org/10.1109/CTS53513.2021.9562850
3. Rehman, W.U., Wang, X., Hameed, Z., Gul, M.Y.: Motion synchronization control for a large civil aircraft's hybrid actuation system using fuzzy logic-based control techniques. Mathematics **11**(7), 1576 pages (2023)

4. Kuznetsov, V.E., Khanh, N.D., Lukichev, A.N.: System for synchronizing forces of dissimilar flight control actuators with a common controller. In: 2020 XXIII International Conference on Soft Computing and Measurements (SCM), St. Petersburg, Russia, pp. 137–140 (2020). https://doi.org/10.1109/SCM50615.2020.9198768
5. Kuznetsov, V.E., Dinh Khanh, N., Lukichev, A.N., Filatov, D.M.: Hybrid steering system's Pid-based adaptive control. In: 2021 IEEE Conference of Russian Young Researchers in Electrical and Electronic Engineering (ElConRus), pp. 979–984 (2021). https://doi.org/10.1109/ElConRus51938.2021.9396303
6. Krstic, M., Kanellakopoulos, I., Kokotovic, P.V.: Nonlinear and Adaptive Control Design. – Энергоатомиздат, 564 с (1995)
7. Nguen, D.Kh., Kuznetsov, V.E.: An algorithm for synchronizing a multichannel steering drive with an adaptive controller. Russian Elec. Eng. **94**(3), 162–168 (2023). https://doi.org/10.3103/S1068371223030112
8. Khanh, N.D., Hung, N.N.: Synthesis of an adaptive controller for movement synchronization in a multi-channel steering system. In: 2024 XXVII International Conference on Soft Computing and Measurements (SCM), Saint Petersburg, Russian Federation, pp. 99–102 (2024). https://doi.org/10.1109/SCM62608.2024.10554163
9. Rehman, W.U., et al.: Motion synchronization in a dual redundant HA/EHA system by using a hybrid integrated intelligent control design. Chin. J. Aeronaut. **29**(3), 789–798 (2016). https://doi.org/10.1016/j.cja.2015.12.018

Software Framework for EEG Signals Analysis Using Machine Learning Methods

N. Shanarova[1]([✉]), M. Lipkovich[1], M. Pronina[3], V. Knyazeva[2],
A. Sagatdinov[2], A. Aleksandrov[2], J. Kropotov[3], and V. Ponomarev[3]

[1] Institute of Problems in Mechanical Engineering, Saint Petersburg,
Russian Federation
nadya.shanarova@gmail.com
[2] Saint Petersburg State University, Saint Petersburg, Russian Federation
[3] N.P. Bechtereva Institute of the Human Brain of Russian Academy of Sciences,
Saint Petersburg, Russian Federation

Abstract. This paper introduces a software framework developed for analyzing EEG signal using machine learning methods. The framework consists of several independent and customizable modules for signal acquisition and preprocessing, feature extraction, model training, evaluation, and interpretation. A unique aspect is the flexibility to tune hyperparameters across all stages of preprocessing and feature extraction.

The framework was applied to two tasks: diagnosis of mental disorders and detection of intention to perform a hand movement. The results demonstrate balanced accuracy rates of 91% for schizophrenia diagnosis, 88% for obsessive-compulsive disorder diagnosis and 77% for movement intention detection. The methodologies employed for both tasks are detailed in the study.

Keywords: electroencephalogram · event-related potentials · machine learning · psychiatric diagnosis · software framework

Introduction

Currently, there is an increasing attention being paid to research focused on the close interaction between neuroscience and computer technologies. One of the areas of such research is the development of brain-computer interfaces (BCIs), which offer promising prospects for augmenting, restoring, and enhancing the functions of the central nervous system [1], as well as for decision support systems in medical practice. These systems can help in refining diagnoses and selecting correction protocols for neurotherapy based on biofeedback methods [2].

Another application of EEG signal analysis is disease diagnosis. In [3] modern views on the problem of diagnosing cognitive impairments are presented, along with various approaches to identifying disease biomarkers. Similar conclusions were drawn from the analysis of EEG complexity in [5]. It is proposed that comparing event-related potentials between an experimental group (diagnosed patients) and a control group (healthy subjects) may allow domain experts to evaluate impairments in cognitive control. These insights could further support disease diagnosis and inform the application of biofeedback techniques to enhance cognitive control mechanisms in neurotherapeutic interventions.

The outlined tasks align with the conventional machine learning workflow: involving data preprocessing, feature extraction, model training, evaluation, and result interpretation. However, standard machine learning libraries typically lack functionality for handling EEG data, while specialized EEG acquisition and processing software, such as "WinEEG"[1], does not offer integrated machine learning methods or developer APIs.

The aim of this study is to develop a software package that includes methods for EEG signal processing, features extraction and training machine learning models on these features for the classification of target EEG patterns. Additionally, the package provides tools for interpretation and validation of results.

The package consists of a set of configurable and extensible modules, which are described in detail in the next section. Python was chosen as the programming language due to its popularity in machine learning. To reuse existing open-source solutions, the developed package integrates with the MNE-Python library[2], which contains numerous methods for reading and preprocessing EEG signals, and is compatible with the machine learning library *sklearn*[3]. The implemented software package was approbated on two tasks from different domains: determining the intention to perform a movement and medical diagnostics. The solutions to these tasks are described in Sect. 2.

1 Framework Structure

The software package is composed of the following modules (Fig. 1):

- Data Reading Module
- Data Preprocessing Module, which includes various types of EEG signal transformations
- Features Extraction Module
- Classifier Training Module, which also includes performance evaluation of the constructed model
- Results Interpretation Module

Data Reading Module. This module is designed to support reading EEG data in various formats and represent them as multidimensional arrays. Currently, it supports raw EEG data, event-related potentials (ERPs), and ERP components obtained via blind source separation (BSS) method. The module can be extended to read other data formats obtained from various research paradigms.

---

[1] http://www.mitsar-eeg.ru/page1.php?id=update, WinEEG v. 2.4, V. A. Ponomarev, Institute of the Brain, RAS.
[2] https://mne.tools/stable/index.html.
[3] https://scikit-learn.org/stable/.

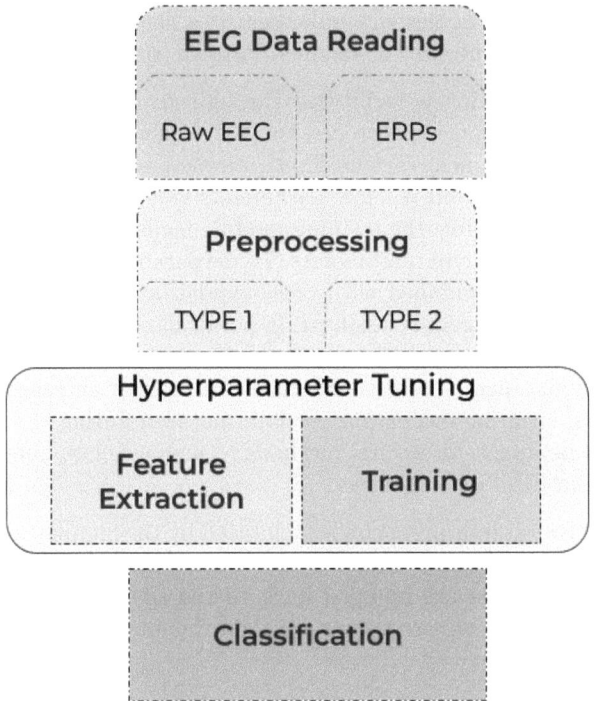

Fig. 1. Block diagram illustrating the stages of the software package

Data Preprocessing. The use of raw EEG data complicates the extraction of indicators associated with specific cognitive processes, underscoring the necessity of signal preprocessing to improve data informativeness. This module incorporates methods for artifact detection and removal, signal filtering, and diverse techniques for segmenting signals into epochs, which are treated as samples for machine learning models. In addition to frequency filtering, spatial Laplacian filters are implemented, which have shown improvements in recognition quality for the tasks under consideration. For ERP data, interval selection using cluster-based analysis tools from WinEEG is supported for subsequent feature extraction.

Feature Extraction. The feature extraction module enables the derivation of features from ERP data by dividing signals into overlapping windows, within which extreme and mean values are calculated, along with identifying the window indices corresponding to these values. Since the characteristics of windows vary based on their size and shift, this feature extraction process is parameterized. For raw EEG, features are derived from both time-domain signals and frequency-domain signals following the application of the Short-Time Fourier Transform (STFT). The features include those based on linear regression, amplitude values, area under the curve calculated from non-overlapping segments of a specified length, as well as more complex features like Common Spatial Pattern (CSP) [6–8] and Discriminative Canonical Pattern Matching (DCPM) [9].

Training Module. Training is facilitated through an interface with the *sklearn* library. The aforementioned preprocessing and feature extraction modules are parameterized; for instance, various filter parameters can be set, signal transformation combinations and window parameters can be defined within the feature extraction phase, while the training module facilitates joint tuning of these parameters and model hyperparameters. Hyperparameter optimization is executed via grid search combined with cross-validation to obtain reliable metric estimates. Model training and classification performance were evaluated through cross-validation on the training dataset and validated on an independent test set. Accordingly, model quality assessment relies on metrics averaged across cross-validation folds, ensuring evaluation on data not seen during training. Furthermore, the module integrates several methods to manage class imbalance, a common challenge in EEG data analysis.

Interpretation. Result interpretation is facilitated by the Shapley method[4], which evaluates the magnitude and direction of feature influence. For ERP data, the most important features are mapped back to the signal segments from which they were derived, and these segments are then visualized on the original signals for expert interpretation.

2 Framework Approbation

The software package described in this work was validated on two tasks: the detection of event-related potentials for determining the intention to move and the classification of psychiatric disorders.

Detecting Movement Intention from Event-Related Potentials. The goal of this task is to train a model to predict, based on a segment of EEG signal, whether movement will occur following the segment. This can be formulated as a binary classification task, where the target classes represent the presence or absence of movement. Such classification outcomes can be used as inputs for a controller to operate electronic devices, offering potential applications in rehabilitation programs for patients with motor impairments [10].

The database comprised of EEG recordings collected from 10 participants (aged 19–43, 7 females) with normal hearing and normal or corrected-to-normal vision. The data were prepared by the Faculty of Biology at St. Petersburg State University. All participants were right-handed and used their right hand for the motor task. Participants gave informed consent to participate in the experiment. The experimental procedures adhered to the principles outlined in the Helsinki Declaration. The experimental design for this task is depicted in Fig. 2.

---

[4] https://pypi.org/project/shapely/.

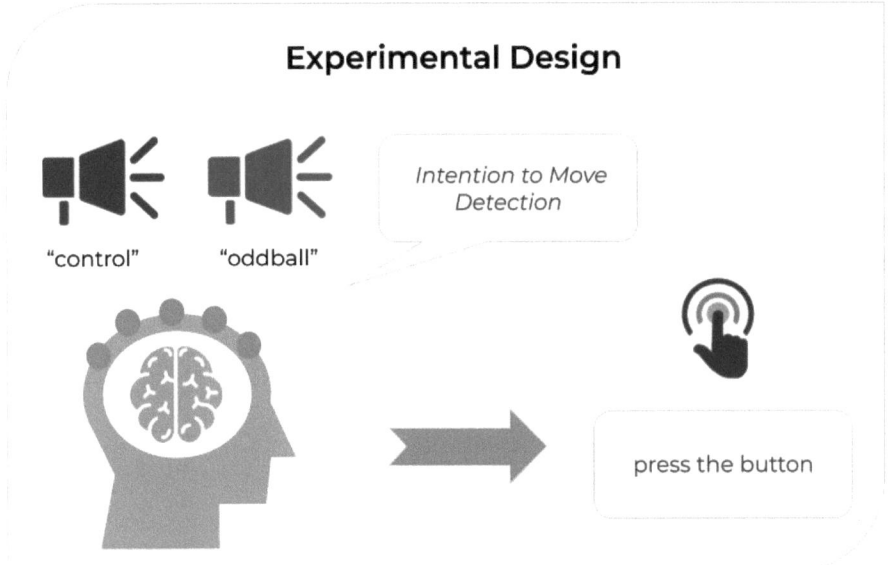

Fig. 2. Experiment Design

Participants were instructed to avoid rhythmic movements, vary the intervals between button presses, and complete a sufficient number of presses, with an average of 104 presses per participant. EEG was recorded using seven electrodes placed at positions F3, Fz, F4, C3, Cz, C4, and Pz according to the international 10–20 system. The reference electrode was located at the tip of the nose, and the ground electrode was placed on the forehead.

Movements were performed under auditory stimulus conditions. Two auditory paradigms were employed: a 'control' paradigm with a 1000 Hz frequency stimulus, and an 'oddball' paradigm, where 1000 Hz standard stimulus was presented with an 85% probability, and a 1200 Hz stimulus was presented with a 15% probability. According to [11], the deviant stimulus in the 'oddball' paradigm exerts a stimulating effect on the motor system, requiring the activation of involuntary attention. Experimental results indicated that this approach improves classification accuracy.

The data processing flow implemented with the framework is presented in Fig. 3:

Signal segments with amplitudes exceeding 100 μV were removed. Additionally, segments where slow waves in the frequency range from 0.16 Hz 1 Hz had amplitudes exceeding 50 μV and fast waves in the range 20 Hz to 35 Hz with amplitudes exceeding 50 μV were also removed.

Features were extracted from both time and frequency domains. Various feature types required distinct signal preprocessing steps, and the framework enabled the creation of parallel data processing pipelines: users could specify the preprocessing steps and parameters required for each pipeline. The output

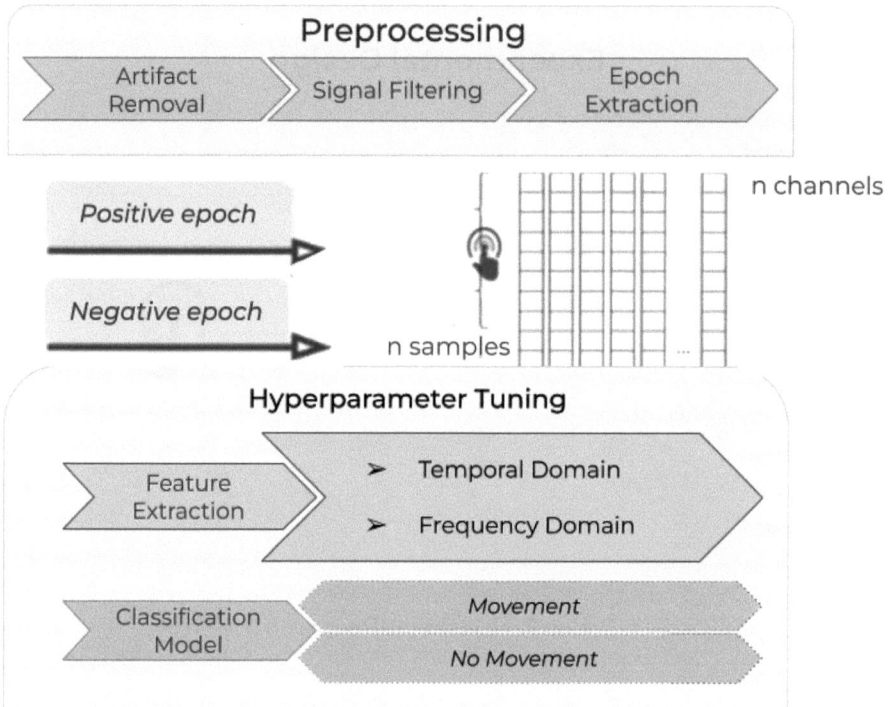

Fig. 3. System workflow for recognizing the intention to move

of each pipeline was then combined into a unified data array for input to the machine learning models. Detailed descriptions of specific preprocessing steps and parameters can be found in [12]. The task was defined as a binary classification problem, with the positive class representing epochs with movement and the negative class representing epochs without movement. Classification models were trained individually for each participant, and samples from different paradigms were not mixed; thus, two models were trained for each participant. Since the number of epochs with and without movements could differ due to the exclusion of artifact epochs, a downsampling approach was applied, where objects from the paradigm with the predominant number of epochs were randomly removed. Therefore, the number of positive and negative classes was balanced for both paradigms.

Accuracy and F-measure are commonly used metrics in binary classification tasks. However, these metrics can be overly optimistic about a model's generalization ability, especially on imbalanced data [13–16]. The primary evaluation metrics in this study were sensitivity and specificity. Sensitivity, or true positive rate (TPR), quantifies the proportion of epochs with a press that were accurately identified by the model as belonging to the positive class. Specificity, or true negative rate (TNR), measures the proportion of epochs without a press that the model correctly classified as the negative class. Decision thresholds for all models were calibrated to achieve a balance between sensitivity and specificity, with balanced accuracy serving as a primary metric due to its ability to reflect this trade-off. The decision threshold determines the classification of each subject: by selecting an optimal threshold from the decision function output, all decision scores below this threshold are assigned to the negative class, while those above are assigned to the positive class. Sensitivity, or true positive rate (TPR), quantifies the proportion of epochs with a press that were accurately identified by the model as belonging to the positive class. Specificity, or true negative rate (TNR), measures the proportion of epochs without a press that the model correctly classified as the negative class. Decision thresholds for all models were calibrated to achieve a balance between sensitivity and specificity, with balanced accuracy serving as a primary metric due to its ability to reflect this trade-off. The decision threshold determines the classification of each subject: by selecting an optimal threshold from the decision function output, all decision scores below this threshold are assigned to the negative class, while those above are assigned to the positive class.

The average balanced accuracy across participants was 74% for the 'control' paradigm and 77% for the 'oddball' paradigm. The improvement in classification accuracy in the 'oddball' paradigm is consistent with the results obtained in [11], suggesting the potential use of the proposed paradigm as a mechanism to enhance movement recognition quality.

Classification of Psychiatric Disorders. The second task chosen for the validation of the developed software package was the classification of patients with psychiatric disorders, specifically schizophrenia and obsessive-compulsive disorder (OCD).

Diagnosing psychiatric disorders at early stages is critically important as it can lead to timely treatment and improved quality of life for patients. One of the main challenges in clinical psychiatry is the lack of objective diagnostics for mental disorders. Emphasis is placed on methods that provide objective information about cognitive functions, enabling timely therapy to reduce the risk of disability in patients [17,18]. Neuropsychological tests are the most common diagnostic tools, yet they often result in inaccurate diagnoses. Machine learning methods can improve diagnostics by creating models that aid clinicians in refining diagnoses and providing useful information for selecting neurotherapy correction protocols based on biofeedback.

Event-related potentials (ERPs) have been extensively studied in the context of sensory and cognitive process disturbances in various psychiatric disorders [19]. ERPs are brain responses triggered by specific events, such as the presentation of a stimulus or the performance of an action. Tasks used to evoke these potentials encompass a wide range of human sensory, motor, and cognitive functions. The Go/NoGo task paradigm is particularly relevant for assessing the brain's action control system [20]. Previous studies of schizophrenia patients using this paradigm have shown reduced amplitudes of P300 and contigent negative (CNV) ERP waves associated with cognitive control processes [21–24]. A recent study employed visual stimuli across three categories: (a) 20 distinct animal images (designated as "A"), (b) 20 distinct plant images (designated as "P"), and (c) 20 images depicting individuals from various professions (designated as "H"). These visual stimuli were presented alongside auditory stimuli. The experimental design included four trial types: A-A, A-P, P-P, and P-H. Tasks were structured into four blocks of 100 trials each, where each block employed a unique set of five "A", five "P", and five "H" stimuli. Each block featured a pseudo-randomized presentation of 100 stimulus pairs, ensuring an equal probability for each stimulus category and trial type.

Before recording, participants practiced the task. After every 200 trials, participants rested for a few minutes. The instruction was to press a button with their right hand as quickly as possible for all A-A pairs and refrain from pressing for other pairs. The experimental design for this task is depicted in Fig. 4.

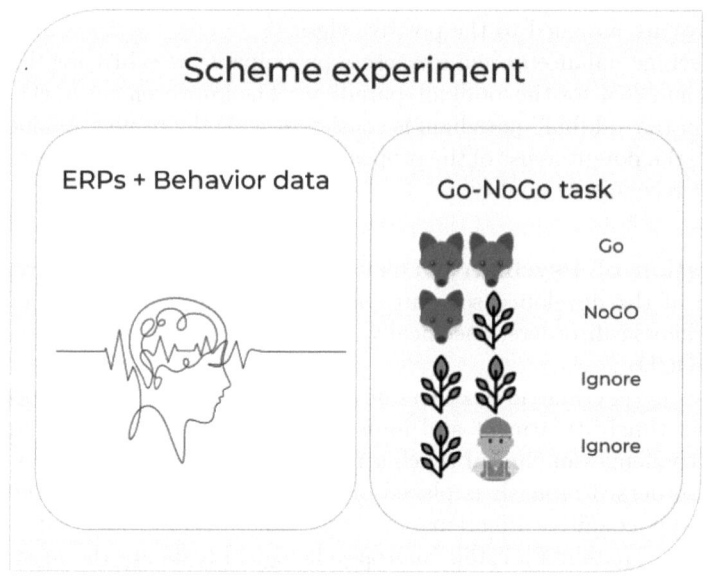

Fig. 4. System workflow for determining the presence of psychiatric disorders based on ERPs

During the described experiment, 19-channel ERPs were recorded. The trial duration was 3000 ms. ERP signals in four conditions were segmented into intervals with an average duration of around 200 ms, corresponding to the main ERP waves associated with cognitive processes, which showed the greatest differences between the participant groups.

The data processing workflow, illustrated in Fig. 5, comprised following stages:

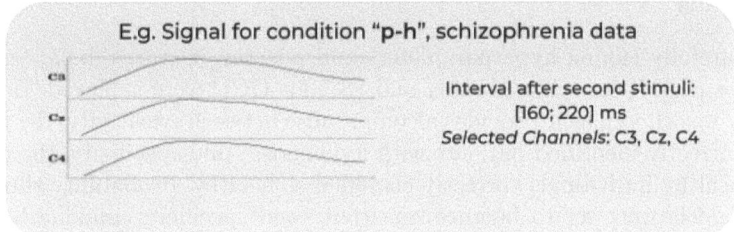

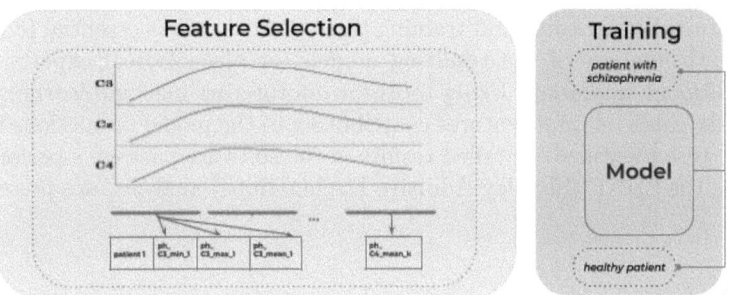

Fig. 5. System workflow for determining the presence of psychiatric disorders based on ERPs

- Data preprocessing, normalization, and the extraction of event-related potential (ERP) intervals demonstrating the most significant contrasts between control and test groups.
- The ERP signals from all channels and conditions were divided into overlapping time windows as per [25]. In each time window, minimum, maximum, and mean values were computed. Additionally, global mean, global maximum, and global minimum values were calculated for each signal, along with the window numbers where these values were reached. Since window characteristics depend on window size and shift, the feature extraction process is parameterized. These parameters are chosen independently for each condition and optimized along with the model hyperparameters.
- Feature selection was applied to combat overfitting and reduce training time. The total number of features can be quite large depending on the chosen window parameters, reaching up to 1443 features for the considered parameter set. The final feature selection method used was sequential feature selection, where the model is trained on the full set of features and features are iteratively removed until the model performance starts to degrade.
- Model training included hyperparameter optimization for models and feature extractors through grid search with cross-validation. Performance met-

rics were calculated as averages across cross-validation folds to ensure robust evaluation

By carefully tuning hyperparameters and selecting features, balanced accuracy rates of 91% for schizophrenia and 88% for OCD were achieved. Balanced accuracy was chosen as it provides a compromise between sensitivity, the proportion of correctly identified patients with a diagnosis, and specificity, the proportion of healthy individuals correctly classified as healthy. Probability thresholds for all models were set to balance sensitivity and specificity, making balanced accuracy an important metric.

Feature selection played a crucial role. It was necessary for two reasons. First, reducing the total number of features helps to combat overfitting and reduces computational complexity and training time. Second, it was essential to conduct an in-depth analysis of the resulting models for physiological experts to provide additional validation. Using tools for interpreting machine learning model results, the most relevant features contributing to the model predictions for each disorder were identified. Detailed results of the software package's performance, including the SHAP (Shapley Additive Explanations) analysis, are presented in [26,27].

Conclusion

We introduced a software framework aimed at advancing research in the application of machine learning techniques to electroencephalogram (EEG) signal analysis. Initially, the framework supported classification tasks; however, it was inherently flexible and could be extended to address a broader range of problem types.

The framework was validated through its application to two specific tasks: the detection of movement intent and the classification of psychiatric disorders in patients. The identified biomarkers associated with movements and disorders were employed in the development of brain-computer interfaces and in diagnosing cognitive control impairments. Furthermore, the results obtained from these studies provided valuable information for designing correction protocols for neurotherapy based on biofeedback. Such protocols trained subjects to regulate brain rhythms to address various issues, such as those encountered in cognitive psychology and in controlling mobile robotic systems.

Future work included plans to adapt the framework for neurofeedback-based control tasks.

References

1. Wolpaw, J., Wolpaw, E.W.: Brain–Computer Interfaces: Principles and Practice. Oxford University Press, Oxford (2012)
2. Juri, D.: Kropotov Quantitative EEG, Event-Related Potentials and Neurotherapy. Academic Press (2009)

3. Babenko, D.V., et al.: Modern Diagnostic Capabilities of cognitive disorders. Russian Family Doctor **24**, 35–44. https://doi.org/10.17816/RFD18986
4. Akar, S.A., Kara, S., Latifoglu, F., Bilgic, V.: Analysis of the complexity measures in the EEG of schizophrenia patients. Int. J. Neural Syst. **26**(2) (2016)
5. Kutepov, I.E., et al.: Complexity of EEG signals in schizophrenia syndromes. In: Proceedings of the 29th International Conference on Computer Graphics and Vision, vol. 2 (2019)
6. Li, Y., Gao, X., Liu, H., Shangkai, S.: Classification of single-trial electroencephalogram during finger movement. IEEE Trans. Biomed. Eng. **51**(6), 1019–1025 (2004). https://doi.org/10.1109/TBME.2004.826688
7. Wang, H.: Optimizing spatial filters for single-trial EEG classification via a discriminant extension to CSP: the Fisher criterion. Med. Biol. Eng. Comput. **49**, 997–1001 (2011). https://doi.org/10.1007/s11517-011-0766-7
8. Wang, H., Zheng, W.: Local temporal common spatial patterns for robust single-trial EEG classification. IEEE Trans. Neural Syst. Rehabil. Eng. **16**(2), 131–139 (2008). https://doi.org/10.1109/TNSRE.2007.914468
9. Minpeng, X., Xiao, X., Wang, Y., Qi, H., Jung, T.-P., Ming, D.: A brain-computer Interface based on miniature-event-related potentials induced by very small lateral visual stimuli. IEEE Trans. Biomed. Eng. **65**, 1166–75 (2018)
10. Plotnikov, S.A., Lipkovich, M., Semenov, D.M., Fradkov, A.L.: Artificial intelligence based neurofeedback. Cybernet. Phys. **8**(4), 287–291 (2025)
11. Knyazeva, V.M., Plakkhin, A.M., Aleksandrov, A.A.: Voluntary movements performance during the involuntary attention activation. J. Evol. Biochem. Physiol. **58**(5), 1604–1612 (2022)
12. Lipkovich, M., Knyazeva, V., Aleksandrov, A., Shanarova, N., Sagatdinov, A., Fradkov, A.: Evoked potentials detection during self-initiated movements using machine learning approach. In: 2023 Fifth International Conference Neurotechnologies and Neurointerfaces (CNN) (2023)
13. Sokolova, M., Japkowicz, N., Szpakowicz, S.: Beyond accuracy, F-Score and ROC: a family of discriminant measures for performance evaluation. In: Sattar, A., Kang, B. (eds.) AI 2006. LNCS (LNAI), vol. 4304, pp. 1015–1021. Springer, Heidelberg (2006). https://doi.org/10.1007/11941439_114
14. Gu, Q., Zhu, L., Cai, Z.: Evaluation measures of the classification performance of imbalanced data sets. In: Cai, Z., Li, Z., Kang, Z., Liu, Y. (eds.) ISICA 2009. CCIS, vol. 51, pp. 461–471. Springer, Heidelberg (2009). https://doi.org/10.1007/978-3-642-04962-0_53
15. Bekkar, M., Djemaa, H.K., Alitouche, T.A.: Evaluation measures for models assessment over imbalanced data sets. J. Inform. Eng. Appl. **3**(10), 27–38 (2013)
16. Akosa, J.S.: Predictive accuracy: a misleading performance measure for highly imbalanced data. In: Proceedings of the SAS Global Forum 2017 Conference, pp. 942–2017. SAS Institute Inc., Cary (2017)
17. Dzhos, Y.S., Kalinina, L.P.: Cognitive event-related potentials in neurophysiology research. https://doi.org/10.17238/issn2542-1298.2018.6.3.223
18. Zueva, I.B., Vanaeva, K.I., Sanets, E.L.: Cognitive evoked potential, P300 component: role in assessment of cognitive function among patient with arterial hypertension and obesity. UDK 616.83/.85:616.1:616-056.52
19. Barry, R.J., Johnstone, S.J.: Clarke AR A review of electrophysiology in attention-deficit/hyperactivity disorder: II. Event-related potentials. Clin. Neurophysiol. **114**, 184–198 (2003)

20. Mueller, A., Candrian1, G., Kropotov, J.D., Ponomarev, V.A., Baschera, G.-M.: Classification of ADHD patients on the basis of independent ERPs components using a machine learning system. Nonlinear Biomed. Phys. **4**(Suppl 1) (2010)
21. Ertekin, E., Üçok, A., Keskin-Ergen, Y., Devrim-Üçok, M.: Deficits in Go and NoGo P3 potentials in patients with schizophrenia. Psychiatry Res. **254**, 126–132. https://doi.org/10.1016/j.psychres.2017.04.052. Accessed 24 Apr 2017. PMID 28460282
22. Sun, Q., Fang, Y., Shi, Y., Wang, L., Peng, X., Tan, L.: Inhibitory top-down control deficits in schizophrenia with auditory verbal hallucinations: a Go/NoGo task. Front. Psychiatry **12**, 544746. https://doi.org/10.3389/fpsyt.2021.544746. PMID: 34149464; PMCID: PMC8211872
23. Kropotov, J.D., Pronina, M.V., Ponomarev, V.A., Poliakov, Y.I., Plotnikova, I.V., Mueller, A.: Latent ERPs components of cognitive dysfunctions in ADHD and schizophrenia. Clin. Neurophysiol. **130**(4), 445–453. https://doi.org/10.1016/j.clinph.2019.01.015. Accessed 1 Feb 2019. PMID: 30769271
24. Oribe N., et al.: Progressive reduction of visual P300 amplitude in patients with first-episode schizophrenia: an ERPs study. Schizophr Bull. **41**(2), 460–470 (2015). https://doi.org/10.1093/schbul/sbu083. Accessed 9 June 2014. PMID: 24914176; PMCID: PMC4332938
25. Mueller, A., Candrian1, G., Kropotov, J.D., Ponomarev, V.A., Baschera, G.-M.: Classification of ADHD patients on the basis of independent ERPs components using a machine learning system. Nonlinear Biomed. Phys. **4**(Suppl 1) (2010)
26. Shanarova, N., Pronina, M., Lipkovich, M., Kropotov, J.: Machine learning based diagnostics of schizophrenia patients. In: 6th Scientific School Dynamics of Complex Networks and their Applications (DCNA), Kaliningrad, Russian Federation, pp. 252–255 (2022). https://doi.org/10.1109/DCNA56428.2022.9923292
27. Shanarova, N., Pronina, M., Lipkovich, M., Ponomarev, V., Müller, A., Kropotov, J.: Application of machine learning to diagnostics of schizophrenia patients based on event-related potentials. Diagnostics (Basel) **13**(3), 509 (2023). https://doi.org/10.3390/diagnostics13030509. PMID: 36766614; PMCID: PMC9913945

A Review on Facial Expression Recognition Approaches, Datasets and Technologies

M. S. Lavanya[1], Vanishri Arun[1], Mayura Tapkire[2], and Yulia Shichkina[3(✉)]

[1] JSS Science and Technology University, Mysuru, India
[2] The National Institute of Engineering, Mysuru, India
[3] Petersburg State Electrotechnical University, Saint Petersburg, Russia
strange.y@mail.ru

Abstract. Face Expression Recognition (FER) has become challenging area in computer vision research due to its wide range of applications. The FER includes comprehending human behaviour, recognizing mental states and facilitating human-computer interaction. FER has become an important field in facial image processing due to its adaptability. The numerous applications of FER highlight the development of facial image processing systems. The FER aids in the comprehension of non-verbal clues, offering perceptions into people's intentions and attitudes in domains like human behaviour understanding. Numerous datasets with various features have been created for the purpose of training and assessing FER systems. This study discusses the datasets that the researchers frequently use, their characteristics and contents and how the data is generated. Additionally, gathering innovative technologies and diverse approaches that are used to datasets in order to achieve the best possible accuracy rate is the main feature of this review. The adaptability and impact of FER highlight the importance of developing advanced facial image processing systems. This review explores the challenges, datasets, and technologies associated with FER, providing a comprehensive overview of the current state and future directions in this evolving field. Every dataset offers insights on distinctive characteristics such subject demographic variety, feelings and annotation techniques. It provides information about the types of datasets that are currently accessible and suggests future paths for the development, application and assessment of datasets in the exciting field of facial expression recognition.

Keywords: FER · CK+ · FER-2013 · Affect Net · RAF-DB

1 Introduction

The face is the most noticeable feature of the human body and it can be used to determine a person's identity, age and even gender. Face recognition research has shown in numerous studies that facial expression identification is becoming more and more popular. Facial emotion detection has a wide range of possible uses, according to scientific studies. The Emotions are a fundamental and inherent aspect of human behaviour, significantly influencing our communication. People use a variety of cues to indicate their emotions, including body language and facial expressions. The most straightforward and significant of these non-verbal communication methods is facial expressions; they act as a universal language that can rapidly express a vast range of emotional states, sentiments and attitudes. They also help with a variety of cognitive tasks. Gaining a more profound comprehension of human behaviour requires the accurate analysis and interpretation of the emotional content of human facial expressions. Indeed, people may convey emotions, knowledge and intentions as well as control relationships and communication with others most effectively and naturally through their facial expressions [1]. Recognizing facial expressions is beneficial across a wide range of systems and applications and is crucial for achieving naturalistic interaction. Facial expressions play a key role in various cognitive tasks, making the accurate reading and interpretation of human emotions essential for a deeper understanding of the human condition. Consequently, the primary goal of facial expression recognition methods and approaches is to enable machines to automatically assess the emotional content of a human face. Endowing computer applications with the capability to recognize human emotional states through facial expressions is a highly important and challenging task with extensive applications [2]. Based on feature representations, FER systems can be classified into two primary groups: dynamic sequence FER and static image FER. The static-based methods [3–5,8] solely encode the feature representation using spatial information from the current single image. While dynamic-based approaches [6–8] take into account the temporal relationship among consecutive frames in the input facial expression sequence, Other modalities, including audio and physiological channels, have also been employed in multimodal systems [9] to aid in the recognition of expression, building on these two vision-based techniques. While facial image-based pure emotion detection can yield encouraging outcomes, integrating additional models into a high-level framework can offer supplementary insights and increase resilience. In recent years, extensive surveys on automatic expression analysis have been published [10–13]. An established set of common algorithmic pipelines for FER has been established by these surveys. But they emphasize conventional techniques and there haven't been many reviews of deep learning. The evolution of deep affect recognition from 2010 to 2017 was examined, with a focus on the merging of audiovisual and physiological sensors, in a very recent assessment on deep learning for human affect recognition [14]. In [15] carries out further focused and in-depth research on deep learning until 2019 for both static and dynamic FER challenges in this study. The goal is to provide an overview of the methodical framework and essential abilities for deep FER to someone who is new to this topic. Even with deep learning's potent feature learning capabilities, issues arise when using it for FER. To prevent over fitting, deep neural networks need a lot of training data. To train the well-known neural network with deep architecture that produced the most promising results in object identification tasks, however, the available facial expression databases are insufficient. Furthermore, there are significant inter-subject differences because of several individual characteristics, including age, gender, ethnicity and expressiveness level. Unrestricted facial expression scenarios frequently exhibit variances in position, illumination and occlusions, in addition to subject identification bias. Due to the nonlinear confusion these disruptions cause with facial expressions, deep networks are more important in order to handle the high intraclass variability and acquire efficient expression-specific representations. Advancements in computer vision technology have significantly improved the accuracy of emotion recognition in images. When images are captured under controlled conditions—such as consistent lighting, standardized backgrounds and predefined camera settings—the accuracy of emotion recognition algorithms can be exceptionally high. This level of precision is achieved because the controlled environment minimizes the variability and noise that can interfere with the algorithm's ability to correctly identify and classify emotions [16]. Despite the advancements in computer vision, emotion recognition in naturalistic conditions still faces significant challenges. One major issue is the high intra-class variation, which refers to the wide range of ways in which the same emotion can be expressed. For example, a smile can vary greatly in intensity, duration and the way it engages different facial muscles. These variations make it difficult for algorithms to consistently recognize the same emotion across different instances. Additionally, there is low inter-class variation, meaning that different emotions can sometimes appear very similar to one another. Subtle differences, such as the slight change in the corners of the mouth or the eyebrows, can be all that distinguishes one emotion from another, such as between a mild smile and a smirk. Furthermore, changes in facial pose—like head tilts or rotations—add another layer of complexity, as they can obscure or alter key facial features that the algorithm relies on to detect emotions. These factors combine to create a challenging environment for emotion recognition algorithms, making it difficult to achieve the same level of accuracy seen under controlled conditions. As a result, recognizing emotions in real-world, naturalistic settings remain an active area of research in computer vision. Numerous businesses that sell a wide variety of consumer goods use facial emotion detection technologies to discover potential clients and recognize customer feedback. The field of machine learning gains greater value from these facial expressions. In the sphere of education, face emotions are used to gauge student's comprehension levels. It is the student's liberty to instruct their student in a way that suits them best. They use both traditional and digital methods to deliver the lessons. The primary goal is for the student is to comprehend the material by the end of each class. Teachers can assess whether or not their students paid attention to the lecture by using facial emotion recognition software. The teacher may alter the way they present the lesson in response to the feedback derived from the facial emotion patterns. The advancement of facial emotion detection technology may also be crucial for hospitals and other healthcare facilities. Patients find it more difficult to seek medical assistance when they become immobile and medical workers must continuously watch them in order to assess how much pain they are experiencing. It is possible to identify and assess the patients' emotions through facial emotion recognition. Studies show that the current state of machine learning models can read six basic emotions: anger, sadness, happiness, disgust, fear, surprise and neutrality. After eliminating the neutral feeling from the list, the five emotions that are left can be classified as either positive (happy and surprise) or negative (anger, sadness, disgust and fear). It is still possible to improve the development of these fundamental feelings to reveal deeper and complex emotions. In healthcare environments, FER technology can serve as an additional tool for medical staff to monitor patient well-being. By analyzing facial expressions, FER systems can help identify emotions that may indicate discomfort, pain, anxiety, or distress. This is particularly useful in scenarios where patients are unable to communicate effectively due to conditions like stroke, dementia, or other neurological impairments. For example, if a patient displays facial expressions associated with pain or discomfort, the system could alert medical personnel to provide timely interventions, potentially improving patient outcomes and comfort. Additionally, FER can be integrated with other patient monitoring systems to provide a more comprehensive view of a patient's condition, contributing to more personalized and effective care. There is significant potential to improve FER technology by developing models that can recognize a broader range of emotions, including more complex and subtle states. Advanced models could integrate contextual information, such as body language or vocal cues, to better interpret the emotional state of a patient. Additionally, improvements in deep learning techniques, such as transformers and multi-modal learning, could enhance the accuracy and reliability of FER systems in diverse healthcare environments. Moreover, integrating FER with other biometric data, such as heart rate, blood pressure, or even brain activity, could provide a more holistic understanding of a patient's emotional and physical state. This multi-modal approach could lead to the development of more sophisticated tools for monitoring and assessing patient well-being. Facial emotion detection technology has the potential to transform patient care by providing real-time insights into patients' emotional states, particularly for those who are unable to communicate effectively. While current systems are capable of recognizing basic emotions, there is room for significant improvement to detect more complex and nuanced emotional states. As this technology evolves, it could become an invaluable tool in healthcare settings, enhancing patient monitoring and leading to more responsive and personalized care. However, the development and implementation of these systems must be carefully managed to ensure they are accurate, reliable, and used ethically. The structure of the paper is organized as follows. In Sect. 2, the various datasets on facial emotion recognition is presented. In Sect. 3, facial emotion recognition approaches and technologies are presented. After that, Sect. 4, Comparative metrics for evaluating various emotion method. Finally, Sect. 5 concludes the paper and draws directions for future work.

2 Facial Expression Databases

2.1 CK+

The CK+ datasets, which contain 593 photos taken from 123 individuals, were made available in 2000 to aid in the automatic recognition of facial expressions. This dataset consists of 593 picture sequences for 123 different people. Thirty-one percent of the images show men and six ninety-nine percent show women between the ages of 18 and 50 [17], which is shown in Fig. 1. The images in the CK + dataset were taken in a controlled laboratory setting. The Facial Action Coding System was used to label the datasets inside (FACS) [18]. Using the Active Appearance Model (AAM) the face and the facial features were traced from a video input. They also account for linear appearance variations. This method works well with images containing the object of interest. AAMs adapt the model to fit unseen images, ensuring accurate alignment. Their ability to capture both shape and appearance variations makes them a strong choice for image analysis tasks.

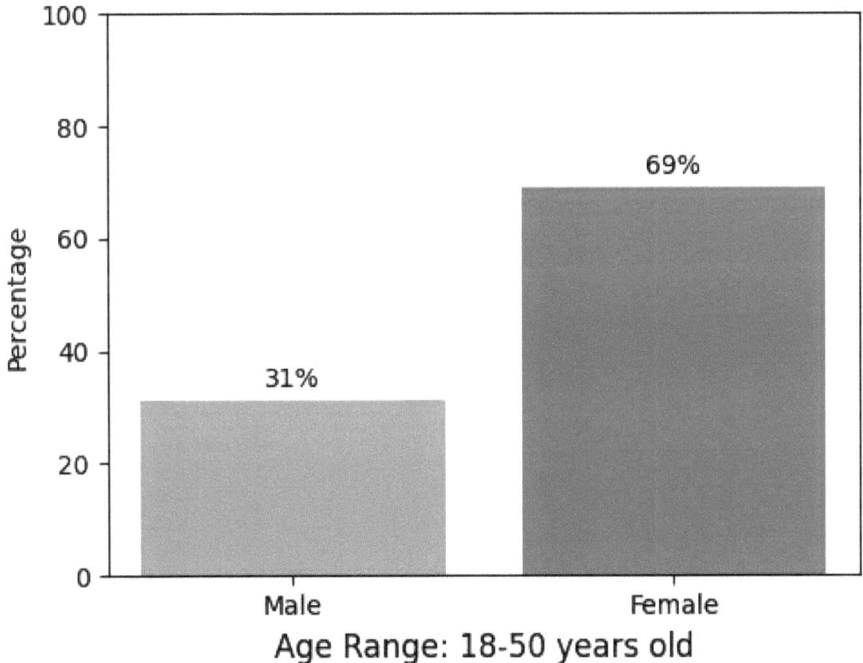

Fig. 1. CK+ Gender Distribution

2.2 MMI

A key tool in the fields of emotional computing, computer vision and facial expression recognition is the MMI (Montreal Multi-lingual Image) dataset, due to its thorough and exhaustive character, it has been widely used and cited across multiple research publications and journals [19].

2.3 JAFFE

The Japanese Female Facial Expression (JAFFE) database contains 213 images that are categorized into seven facial expressions plus one neutral inserted image shown in Fig. 2. Ten Japanese female models provided the data. Six emotive adjectives were used to rate each image by 60 Japanese participants. Planning and assembly of the database was done by Jiros Gyoba, Miyuki Kamachi and Michael Lyons. Anger, disgust, fear, happiness, sadness, surprise and disdain are among the feelings. The JAFFE database was meticulously planned and assembled by Jiros Gyoba, Miyuki Kamachi, and Michael Lyons. Their work involved capturing the images, ensuring consistent and accurate representation of each expression, and organizing the data for research purposes. The JAFFE database is widely used in emotion recognition research, psychological studies, and the development of facial expression analysis systems. Its well-defined set of expressions and standardized ratings provide a reliable basis for training and testing models in emotion recognition tasks. The database's focus on a specific

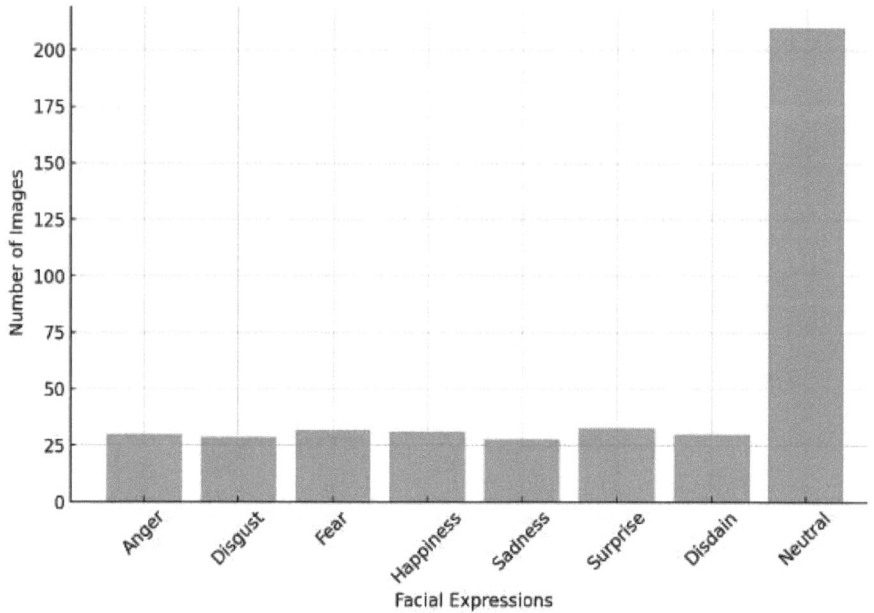

Fig. 2. Distribution of Facial Expression in JAFFE database

demographic (Japanese females) and the inclusion of both emotional and neutral expressions make it a valuable tool for studying cultural and gender-specific differences in facial expression and emotion perception. The JAFFE database, with its detailed and carefully rated images, offers a rich resource for exploring the nuances of facial expressions and advancing the understanding of emotional expression in various applications [20].

2.4 FER2013

Facial Expression Recognition (FER) is one of the most used datasets. The 35,887 images that make up the collection are sized to 48 by 48. There are 28,709 training images, 3589 validation images and 3589 testing images in FER2013. Three columns in FER2013 define each image. These columns are titled "Emotion type" and are numbered 0–6, with each column describing a different emotion: anger, disgust, fear, happiness, sadness, surprised and neutral [21]. The second column is an array of numerical values representing the images. The last column indicates the type of image whether a training or a testing data. The Fig. 3 represents the number of images in the respective dataset category (training, validation and testing) for each facial expression.

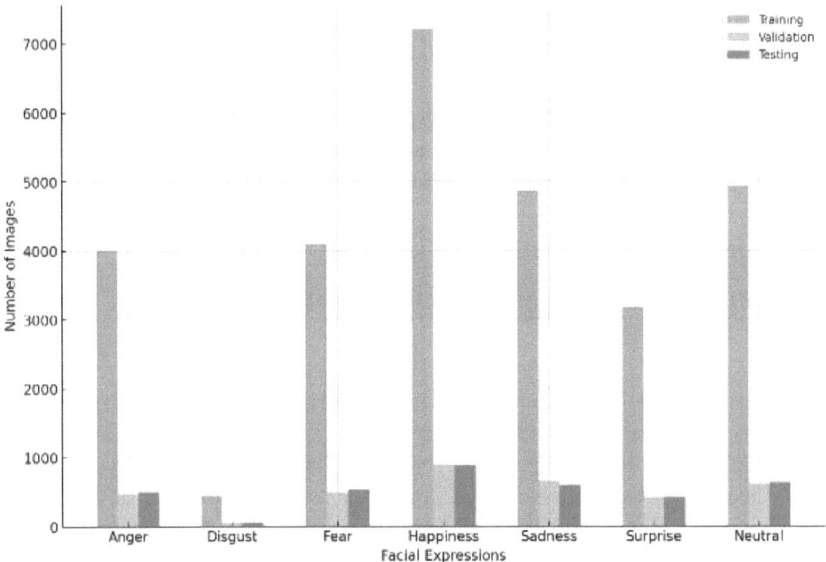

Fig. 3. Distribution of Facial Expression in FER2013 Dataset

2.5 RAF-DB

RAF-DB is a facial image database including 29,672 images. The images were personally annotated by forty individuals. The features of the image are varied with regard to gender, ethnicity and age. As suggested by the term RAF-DB, this database's objective is to classify complicated facial expressions from the actual world based on their facial expression type. In addition to fundamental emotions, RAFDB also takes into account compound emotions [22].

2.6 AffectNet

The shortcomings of earlier image datasets, which frequently disregarded the significance of valence and arousal in characterizing images within a continuous dimensional model, led to the creation of AffectNet. The Arousal denotes the degree of an emotion, whereas valence describes its positive or negativity, it offers a more sophisticated comprehension of emotions and facial expressions by integrating various elements, With more than a million facial images, AffectNet is the biggest database of its kind. Human annotators have carefully classified around half of these images with seven different face expressions: neutral, happy, sad, surprised, fearful, disgusted and angry. The dataset additionally contains annotations for 68 facial landmarks, which are significant areas of the face that are used to identify. It stands as the largest dataset of its kind, featuring over a million facial images. Of these images, approximately half have been meticulously annotated by human experts with seven distinct facial expressions: neutral, happy, sad, surprised, fearful, disgusted, and angry.

Additionally, AffectNet includes annotations for 68 facial landmarks. These landmarks are crucial for identifying and analyzing key facial regions and features, enhancing the accuracy of emotion recognition by providing detailed spatial information about the face. By integrating both expression labels and continuous emotion dimensions, AffectNet offers a more sophisticated tool for studying and understanding facial emotions [23].

3 Facial Emotion Recognition Approaches and Technologies

The process of recognizing facial emotions involves several distinct steps, starting from face detection and culminating in emotion classification. Face detection identifies and isolates the face by determining key features like eyes, nose and mouth. Once detected, specific facial landmarks are extracted to understand the face's geometric structure. The data may then undergo preprocessing, such as normalization and alignment, to enhance quality. Finally, Deep Learning algorithms analyze the preprocessed features to classify the emotion into categories like happiness, sadness and anger. Once the face is detected, facial landmarks are extracted to understand the geometric structure of the face. Facial landmarks are specific points on the face that correspond to key facial features and their relationships. This step involves: Landmark Detection: Algorithms such as the Dlib facial landmark detector or the MediaPipe Face Mesh are used to identify and map landmarks on the face. Commonly, 68 or 5 key landmarks are detected, including points around the eyes, eyebrows, nose, and mouth. Geometric Analysis: The positions and distances between landmarks are analyzed to capture the face's geometric configuration. This information is crucial for understanding facial expressions and movements [24].

3.1 Face Detection

The ability of a computer to recognize and find human faces in an image or video is known as face detection. A widely used and highly successful technique for face detection is the application of cascade classifiers based on Haar features. Due to their well-known simplicity and resilience, these classifiers are frequently used in face detection applications. By analyzing patterns of intensity differences between neighboring rectangular parts of an image, the Haar cascade classifier is able to recognize the existence of facial features like the mouth, nose and eyes. Due to its rapid image processing speed and accuracy in identifying facial regions, it is widely used in a variety of computer vision tasks [25]. The Haar Cascade was used to identify the mouth and the eyes only [26]. Viola Jones algorithm is also a well-known face detection model. It uses rectangular features to recognise the human face in the image [27]. Haar Features: Haar features are essentially simple rectangular features that measure the intensity differences between adjacent regions of an image. For example, one common Haar feature might measure the difference in pixel intensity between the area above the eyes and the area below the eyes. These features are designed to capture the presence of facial structures, such as the eyes, nose, and mouth.

Training with Positive and Negative Samples: The Haar cascade classifier is trained using a large number of images containing faces (positive samples) and images without faces (negative samples). During training, the classifier learns to recognize patterns associated with facial features by adjusting weights on the Haar features to improve detection accuracy.

Cascade of Classifiers: The "cascade" refers to a series of increasingly complex classifiers arranged in a sequence. Each stage of the cascade performs a specific check, starting with simple and fast calculations and progressing to more detailed evaluations. If a face is detected in a stage, the process moves to the next stage. If a face is not detected, the region is quickly discarded, improving processing speed.

Real-Time Detection: Haar cascade classifiers are known for their fast processing speed, making them suitable for real-time applications. The cascade structure allows for rapid rejection of non-face regions, focusing computational resources on areas more likely to contain faces. Face detection technologies, such as Haar cascade classifiers and the Viola-Jones algorithm, have significantly advanced the ability to identify and locate human faces in images and video streams. Haar cascade classifiers utilize simple rectangular features and a cascade structure to achieve fast and accurate face detection. The Viola-Jones algorithm, leveraging Haar-like features, integral images, and AdaBoost, provides a robust framework for real-time face detection. Both techniques have made substantial contributions to the field of computer vision and continue to be integral in various practical applications.

3.2 Face Extraction

The crucial areas of a face that define the form and location of vital biometric features like the mouth, nose and eyes are referred to as features. For a variety of facial analysis tasks, such as recognition and expression analysis, these traits are essential. By using clustering technologies, related features can be grouped together to improve the efficiency and accuracy of facial analysis. K-means algorithm is one of the most used clustering approaches. This technique effectively groups comparable facial features together by partitioning the data into K separate clusters based on feature similarity. For tasks like face identification, recognition and emotion classification, K-means helps to organize and analyze facial features more effectively by maximizing the variation across clusters and limiting the variance within each cluster. Facial features play a pivotal role in defining the unique identity and expressions of an individual. For instance: Eyes: The shape, size, and spacing of the eyes are unique to each person and are key indicators of identity and emotional state. Nose: The structure and position of the nose help in differentiating individuals and contribute to the overall geometry of the face. Mouth: The shape and movements of the mouth are crucial for recognizing speech, emotions, and expressions [28]. In facial analysis, clustering techniques are often employed to group similar features together, which can enhance the efficiency and accuracy of the analysis. Clustering helps in organizing the data in such a way that similar features are grouped into clusters, making it easier for algorithms to process and analyze the face.

3.3 Data Augmentation

Data augmentation strategies are used to increase the diversity and quality of the training dataset in order to produce the best possible results in emotion categorization. Increasing the contrast and brightness of the images in the dataset is a typical strategy that helps the model become more resilient to varying lighting conditions. Resizing the images to a standard size also guarantees consistency throughout the dataset, which contributes to better model training. Researchers additionally employ diverse filtering methodologies and edge detection methods to augment the characteristics present in the images. Sobel edge detection is a well-liked technique that makes it simpler to recognize and examine the facial features essential for precise emotion classification by emphasizing the edges through intensity fluctuations. These data enhancement techniques collectively contribute to the creation of a more comprehensive and diverse dataset, leading to improved performance of emotion recognition models. One specific application of data augmentation is the generation of synthetic facial expressions. This involves creating new, realistic facial expressions based on the existing data to enrich the training dataset. By doing so, models are exposed to a broader range of facial variations, helping them learn to recognize and interpret subtle nuances in expressions that might not be well-represented in the original dataset. Synthetic facial expressions can be generated through various methods, such as geometric transformations (like rotation, scaling, and flipping), adding noise, or using more advanced techniques like generative adversarial networks (GANs) and facial morphing. These methods can produce new facial expressions that maintain the original emotional content but vary in aspects like angle, lighting, or intensity. This variation is crucial for training models that are more resilient to real-world variability, such as different lighting conditions, camera angles, and individual facial characteristics [29].

3.4 Emotion Classification

The Convolutional neural networks (CNN) facilitate and expedite the analysis of facial emotions in a manner similar to that of human brains [30]. CNN learns the data patterns. It could be colored, brilliant, or black. CNN is a multi-layer system that can identify facial landmarks as well. All that is visible to the computer is the two-dimensional array of a pixel-based image. After numbering each pixel, they are compared to see if they match or not. CNN then compares groups or sections of the photos instead of the entire image. CNN uses Rectified Layer Units (ReLU) for filtering, pooling, normalizing and layering. CNN is regarded as a very reliable method for a range of image and object recognition tasks due to its strong feature extraction capabilities from pictures and hierarchical structure.

3.5 State-of-the-Art Technologies in FER
3.5.1 Self-attention and Transformers

These mechanisms focus on the most relevant parts of an image or a sequence, allowing models to pay attention to critical regions of a face. Transformers have started to outperform traditional CNNs in several tasks, including FER and Spatial Attention method highlights important facial regions such as eyes, mouth and eyebrows, which are crucial for recognizing expressions. In FER, transformers have started to outperform traditional CNNs because of their ability to integrate information from across the entire image, rather than just local patches. This capability is particularly useful for recognizing facial expressions, where the context provided by different facial features (e.g., how the mouth's shape interacts with the eyes' squint) is crucial for accurate interpretation. Spatial Attention is a technique that can be incorporated within transformer models to further enhance their focus on important facial regions. In the context of FER, spatial attention mechanisms highlight areas like the eyes, mouth, and eyebrows. These regions are critical for distinguishing between different expressions, as they tend to exhibit the most significant variations in facial expressions. By integrating spatial attention, transformers can better emphasize these crucial facial areas, improving their ability to recognize and classify expressions accurately. This targeted focus allows the model to ignore less relevant background information or less informative regions of the face, leading to more precise and reliable emotion recognition [31].

3.5.2 GANs

These are used for data augmentation, generating synthetic facial expressions to enrich training datasets and for creating more robust and diverse training examples. The generator's role is to create synthetic data that resembles real data. It starts by taking random noise as input and tries to transform it into something that looks like the real data (e.g., an image of a face). The two networks are trained simultaneously in a zero-sum game. The generator improves by learning to create more realistic data that can fool the discriminator, while the discriminator improves by getting better at distinguishing real from fake data. Over time, the generator becomes adept at producing highly realistic synthetic data that is almost indistinguishable from the real data. GANs are used to create synthetic data to augment training datasets, especially in scenarios where collecting real data is expensive or difficult. For example, in facial recognition or emotion detection, GANs can generate diverse facial expressions, poses, and lighting conditions to improve model training [32].

4 Comparative Metrics for Evaluating Various Emotion Methods

4.1 Accuracy

A model may be constructed that cannot be replicated using a different technique for a given emotion, dataset, training set and testing set [12]. There are several aspects that affect a model's accuracy, especially when it is evaluated using a confusion matrix. These considerations include data quality, model selection and more. The dataset's size, quality and balance are all very important; well-balanced, large enough and high-quality datasets guarantee improved model performance. The relevance of the data given into the model is improved by feature engineering and selection as well as thorough data pretreatment. Accuracy can be greatly impacted by selecting the right model and optimizing its hyperparameters because different models and configurations work better in various situations. Regularization techniques penalize overly complex models, hence preventing overfitting. Important factors include the size of the batch, the number of epochs, the training procedures and the application of ensemble methods. A more trustworthy appraisal of the model's performance is ensured by evaluation strategies including cross-validation and the use of a distinct validation set. Understanding and fixing the model's faults is made easier with a thorough examination of the confusion matrix, adding domain expertise, using data augmentation and utilizing transfer learning are further strategies that help reach greater accuracy. As a result, increasing model accuracy is a complex process that calls for the careful optimization of a number of variables pertaining to the model, dataset, training protocols and assessment techniques.

4.2 Average Processing Time

Real-time emotion recognition plays a vital role in determining the effectiveness of a model. To fully assess the model's performance, it's important to consider not just its accuracy but also the average processing time it requires. These two factors together provide a comprehensive measure of the model's efficiency and reliability. In practical applications, especially in challenging environments like outdoor settings or remote locations, hardware such as the Raspberry Pi is commonly used due to its affordability, portability, and energy efficiency. However, these devices have limited processing power, which necessitates a careful balance between maintaining high accuracy and ensuring fast processing times. This compromise is essential to ensure that the model can perform well under real-world conditions where quick, reliable emotion recognition is crucial. Successfully navigating this trade-off is key to deploying effective emotion recognition systems in scenarios where both speed and accuracy are critical.

4.3 Quality of Datasets

Datasets are foundational to the development and evaluation of machine learning models, particularly in tasks like emotion recognition, where accuracy is paramount. The quality and accuracy of the data used for training and testing directly influence the model's performance. Properly labelled and well-curated datasets are essential for ensuring that the model can learn and generalize effectively to new data. However, not all datasets meet these standards. For example, FER2013, despite being a widely used and large dataset for facial expression recognition, has several shortcomings that complicate its use. One of the primary issues is its unbalanced data distribution, where certain emotions are overrepresented while others are underrepresented. This imbalance can bias the model, making it more likely to correctly predict the dominant classes while struggling with the less frequent ones. Moreover, FER2013 contains a significant number of invalid samples, which further hinders model training and evaluation. These invalid samples include non-face images that are erroneously included in the dataset, poorly cropped faces where essential facial features are not fully visible, and mislabelled expressions that do not accurately represent the intended emotion. These inaccuracies can mislead the model during training, leading to poorer generalization and reduced performance in real-world applications. The presence of these issues in datasets like FER2013 highlights the importance of careful data curation. Ensuring that datasets are well-defined, balanced, and free from errors is crucial for developing robust and accurate models. Without addressing these challenges, the effectiveness of emotion recognition systems can be significantly compromised, leading to unreliable results in practical applications [33]. It is common to get confused when classifying facial emotions, especially when emotions like surprise and terror are involved. These emotions are thought to be opposites, yet because of their comparable facial expressions, they frequently cause classification errors. It is difficult for models to reliably discriminate between them because of their similar traits [34].

5 Conclusion and Future Work

Natural human expressions are blends of underlying emotions, and models designed for automated emotion processing should account for this complexity. Traditional classification techniques typically assign a single emotion label, but this can be inaccurate when the emotional content is ambiguous or mixed. Instead of relying on single-label assignments, these expressions should be described using methods that generate multiple emotion hypotheses. This approach better captures the nuanced nature of emotional expressions. Research on facial emotion identification has come a long way, with a variety of datasets, face detection algorithms and classification strategies available that are customized to meet the unique goals of individual studies. Selecting a trustworthy dataset is essential since it has a direct impact on the models' accuracy. The nature and size of the data are important factors to take into account; larger datasets typically offer more thorough coverage and result in higher accuracy. Image augmentation techniques are frequently used to further enhance the visibility of face characteristics. By producing variations of the current images, these strategies improve the dataset and aid in the model's improved generalization. For the classification of facial emotions, CNNs are widely used due to their effectiveness in image recognition tasks. The development of robust classifiers within these networks is a focal point to enhance accuracy levels. Researchers aim to fine-tune these classifiers to distinguish between subtle differences in facial expressions more accurately. It is recommended that future research concentrate on mapping the different components of facial recognition, such as datasets, face detection algorithms and classification techniques, in order to fully comprehend their interdependencies and effects. This all-encompassing method would clarify how the features of datasets impact each stage's performance in the face emotion recognition pipeline and vice versa. Through this process, scientists can pinpoint optimal methodologies and possible avenues for enhancement, ultimately resulting in enhanced efficaciousness and precision of emotion detection systems.

Conflicts of interest. The authors have no conflicts of interest to declare.

References

1. Pantic, M.: Facial expression recognition. In: Encyclopedia of Biometrics, pp. 400–406. Springer, New York (2009)
2. Liebold, B., Richter, R., Teichmann, M., Hamker, F.H., Ohler, P.: Human capacities for emotion recognition and their implications for computer vision. i-com **14**(2), 126–137 (2015)
3. Shan, C., Gong, S., McOwan, P.W.: Facial expression recognition based on local binary patterns: a comprehensive study. Image Vis. Comput. **27**(6), 803–816 (2009)
4. Liu, P., Han, S., Meng, Z., Tong, Y.: Facial expression recognition via a boosted deep belief network. In: Proc. IEEE Conf. Comput. Vis. Pattern Recognit., pp. 1805–1812 (2014)

5. Mollahosseini, A., Chan, D., Mahoor, M.H.: Going deeper in facial expression recognition using deep neural networks. In: Proc. IEEE Winter Conf. Appl. Comput. Vis., pp. 1–10 (2016)
6. Zhao, G., Pietikainen, M.: Dynamic texture recognition using local binary patterns with an application to facial expressions. IEEE Trans. Pattern Anal. Mach. Intezlligence **29**(6), 915–928 (2007)
7. Jung, H., Lee, S., Yim, J., Park, S., Kim, J.: Joint fine-tuning in deep neural networks for facial expression recognition. In: Proc. IEEE Int. Conf. Comput. Vis., pp. 2983–2991 (2015)
8. Zhao, X.: Peak-piloted deep network for facial expression recognition. In: Leibe, B., Matas, J., Sebe, N., Welling, M. (eds.) ECCV 2016. LNCS, vol. 9906, pp. 425–442. Springer, Cham (2016). https://doi.org/10.1007/978-3-319-46475-6_27
9. Corneanu, C.A., Simon, M.O., Cohn, J.F., Guerrero, S.E.: Survey on RGB, 3D, thermal and multimodal approaches for facial expression recognition: history, trends and affect-related applications. IEEE Trans. Pattern Anal. Mach. Intell. **38**(8), 1548–1568 (2016)
10. Zeng, Z., Pantic, M., Roisman, G.I., Huang, T.S.: A survey of affect recognition methods: audio, visual and spontaneous expressions. IEEE Trans. Pattern Anal. Mach. Intell. **31**(1), 39–58 (2009)
11. Sariyanidi, E., Gunes, H., Cavallaro, A.: Automatic analysis of facial affect: a survey of registration, representation and recognition. IEEE Trans. Pattern Anal. Mach. Intell. **37**(6), 1113–1133 (2015)
12. Pantic, M., Rothkrantz, L.J.M.: Automatic analysis of facial expressions: the state of the art. IEEE Trans. Pattern Anal. Mach. Intell. **22**(12), 1424–1445 (2000)
13. Fasel, B., Luettin, J.: Automatic facial expression analysis: a survey. Pattern Recognit. **36**(1), 259–275 (2003)
14. Rouast, P.V., Adam, M., Chiong, R.: Deep learning for human affect recognition: insights and new developments. IEEE Trans. Affective Comput. (2019, to be published). https://doi.org/10.1109/TAFFC.2018.2890471
15. Valstar, M.F., Mehu, M., Jiang, B., Pantic, M., Scherer, K.: Metaanalysis of the first facial expression recognition challenge. IEEE Trans. Syst. Man Cybern. B Cybern. **42**(4), 966–979 (2012)
16. Sariyanidi, E., Gunes, H., Cavallaro, A.: Automatic analysis of facial affect: a survey of registration, representation and recognition. IEEE Trans. Pattern Anal. Mach. Intell. **37**(6) (2015). https://doi.org/10.1109/TPAMI.2014.2366127
17. Kalsum, T., Majid, M., Anwar, S.: Emotion recognition from facial expressions using hybrid feature descriptors, The Institute of Engineering and Technology, pp. 1003–1012 (2019)
18. Lucey, P., et al.: The extended Cohn-Kanade DATASET (CK+): a complete dataset for action unit and emotion-specified expression. In: 2010 IEEE Computer Society Conference on Computer Vision and Pattern Recognition - Workshops, San Francisco, CA, pp. 94–101 (2010)
19. Abiodun, R., Makanju, A.O.: A survey of approaches and challenges in real-time facial expression recognition in video. J. Vis. Commun. Image Represent. (2020)
20. Zhongzhao, X., Li, Y., Wang, X., Liu, Z.: Convolutional neural networks for facial expression recognition with few training samples. In: 37th Chinese Control Conference (CCC), China, pp. 1–6 (2018)
21. Fan, Y., Lam, J.C.K., Li, V.O.K.: Multi-region ensemble convolutional neural network for facial expression recognition. In: Kůrková, V., Manolopoulos, Y., Hammer, B., Iliadis, L., Maglogiannis, I. (eds.) ICANN 2018. LNCS, vol. 11139, pp. 84–94. Springer, Cham (2018). https://doi.org/10.1007/978-3-030-01418-6_9

22. Jyoti, S., Sharma, G., Dhall, A.: Expression empowered residen network for facial action unit detection. In: IEEE International Conference on Automatic Face and Gesture Recognition 2019, Lille, France, pp. 262–269 (2019)
23. Tautkute, I., Trzcinski, T., Bielski, A.: I know how you feel: emotion recognition with facial landmarks. In: Proceedings of the IEEE Conference on Computer Vision and Pattern Recognition Workshops, pp. 1878–1880 (2018)
24. Giannopoulos, P., Perikos, I., Hatzilygeroudis, I.: Deep learning approaches for facial emotion recognition: a case study on FER-2013. In: Advances in Hybridization of Intelligent Methods, Smart Innovation, Systems and Technologies, vol. 85
25. Aljaloud, A.S., Ullah, H., Alanazi, A.: Facial emotion recognition using neighborhood. Int. J. Adv. Computer Sci. Appl. **11**(1), 299–306 (2020)
26. Yang, D., Alsadoon, A., Prasad, P.W.C., Singh, A.K., Elchouemi, A.: An emotion recognition model based on facial recognition in virtual learning environment. Procedia Computer Sci. **125**, 2–10 (2018)
27. Dandil, E., Ozdemir, R.: Real-time facial emotion classification using deep learning data science and applications **2**(1), 13–17 (2019)
28. Farhang, Y.: Face extraction from image based on K-means clustering algorithms. Int. J. Adv. Comput. Sci. Appl. **8**(9) (2017)
29. Yang, K., Ahuja: Detecting faces in images: a survey. IEEE Trans. Pattern Anal. Mach. Intell. **24**(1), 34–58 (2002)
30. Ozdemir, M.A., Elagoz, B., Alaybeyoglu, A., Sadighzadeh, R., Akan, A.: Real time emotion recognition from facial expressions using CNN architecture. In: Medical Technologies Congress (TIPTEKNO) Izmir, Turkey, pp. 1–4 (2019)
31. Ma, F., Sun, B., Li, S.: Facial expression recognition with visual transformers and attentional selective fusion. IEEE Trans. Affect. Comput. **14**(2), 1236–1248 (2023)
32. Zhang, X., Zhang, F., Xu, C.: Joint expression synthesis and representation learning for facial expression recognition. IEEE Trans. Circuits Syst. Video Technol. **32**(3), 1681–1695 (2022)
33. Nguyne, H.-D., Yeom, S., Lee, G.-S., Yang, H.-J., Na, I.-S., Kim, S.-H.: Facial emotion recognition using an ensemble of multi-level convolutional neural networks. Int. J. Pattern Recognit Artif Intell. **33**(11) (2019)
34. Tian, Y.-I., Kanade, T., Cohn, J.F.: Recognizing action units for facial expression analysis. IEEE Trans. Pattern Anal. Mach. Intell. **23**(2), 97–115 (2001)

An Open Source Laboratory Information Management System for Biobank Optimization in a Robotic Environment

Anna Nozdracheva[1], Alexey Nozdrachev[2], Larisa Rybak[4(✉)],
Vladislav Cherkasov[3], and Dmitry Malyshev[3]

[1] Gamaleya National Research Center for Epidemiology and Microbiology,
18 Gamaleya Street, 123098 Moscow, Russia
[2] Institute of Precision Mechanics and Computer Science S.A. Lebedev,
51 Leninsky Prospekt, 119991 Moscow, Russia
[3] Belgorod State Technological University n.a. V.G. Shukhov, 46 Kostyukova Street,
308012 Belgorod, Russia
[4] Federal Research Center "Computer Science and Control" of the Russian Academy
of Sciences (FRC CSC RAS), 44/2, Vavilova Street, 119333 Moscow, Russia
rlbgtu@gmail.com

Abstract. Population and cohort studies have become one of the main directions of biomedical research, the resource support of which is associated with the active development of a network of biobanks and extensive collections of biosamples. The key task of information support of modern biobanks is the effective management of information about biosamples and their donors, results of laboratory analyses, as well as optimization of human-robotic interaction while complying with ethical requirements and industry standards for quality assurance of the pre-analytical stage of research. This research applies the concept of "specimen life cycle traceability" to build a comprehensive biobank information management system based on analyzing user requirements, ethical and legal principles, and the requirements of industry standards for biosample quality. A laboratory information management system (LIMS) that integrates the management of research projects, biosamples, clinical information, quality control, and multidimensional information requests is established and demonstrated.

Keywords: Laboratory Information Management System · Biobank · Biobanking · Database · Aliquoting · Robotic System

1 Introduction

Biobanking is one of the spheres of human activity undergoing rapid growth and is a source of biomaterial for various areas of research - from genomic technologies, to the creation of highly effective means of diagnosis and treatment of pathological conditions in all areas of medicine. A biobank is an organization that receives, stores, processes and/or distributes samples according to available requests. It contains data not only on the place of actual storage of biosamples, but also information on biomaterial donors and a full range of operations of the pre-analytical stage of medical research [1]. Most biobanks store human samples (all available tissue types), but there are banks of samples obtained from animals and from environmental objects (e.g., wastewater) [2]. Biobanks have become widespread as a material base for medical research, including infectious pathology. The availability of passported collections makes it possible to organize relevant studies in a relatively short time and guarantee their reliability and reproducibility [3].

Research of herd immunity to various infectious agents, including those potentially associated with biothreats [4], for example COVID-19 [5], have become widespread. There are both small and quite extensive collections of biological samples collected either for specific nosocomial or population studies or due to diagnostics in medical and preventive organizations [3]. Technical means of automation and information-analytical systems that allow data management are being introduced everywhere to optimize the work of biobanking and biosample collections. This helps to improve the quality of the pre-analytical phase of the research and minimize the risk of errors [6].

The Laboratory Information Management System (LIMS) allows the implementation of a quality management system and is the center of information management within the biobank. This meets the requirements of international organizations (including ISBER) for biobank operation [1,7–9]. However, most LIMS used in a biobank for data management are simple sample accounting systems with separate functions [1,6]. The ability of these programs to integrate new robotic systems for biobanking process optimization into the operation is significantly limited, the function of providing control of biomaterial storage conditions, and the implementation of coordination of research projects is missing. It should be noted that unified requirements for the management system and IT infrastructure that underlies the biobank have not been developed to date [10]. There are many commercial open source LIMS, each of which has its own specificity and is primarily focused on the clinical laboratory. Significantly fewer solutions are offered for the needs of biobanks. An important limitation of the introduction of specialized software tools is that the cost of LIMS for biobanks is often higher than what can be afforded by an individual laboratory collecting samples for research.

In addition, the vast majority of biobank management systems on the market are mass-produced, and research users cannot adjust the administrative approval process and technological workflow in time to meet the needs of research. This quickly renders such systems obsolete, causes inconvenience and loss of money [6]. An interesting technological solution in the IT of biobanking is the creation of digital copies of a real biobank or virtual biobanks that contain all the information accompanying the actual storage of biosamples [11]. This approach allows scaling up the research conducted in the biobank by including data from other interested laboratories and collaborating clinical centers, analyzing the array of data collected from different sources, organizing access to the stored data for other researchers, etc. The article proposes a technological solution for information support of the population biobank of infectious pathology, including the use of a robotic system for aliquoting biomaterial. The materials of such a biobank are used to conduct studies of herd immunity. The main type of biomaterial is serum and blood plasma, because marker molecules, including specific antibodies, are preserved in it longer than in whole blood during low-temperature storage [12]. At the same time, the process of serum separation from the blood clot and the subsequent aliquoting are the stages most exposed to various risks: from contamination of equipment and contamination of personnel, to the occurrence of errors in determining the correspondence of the biomaterial to its donor. To automate this process, an aliquoting system can be used, which requires the organization of automatic entry of aliquoting results into LIMS.

At the present stage, aliquoting system have been developed that allow aliquoting of serum that has undergone primary processing in the form of separation from the blood clot. In addition, the software of such stations has a closed architecture, users have limited access to program components. To unify the technological process of aliquoting and improve its quality, the creation of an intelligent robotic system (RS), taking into account the variation of the volume of initial samples and the presence of different phases of liquid in them, it is characterized in more detail in other papers [13–15].

The work describes in detail the created laboratory information management system (LIMS) of the biobank, which allows to take into account the peculiarities of the organization of the study of herd immunity in relation to actual infectious agents, as well as to use the developed RS of automated aliquoting of serum or blood plasma from tubes with whole blood.

The aim of the research is to develop an effective, ergonomic, compliant with ethical principles and industry quality standards information system of the population biobank under the conditions of using an aliquoting system to improve the quality of the pre-analytical stage of medical research.

2 Overview About the Technological Process of Biobanking

The following basic principles of organization of information system to ensure the technological process of population biobank are formulated:

- functional flexibility and compliance with industry standards. The system should take into account the regulatory requirements with the possibility of editing when they change and updating in the future. In addition, there should be an option to connect the software of high-tech equipment used in the practice of biobanking (ELISA-analyzers, optical density scanners, nucleic acid amplifiers, etc.), including aliquoting system.
- ensuring traceability of the biosample life cycle within the biobank. To ensure the reliability and reproducibility of studies using biosamples, each biosample must be selected and stored in compliance with the quality requirements of

the pre-analytical stage of the research. For this purpose, the information system should be able to trace all conditions of storage and processing of the biosample from the moment of receipt to disposal, which can be realized by ensuring the continuity and stage of biosample data entry, as well as the option to view related data.
- accordance of the system with ethical and legal requirements. Storage of biosamples in the biobank should comply with the requirements of legislation in the field of personal data protection of biosample donors. Entering data on biomaterial donors into the information system should be accompanied by entering details of documents ensuring the legality of information collection. The system should block the possibility of using unauthorized personal data, as well as unauthorized access and copying of stored data.
- ensuring automatic exchange of data and duplication of channels for their input into LIMS by the operator (in manual mode) and by the RS (in automatic mode). LIMS should store and structure all information in the biobank regardless of the source of its receipt and the method of its introduction (manually or with the help of high-tech equipment). Introduction of biomaterial aliquoting results into several serum aliquots from one blood sample can be performed both manually by the operator through the user interface and by means of the MS in automatic mode without the operator's participation.

The technological process of biobanking includes the following main **stages**:

Stage 1 - receiving biosamples from external organizations, entering data about them into the information system. The data on each donor of biosample can be different and includes the following main headings: living place, age, gender, information about infectious and somatic diseases, immunization status, etc. The entering of these data is inherently connected with ethical and legal principals. According to the requirements of the Declaration of Helsinki of the World Medical Organization, biomaterial can be collected only in case of compliance with the legislation in the field of personal data protection and informed voluntary consent of the donor to the collection of biomaterial. In this regard, the receipt of human blood samples to the biobank can be carried out only in the presence of documents confirming the legality of the collection. These may include an agreement or contract on scientific cooperation with the organization that directly collects the biomaterial, the decision of the local ethical committee on the compliance of the research with ethical standards, the presence of voluntary informed consent of the donor for the collection and storage of biomaterial.

Stage 2 - aliquoting the biomaterial into smaller volume portions. The necessity of this stage is associated with the detrimental effect of repeated freeze-thaw cycles on the quality of the biomaterial. In this regard, different aliquots of the same sample can be used for different studies, thus reducing costs. The results of aliquoting (number and volume of aliquots) should be entered into the LIS either manually or automatically. In accordance with the first principle of LIS organization, the development of the technological process of biomaterial aliquoting by means of MS within the biobank includes high-precision preparation of the sample for storage in accordance with international requirements of modern biobanking (ISO 20387 "Biotechnology. Biobanking. General requirements", ISO/TR 22758 Biotechnology - Biobanking - Implementation guide for ISO 20387, ISO 21899 Biotechnology - Biobanking - General requirements for the validation and verification of processing methods for biological material in biobanks).

The tubes are filled with dark red blood clot in the lower part and translucent blood serum in the upper part. An important difference of the designed RS is the possibility of aliquoting from these tubes by taking only blood serum without blood clot particles (erythrocytes). The design of the RS [14,15] is shown in Fig. 1. The PC is equipped with a vision system based on the YOLO neural network, which allows you to automatically move tripods and take serum from test tubes, as well as determine the amount of serum available for digging.

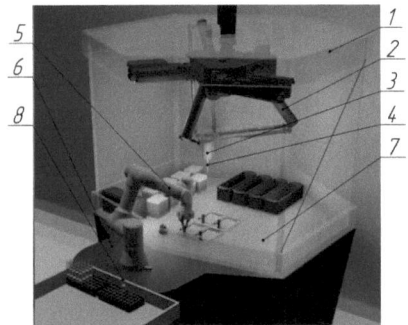

Fig. 1. A robotic system for aliquoting biological material.

The RS includes: housing 1, in which a parallel Delta robot 2 is located, moving a dosing device 3 [13], which performs dosing of biological fluid into equal aliquots. The replaceable tip 4 on the remote control is fixed with an O-ring. The 6-DOF serial manipulator 5 is mounted on a fixed base 6 and provides movement of the rack with tubes 8 using a gripper device within the clean zone 7.

Since each blood sample is taken from a specific donor, the identity of the aliquots formed by the RS is to be accounted for. In some cases, blood samples are received from external organizations in the form of numbered tubes with biomaterial and a list of persons with relevant information about them (on paper or electronic media). To keep track of tubes and associated aliquots, as well as to prevent logical errors, a rigid sequence of filling the rack with a fixed point of placing the first tube, which corresponds to the list of donors recorded in the database, has been introduced. For this purpose, a rack with a capacity of 12 tubes (rack № 1) with contrasting inscriptions for automatic reading of the place-position of each sample is designed for RS operation. With the help of an integrated convolutional neural network, it determines the presence of serum in the tube, its volume and possible deviations in serum quality, which allows the RS to aliquot the biomaterial. If critical deviations are found (poor serum segmentation), the operator will receive an error message. We have described the issues of RS creation in more detail in earlier papers [13–15].

Aliquoting is the manipulation of pipetting serum and transferring it into aliquot tubes while maintaining a link to the biomaterial donor data. The following aliquot volumes (μL) are provided: 200, 400, 600, 800, 1000, 1200, 1400. The number of aliquots and the preferred volume of each of them is set by the operator at the first stage of work. The volume of the last aliquot cannot be calculated precisely because the volume of the blood sample inevitably varies (up to 12 ml). Therefore, the volume of the last aliquot is formed based on the results of RS operation.

Stage 3 - storage of the obtained aliquots. Aliquots are subject to long-term storage at low temperatures ($-80\,°C$) in a biobank. Data on aliquots include the date of their formation and volume, as well as the indication of the places of their further storage (number/code of the rack (rack № 2), number of the cell in the rack). This information is automatically entered by the RS into the information system during the aliquoting process and linked to the serum donor data. The rack is coded, the cipher reflects the number of occupied cells, the number of individual aliquots of each patient, the date of filling and technical information. It is possible for the operator to remove the required aliquots from several donors from the racks for further testing after a period of time, after which the thawed aliquots are disposed of and the rack space is cleared.

Stage 4 - unloading aliquots that meet the search criteria from the biobank for further research. Using the information system, the operator is able to select from the list of samples stored in the biobank those that meet the search criteria, track their location and take them out of the low-temperature storage for further research.

Stage 5 - entering the results of biosample research into the biobank information system.

The results of research should be entered into the information system in such a way that the operator will be able to form representative samples and conduct population studies related to the assessment of the prevalence of infectious disease markers among the population.

3 Hardware and Software of Biobank Operation Using LIMS

3.1 Structure of Information System (LIMS)

For information support of the technological process of biobanking with the use of RS, an information system LIMS was created, which takes into account the peculiarities of the organization of the population biobank of blood serum samples. This system includes a database (DB) (about biomaterial donors, results of RS work, places of storage of obtained aliquots in low-temperature storage) and protocols of interaction between RS, DB and operator. In its turn, the database consists of two parts - the user part, with which the operator's software interacts, and the system part, which determines the RS operation and is inaccessible for user intervention.

The structure of the user part of the database corresponds to the main technological stages of biobanking and aliquoting of biomaterial using RS, and takes into account their functional characteristics (see Fig. 2).

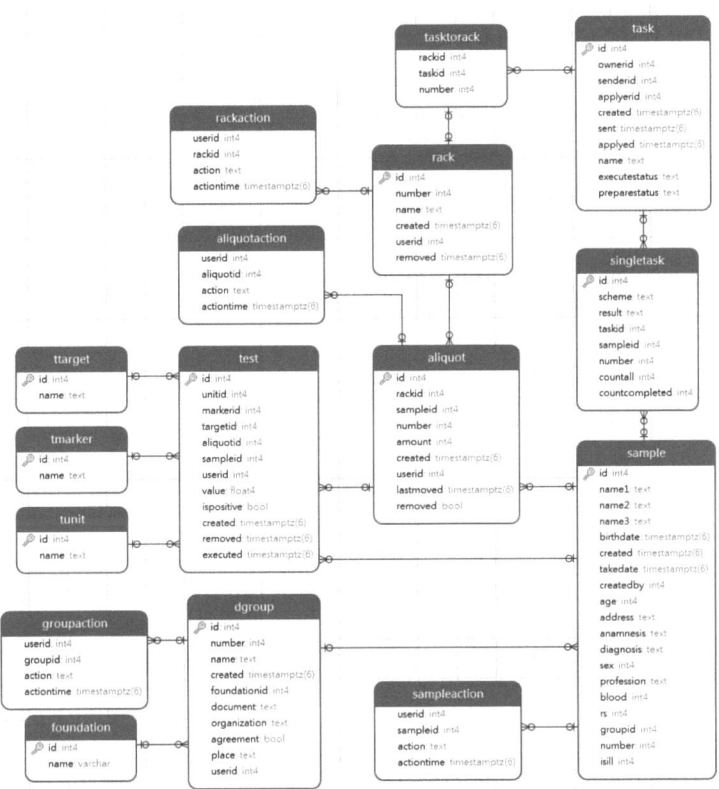

Fig. 2. Relational data model (cuser table is hidden for clarity).

A database under the control of PostgreSQL DBMS was used to store the user and system parts of the database. The database can be located either on the same machine with the client program, or on a separate server. Maintenance of the database (including its backup) is carried out using DBMS tools.

The user part of the database stores data on the receipt of sample parties (including the source and reason for receipt). For each sample in the party, data on the donor of this sample (sex, age, anamnesis, etc.) are specified. The database structure allows storing data on typical manipulations with biosamples, namely, creation and writing off aliquots, storing biosamples, conducting tests, structuring and analyzing the stored data. The database contains tables for storing excavation tasks for RSs and the results of their fulfillment, as well as for ensuring automatic entry of aliquots created by RSs into the database. The information on all operator's actions leading to changes in the database can be saved. To start working with the database from the operator's workplace, the user must be authorized in the system.

3.2 Program Implementation of the Technological Process of Biobanking Using RS

A special client program has been developed to access the biobanking information system. It is a Windows Forms Application with DevExpress UI Controls developed from scratch in several stages as new capabilities were needed to be added to it, as well as in order to provide a more clear and intuitive presentation of data to the operator. It allows the operator to solve the following tasks:

- setting data on objects from which samples are obtained;
- selection of samples for excavation;
- setting the location of tubes with samples in the rack;
- setting the number and volumes of aliquots to be obtained from each sample;
- automatic entry of information on the results of the robotic system operation into the database;
- reflection of manual work with available racks (adding manually excavated aliquots, deleting existing ones; adding, deleting or changing rack numbers; entering information on reaction results);
- selection and formation of an order for unloading aliquots that meet certain requirements for further laboratory tests.

Interaction between the information system and the robotic system is carried out through the database. When a task generated and sent for execution by the operator appears in the database, the RS detects it and starts dripping. The dripping result saved by the RS becomes available to the operator. The operator either sends a command to apply the result or cancels the task. The results of the dripping task are automatically entered into the database, the operator only needs to assign a number to each of the rack received at the RS output, under which it will be stored in the biobank.

Storage of information about donors, aliquots and rack in the database is carried out without actual deletion, which may allow in the service mode to restore records erroneously deleted by the operator.

The main window of the program is an interface for viewing the list of parties, samples in each of them, aliquots and conducted tests for each sample (see Fig. 3). It allows access to the following windows: Viewing the list of racks stored in the low-temperature storage ("Racks"), preparation and launching of RS task for sample excavation, registration of excavation results ("RS Tasks"), search for samples by marker ("Search").

Information Management System for Biobank Optimization   77

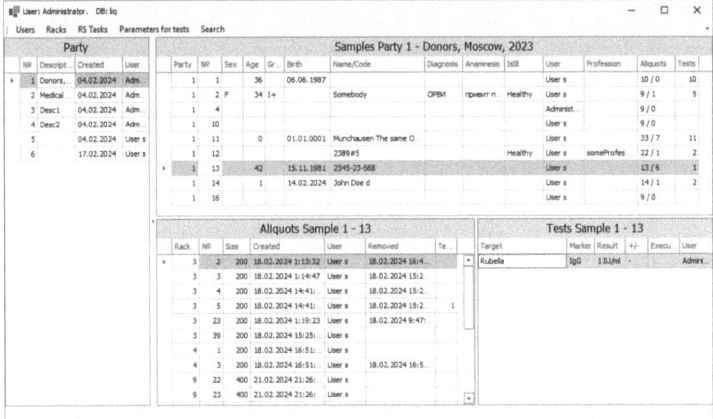

Fig. 3. Example of the database view when simultaneously displaying the list of parties, the list of samples in party №1, all aliquots and conducted test for sample №13

The window for viewing the list of racks is an interface for viewing the list of racks stored in the low-temperature storage, the layout of aliquots in each of them, and the results of laboratory tests. The window allows to perform aliquoting in "manual mode", i.e. without using the RS, and to enter data on laboratory tests (see Fig. 4).

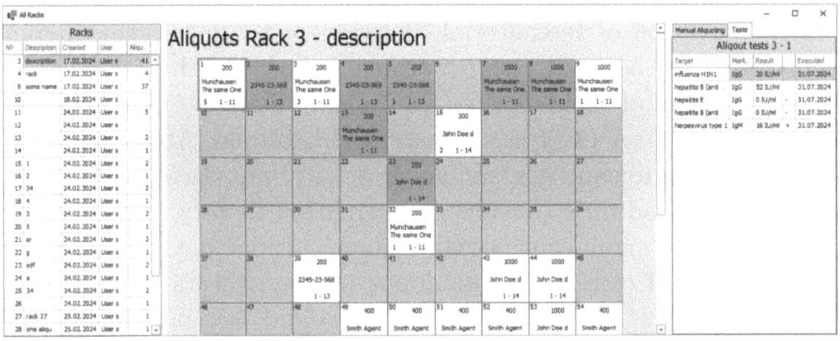

Fig. 4. Viewing the list of racks stored in the database, the scheme of aliquots location in rack №3 in mode of viewing data.

In accordance with the second principle of LIMS organization, to ensure traceability of the biosample life cycle within the biobank (receipt/schedule of biosamples, number and volume of aliquots, places of their low-temperature storage in racks in cold rooms, results of subsequent laboratory tests, etc.), the possibility of both element-by-element viewing of the list of stored racks and aliquots and using summary tables is implemented. Two ways of viewing the contents of racks are available to the operator: aliquots in them can be displayed either in the form of a list or in the form of a 9 × 9 grid, in the cells of which the tubes with serum are located, which imitates the physical location of aliquots in the rack.

The database provides a list of markers of the most relevant infectious diseases for which samples in the biobank can be tested by enzyme-linked immunosorbent assay (ELISA) and hemagglutination inhibition reaction (HIA) methods, the possibility of entering the results of these tests in the database by the operator, structuring and analyzing the data available in the system. The list of infectious diseases and corresponding markers is customizable, the program provides the possibility to expand it as needed (main menu item "Parameters for tests"). It is possible to upload an anonymized part of information to an external medium after structuring and primary data analysis. In order to access the database, the operator must be authorized. At startup, the program asks for a user name and password.

3.3 Operation of LIMS at Each Stage of the Technological Process

At the first stage of the technological process, in order to minimize the risk of errors, loading of the rack with samples and their aliquoting by the RS can be performed only after the operator has entered the data on each storage object, referred to in the graphical user interface as "donor".

In accordance with the third principle of LIMS organization, in order to comply with the requirements of the legislation in the field of personal data protection, the DB implements the protocol of entering the details of documents (agreement or contract on scientific cooperation with the organization performing the direct selection of biomaterial, decision of the local ethical committee on the compliance of the research with ethical standards, the presence of voluntary informed consent of the donor for the collection and storage of biomaterial) in the formation of a set of data on each blood sample. After that, the possibility of filling in significant epidemiologic data about each biomaterial donor (sex, age, residential address (locality), vaccination history, presence of comorbidities, employment, significant notes) becomes available. Only in the presence of voluntary informed consent of the donor to the collection of biomaterial is it possible to enter his/her surname, first name and patronymic. If there is an agreement/contract between the biobank (collection of biosamples) and the organization directly carrying out the selection of biomaterial and its anonymization for depositing in the biobank, it is not allowed to enter the full name or other personal data of donors in the database. In this case, the patient's outpatient card number or other numeric designation shall be entered or the field shall be left blank.

At the second stage of the technological process, upon completion of data entry on biomaterial donors by the operator into the database, biomaterial aliquoting is available in one of two modes: directly by the operator (laboratory technician) in the so-called "manual mode" and with the help of RS - "automatic mode", which corresponds to the fourth principle of LIMS organization.

This allows to use the information system autonomously from the robotic equipment, which significantly expands the area of practical application of the created LIMS. In the manual aliquoting mode, the operator enters data on the volume and position in the rack of each of the obtained aliquots. For each aliquot, its serial number in the rack, brief data of the donor, the volume of material, and the number of laboratory tests performed with this aliquot are displayed (see Fig. 5).

In an automated aliquoting scenario using a RS, a rigid sequence of filling a specially designed rack with a card number or the donor's full name linked to the number of his or her sample in the rack is set to account for the tubes from each storage object and their associated aliquots, and to prevent logical errors. Thus, the operator fills a sheet in the program interface with data on storage objects in the form of a list (from 1 onwards) and arranges blood samples in the same sequence in the rack (see Fig. 5).

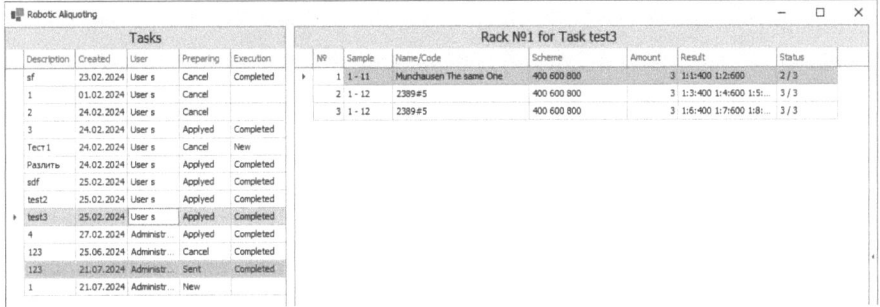

Fig. 5. Example of data output of aliquoting results in «automatic mode» by means of a RS.

To manage and provide information support for the RS operation, a database system component is provided, the operation of which takes place without operator's participation to prevent the risk of errors related to the "human factor". Thus, the work of the RS, the database and the operator is harmonized.

At the third stage of the technological process, to ensure the convenience of viewing the contents of the rack, sent or already in low-temperature storage, as well as entering the results of laboratory tests and search for aliquots that meet certain requirements, the rack viewing window to the right of the image of the rack displays a list of laboratory tests that have been carried out with the currently allocated aliquot on the rack (see Fig. 4).

At the fourth stage of the technological process of biobanking provides the opportunity to remove from the racks the necessary aliquots from several storage objects for their further laboratory examination. After the research, the thawed aliquot is disposed of and the spaces in the racks are released. The operator enters information about aliquot removal from the racks and the results of the tests performed into the database.

A common task for the operator is to select from the database, the results of laboratory tests, those samples that meet certain conditions for further research, which corresponds to the fifth stage of the technological process of biobanking. To create the possibility of primary statistical analysis of data and the use of information system to ensure the entire process of biobanking in the database implemented the possibility of loading data into tables with built-in functions of grouping, filtering and sorting. Figure 6 shows an example of using such a table to search for samples tested for the presence of hepatitis B markers.

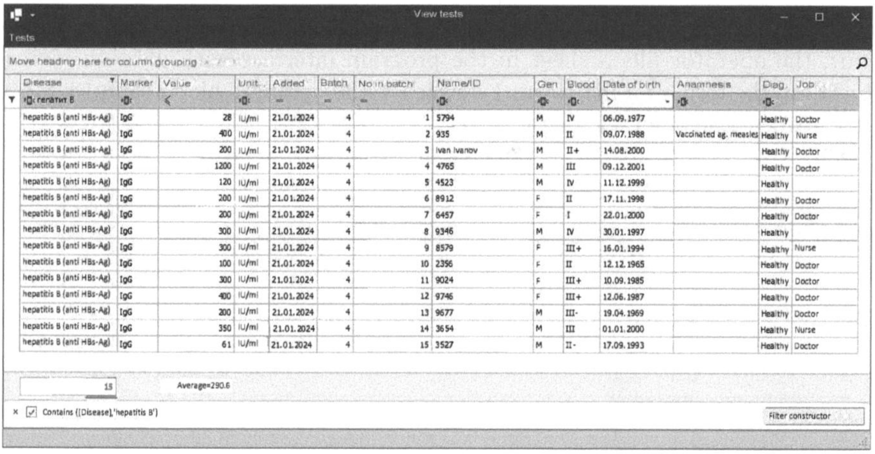

Disease	Marker	Value	Unit	Added	Batch	No in batch	Name/ID	Gen	Blood	Date of birth	Anamnesis	Diag.	Job
hepatitis B (anti HBs-Ag)	IgG	28	IU/ml	21.01.2024	4	1	5794	M	IV	06.09.1977		Healthy	Doctor
hepatitis B (anti HBs-Ag)	IgG	400	IU/ml	21.01.2024	4	2	935	M	II	09.07.1988	Vaccinated ag. measles	Healthy	Nurse
hepatitis B (anti HBs-Ag)	IgG	200	IU/ml	21.01.2024	4	3	Ivan Ivanov	M	II+	14.08.2000		Healthy	Doctor
hepatitis B (anti HBs-Ag)	IgG	1200	IU/ml	21.01.2024	4	4	4765	M	III	09.12.2001		Healthy	Doctor
hepatitis B (anti HBs-Ag)	IgG	120	IU/ml	21.01.2024	4	5	4523	M	IV	11.12.1999		Healthy	
hepatitis B (anti HBs-Ag)	IgG	200	IU/ml	21.01.2024	4	6	8912	F	II	17.11.1998		Healthy	Doctor
hepatitis B (anti HBs-Ag)	IgG	200	IU/ml	21.01.2024	4	7	6457	F	I	22.01.2000		Healthy	Doctor
hepatitis B (anti HBs-Ag)	IgG	300	IU/ml	21.01.2024	4	8	9346	M	IV	30.01.1997		Healthy	
hepatitis B (anti HBs-Ag)	IgG	300	IU/ml	21.01.2024	4	9	8579	F	III+	16.01.1994		Healthy	Nurse
hepatitis B (anti HBs-Ag)	IgG	100	IU/ml	21.01.2024	4	10	2356	F	II	12.12.1965		Healthy	Doctor
hepatitis B (anti HBs-Ag)	IgG	300	IU/ml	21.01.2024	4	11	9024	F	III+	10.09.1985		Healthy	Doctor
hepatitis B (anti HBs-Ag)	IgG	400	IU/ml	21.01.2024	4	12	9746	F	III+	12.06.1987		Healthy	Doctor
hepatitis B (anti HBs-Ag)	IgG	200	IU/ml	21.01.2024	4	13	9677	M	III-	19.04.1969		Healthy	Doctor
hepatitis B (anti HBs-Ag)	IgG	350	IU/ml	21.01.2024	4	14	3654	M	III	01.01.2000		Healthy	Nurse
hepatitis B (anti HBs-Ag)	IgG	61	IU/ml	21.01.2024	4	15	3527	M	II-	17.09.1993		Healthy	Doctor

Fig. 6. Example of the output of a database search for samples tested for hepatitis B markers.

4 Conclusion

The proposed LIMS was created using a publicly available software environment, which, nevertheless, allows us to perform the entire set of tasks included in the technological process of biobanking and, to some extent, to expand the analytical capabilities. It is demonstrated that the created LIMS fully fulfills the tasks common for all LIMS and stated in the scientific literature [7–9], as well as those specific for blood serum banking, including the analysis of population immunity.

The proposed solution can be used both in a small laboratory and for a large biobank, and meets the following requirements: compliance with the specifics of population biobanking and the requirements of international standards for biobanking, possibility to trace the whole life cycle of a sample within the biobank, functional flexibility in use, the ability to change the used volume of data on samples and processes, possibility to connect technological tools for biobank optimization (aliquote system, ELISA-analyzers, etc.), convenient, intuitive interface, ease of implementation and maintenance.

The demonstrated functional capabilities of the database allow to flexibly customize the criteria for selection of aliquots or donors of biomaterial in accordance with the task at hand. The proposed solution is innovative for robotic complexes for biomaterial aliquoting and significantly expands the boundaries of its applicability in practice. Thus, for small laboratories and organizations that have no possibility to purchase large low-temperature storage facilities equipped with specialized LIMS, our proposed solution has a number of advantages and allows organizing the process of biobanking at a high scientific and technological level.

At the moment, the user application works with the database, connecting directly to the DBMS server. This solution is good from the point of view of ease of development, but from a security point of view, a better architectural solution would be a solution with a separately developed server application running on the DBMS side, to which the user application connects. This architecture protects the database from unauthorized changes, because the user application does not have a DBMS user login and password. In addition, it is better suited for implementing a fully multi-user system with the possibility of simultaneous access by several users.

5 Conclusions

The database created for informational support of the RS work, oriented to the needs of the biobank (or collections of blood serum samples) takes into account the shortcomings of existing laboratory information systems (LIMS), meets the stages of the technological process of biobanking, the requirements of ethics and ergonomics, as well as has functional flexibility, traceability of biosample movement and allows to structure, analyze and upload the information necessary for the operator. In addition, the database contains all information about the life cycle of biosamples within the biobank and ensures the implementation of quality management system for biobanks and biosample collections, compliance with international quality standards for biobanking: ISO 9001 "Quality Management Systems. Requirements", ISO 27001 "Information Security Systems. Requirements", ISO 20387 "Biotechnology. Biobanking. General requirements", ISO/TR 22758 Biotechnology—Biobanking—Implementation guide for ISO 20387, ISO 21899 Biotechnology—Biobanking—General requirements for the validation and verification of processing methods for biological material in biobanks.

Acknowledgments. This work was supported by the state assignment of Ministry of Science and Higher Education of the Russian Federation under Grant FZWN-2020-0017. This work was carried out using equipment of High Technology Center at BSTU named after V.G. Shukhov.

Disclosure of Interests. The authors have no competing interests to declare that are relevant to the content of this article.

References

1. Campbell, L.D.: Best Practices for Repositories: Collection, Storage, Retrieval and Distribution of Biological Materials for Research International Society for Biological and Environmental Repositories, 3rd edn. (2012)
2. Hewitt, R., Watson, P.: Defining biobank. Biopreserv. Biobank. **11**(5), 309–15 (2013). https://doi.org/10.1089/bio.2013.0042
3. Anisimov, S.V., Granstrem, O.K., Meshkov, A.N., et al.: National association of biobanks and biobanking specialists: new community for promoting biobanking ideas and projects in Russia. Biopreserv. Biobank. **19**(1), 73–82 (2021). https://doi.org/10.1089/bio.2013.0042
4. Haselbeck, A.H., Im, J., Prifti, K., et al.: Serology as a tool to assess infectious disease landscapes and guide public health policy. Pathogens **11**(7), 732 (2022). https://doi.org/10.3390/pathogens11070732
5. He, S., Han, J.: Biorepositories (biobanks) of human body fluids and materials as archives for tracing early infections of COVID-19. Environ. Pollut. **274**, 116525 (2021). https://doi.org/10.1016/j.envpol.2021.116525
6. Zheng, L., Wang, L.: Comprehensive information management system for a medical research cohort biobank based on quality by design. BMC Med. Inform. Decis. Mak. **23**(1), 222 (2023). https://doi.org/10.1186/s12911-023-02318-w
7. Elliott, P., Peakman, T.C., UK Biobank.: The UK Biobank sample handling and storage protocol for the collection, processing and archiving of human blood and urine. Int. J. Epidemiol. **37**(2), 234–44 (2008). https://doi.org/10.1093/ije/dym276
8. Nagai, A., et al.: Overview of the Biobank Japan Project: study design and profile. J. Epidemiol. **27**, S2–S8 (2017). https://doi.org/10.1016/j.je.2016.12.005
9. Cicek, M.S., Olson, J.E.: Mini-review of laboratory operations in biobanking: building biobanking resources for translational research. Front. Public Health **28**(8), 362 (2020). https://doi.org/10.3389/fpubh.2020.00362
10. Bendou, H., Sizani, L., Reid, T., et al.: Baobab laboratory information management system: development of an open-source laboratory information management system for biobanking. Biopreserv. Biobank. **15**(2), 116–20 (2017). https://doi.org/10.1089/bio.2017.0014
11. Reijs, B.L., Teunissen, C.E., Goncharenko, N., Betsou, F., et al.: The central biobank and virtual biobank of biomarkapd: a resource for studies on neurodegenerative diseases. Front. Neurol. **15**(6), 216 (2015). https://doi.org/10.3389/fneur.2015.00216
12. Nozdracheva, A.V., Semenenko, T.A.: Assessment of the reliability of serological monitoring results based on biobank materials. Cardiovasc. Ther. Prev. **22**(11), 3709 (2023). https://doi.org/10.15829/1728-8800-2023-3709
13. Rybak, L., Carbone, G., Malyshev, D., Voloshkin, A.: Design and optimization of a robot dosing device for aliquoting of biological samples based on genetic algorithms. Machines **12**(3), 172 (2024). https://doi.org/10.3390/machines12030172
14. Voloshkin, A., Rybak, L., Cherkasov, V., Carbone, G.: A novel design of a robotic system for biological fluid aliquoting. In: Proceedings of SYROM 2022, ROBOTICS. IISSMM. Mechanisms and Machine Science, vol. 127, pp. 271–279. Springer, Cham (2023). https://doi.org/10.1007/978-3-031-25655-4_28
15. Rybak, L.A., Cherkasov, V.V., Malyshev, D.I., Carbone, G.: Blood serum recognition method for robotic aliquoting using different versions of the YOLO neural network. In Advances in Service and Industrial Robotics. RAAD 2023. Mechanisms and Machine Science, vol. 135, pp. 150–157. Springer, Cham (2023). https://doi.org/10.1007/978-3-031-32606-6_18

Methods of Objects Recognition System Enhancement in Computer Vision

Andrey N. Kokoulin(✉) [iD] and Rostislav A. Kokoulin

Perm National Research Polytechnic University, Perm, Russian Federation
`a.n.kokoulin@pstu.ru`

Abstract. This article presents methods that allows to increase the accuracy of objects recognition systems and achieve better robustness of neural networks. The first proposed method of neural networks enhancement is the two-stage scheme of object recognition with two models trained on different sets of classes.

The second method uses the new augmentation approach based on FGSM and JSMA algorithms to prepare the training dataset for model training. Combination of these methods allows us to achieve the ~ 20% increase of accuracy on complicated test images where false recognition occurred for original neural network model.

Keywords: object detection · YOLO · classification · two-stage scheme

1 Introduction

The development of machine learning methods (AI/ML) makes it possible to solve a wide class of computer vision problems. In particular the problems of object detection and recognition [1]. But those methods can't solve all of the problems. For example, recognition of different sized objects, especially small ones.

Another kind of problems is that accuracy strongly depends on the quality of image or video. Different kinds of external influences can negatively affect the object recognition. For example, bad or strong illumination, interference, blur.

There is an example of tooth and holes recognition on the Fig. 1. Large objects like tooth were recognized well. But small objects like holes were found in random places where they cannot be. The real holes in tooth weren't detected.

The problem is placed in uncommon object representation of natural scenes, for example, from surveillance cameras. On those videos objects at different distances are projected onto the two-dimensional space of the photograph at different scales relative to each other and other objects. That leads to a difference between sizes of different objects by several times. Also, the ideal conditions to take a picture cannot be provided. And there is always some loss of quality due to external impact. Many false positive and false negative errors (FP, FN) occur, and many very distant objects are recognized as background.

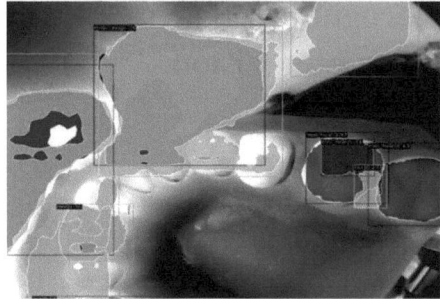 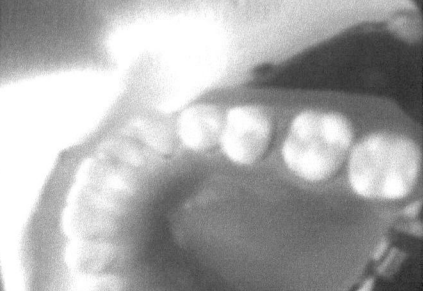

Fig. 1. Tooth and holes recognition

There are some ways to solve this problem of recognition accuracy and network robustness. First way is to use combination of a few networks, trained on the same classes but with different object size. For example, one will detect big cars, which are close to the camera, and another will detect small cars, which are in the background.

Second way is to use SAHI. Slicing Aided Hyper Inference uses slicing algorithm which split image into manageable slices [4]. Slice is much easier to handle than original image due to its lower size. Detection applies to each slice and the output result is the combination of objects detected on the slices. But this method has low object detection performance, because all of the slices should be checked, including those where there are no objects of interest.

Third way is to use FPN. Feature Pyramid Network has top-down architecture with lateral connections is developed for building high-level semantic feature maps at some scales of objects [5]. It helps to find objects that has different sizes, but you must have a layer for each scale. Usually there are 3–4 of them. And if the difference between the biggest object and the least one is huge, objects with medium sizes may be skipped.

None of these approaches can solve all of the problems.

Another problem are adversarial impacts of different factors on the image or video. Due to uncommon image shooting conditions or adversarial attacks some interference may be added on the image. For human the image may looks the same as before the attack, or looks like image in lower quality. But for neural network it will be completely different image [2]. Those attack must have some bounds in order to be undetected by humans or other programs. The first bound is L_0. Attack with this bound includes changing of some features of input data. For example, changing a few pixels. The small number of changes makes them realistic. The bound is L_1. This bound sets the maximum sum of perturbation values. The third bound is L_2. It sets the maximum number of changes using Euclidean distance/Pythagorean distance. And the last bound is l_∞. This bound sets the maximum change of a pixel [3].

Among the methods for increasing the resistance of an CNN to data distortion and adversarial attacks, three groups of methods can be identified [4, 5]:

- Improving the model (creating sustainable networks);
- Elimination of harmful properties of processed data;
- Hiding key information about the CNN model and architecture.

The first group includes methods for distillation and gradient masking, methods for using malicious images for training, and methods for learning to recognize out-of-distribution (OoD) data.

The second group includes various methods of data preprocessing, including data normalization and compression, removal of noise and data anomalies using filters, as well as statistical methods for identifying malicious images. These methods are widely used in most computer vision systems, and it is difficult to create a new universal algorithm for preprocessing and improving the input data. Basically, it requires binding to a specific task and to the properties of the input data. Therefore, this group of methods will not be considered in this work.

Methods for hiding and masking information about an CNN and computer vision systems in general include many methods that control the execution of requests and receiving feedback from the point of view of information security of computer vision systems, including using methods related to information security. It is possible to draw conclusions about the non-universality of the group of methods, since their implementation will depend on the specifics of the computer vision system.

We propose two different approaches, that can help to solve those problems. First one is similar to SAHI. But instead of making a detection on all of the slices in our two-stage algorithm only those where the probability of finding an object is highest are checked. Second one is a special data augmentation. It allows to add the most complicated data samples into the train dataset to enhance the neural network robustness. It works together with processing the image significance map to exclude areas of image, that do not have any useful information from the classification decision process.

2 Two-Stage Object Detection Approach

2.1 Differences from Single-Stage Approach

Multi-stage image processing schemes were studied by the authors from the point of view of increasing the accuracy of object recognition and eliminating false positive recognition errors (False Positive, FP). This method is based on the assumption that the entire set of object classes can be divided into hierarchical groups according to the principle of structural (geometric) nesting. At the top of the hierarchy there are "superobjects" that define the permissible geometric boundaries of the area of interest; on the child branches of the hierarchy there are the required classes of objects that do not geometrically extend beyond the boundaries of the area of interest. Thus, the entire process of object recognition can be divided into two stages: at the first stage, one or more "superobjects" are detected, a "superobject" mask is selected, and the background is cut off using the mask; at the second stage, objects are recognized within the area of interest limited by the mask. An important feature of the method is the division of many classes into two disjoint groups: the CNN of the first stage is trained only to search for "superobjects", the CNN of the second stage is trained to search for the desired objects. Multi-stage schemes made it possible to achieve good results in a project to recognize weapons in a crowd and in a project to create an anthropomorphic dental simulator. But it should be noted that this approach is not universal, since a necessary condition must be met - nesting of objects.

In two-stage approaches (Fig. 2B), the model defines a set of regions of interest (ROIs). Only the candidates from this set are proceed on the next stage. Other areas of the image are considered as a background. In single-stage approaches (for example, SSD and YOLO architectures), the ROI selection step is absent. Object detection is performed directly in possible locations, which are determined by the neural network architecture.

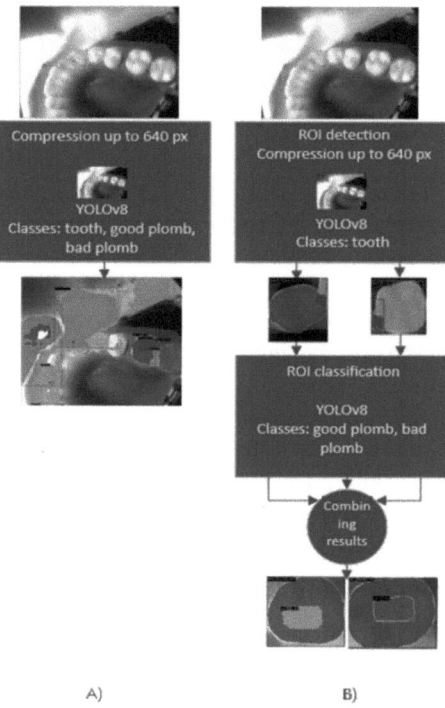

Fig. 2. One-stage (A) and two-stage (B) approaches

2.2 Advanced Two-Stage Approach

But common two-stage scheme has a disadvantage. Neural networks compress images before the start of detection. It is needed to optimize input data. For example, YOLO compress it up to 640x640 px. It is not a problem for one-stage approach, since it applies only one detection. But in two-stage approach we use output data from the first stage as an input data for the second one. It leads to some problems.

First of all, during compression, image proportions often become distorted, because original image didn't have square form. But not only original image can have different shape. If ROI has different shape, it will also be distorted before the start of detection on the second network. It means that object will be distorted two times and it may cause detection mistakes.

Second problem is that compression leads to an information loss. It means that some details from original picture can be missed on compressed one. Since we have to compress image, this problem also appears in one-stage approach. But in two-stage approach we can lower the compression as it is in advanced scheme.

The advanced scheme (Fig. 3) was made to solve those problems of original approach. As we said YOLO compresses the original input image in high quality into a 640 x 640 px before the start of detection. Then we send all the detected ROIs to the second stage in low quality. And this is the problem area. To solve the first problem we have changed the scheme. We could improve the quality of ROI, but there is no need in this, because we already has it in high quality. We can crop it from the original image using mapping algorithm. To process mapping we project coordinates of ROI on the input image and crop the resulting area. This area is ROI in high quality. Then the area is used as input data on the second stage instead of ROI.

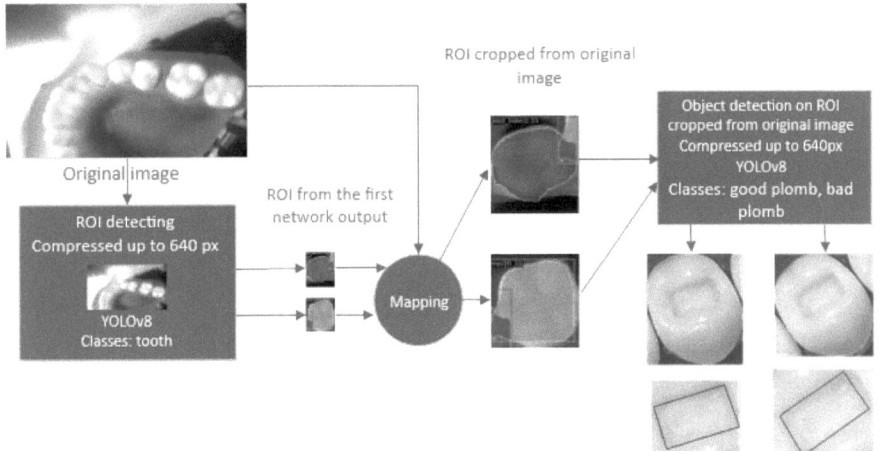

Fig. 3. Advanced two-stage scheme

The main advantage of proposed recognition approach with two levels of classification is recognition with two independently trained models. The first model detects all "superclasses" in the photo and selects them in the image. In our case superclasses include different teeth types. In other computer systems classes may include people, vehicles, various objects (for example, you can take the standard YOLO model). The second network uses these regions of interest (ROI) to recognize objects that belongs to superclasses, for example, the tooth plombs classes. The second network was trained only on this class using a large set of photographs (we used a set of 10 thousand photographs of different tooths and plombs).

3 Augmentation of the Train Set

The most common and effective principle of model improvement is the principle of augmentation. The importance of increasing the variability of the training set to reduce the factor of "CNN overtraining" and susceptibility to "data poisoning" is substantiated in many studies [5].

Among the data transformation algorithms used in augmentation, two groups can be distinguished:

- Algorithms that use knowledge about the CNN model and the features of the learning process;
- Algorithms that use knowledge about the specific input data for a particular task.

Let us limit our study to algorithms that use knowledge about the CNN model. The first step is to look at the principles behind malware attacks, looking at the common white-box attacks FGSM, JSMA and C&W. This analysis will help you choose a strategy for generating training samples that ensure the CNN's resistance to attacks.

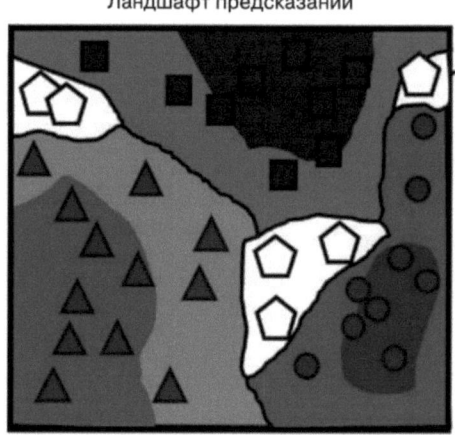

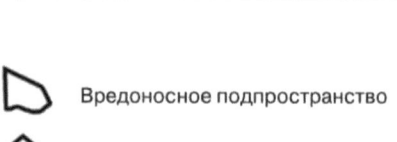

Fig. 4. Predictions landscape of neural network model

It is necessary to analyze the influence of the features of data processing by neural networks on the results of data detection and classification in order to filter out false image features that lead to errors. Understanding the results of the analysis will make it possible to create new methods for augmenting images of the training set, which will increase the robustness of the ANN.

Taking into account the fact that malicious input images are located in subspaces - continuous areas of erroneous classification where there is uncertainty (see Fig. 4), one of the augmentation methods will be to synthesize a large number of malicious images and train an ANN model based on them.

3.1 Feature Significance Maps

Feature significance maps help to visualize which features or variables are most influential in predicting an outcome or in describing patterns within data. They show which pixels or regions of the image affect the prediction the most. Understanding which parts of the input data are important for predictions can help validate and debug models, ensuring they make decisions for the right reasons.

There are some ways to build the map. Vanilla Gradient is the most basic form. It involves calculating the gradient of the output class score with respect to the input image pixels. This provides a rough idea of pixel importance. Grad-CAM uses gradients flowing into the last convolutional layer to produce a coarse localization map highlighting important regions in the image. Guided Backpropagation enhances the gradient information by only allowing positive gradients to propagate back through the network, leading to sharper and more informative saliency maps. Keras-vis is a fast way to build a feature significance map. To increase its speed gradients are assumed to be linear. It can negatively affect the quality of the map of some neural networks, but it works well in the most cases. LIME gives multiple variations of the input data to the neural network and evaluates the impact that each change has on the output. By analyzing the results, LIME can determine which aspects of the image are most important to the classifier [6, 7].

3.2 Using of Feature Significance Maps for Augmentations

Feature significance maps shows the most important pixels. Using it we can check which one of them were chosen correctly of incorrectly. Our approach includes 7 steps and uses neural network for segmentation.

1. First step is to create significance map for an object and perform object recognition;
2. apply mask obtained during segmentation onto a significance map and find areas of the significance map that are outside the borders;
3. blur or interfere those areas to decrease their impact on the training process;
4. improve the quality of the areas which are inside the borders to increase their impact on the process;
5. add image to a train dataset.

3.3 Adversarial Attack

Sometimes neural networks make wrong decisions. It happens because during the training session some features were selected incorrectly. Those adversarial examples are very important to found. Adversarial attacks are used to do this. All of those examples have something similar and those methods are used to find an influence that changes a normal example to an adversarial. The addiction of the adversarial examples to the train set can improve its quality of recognition. There are some approaches to implement this. They vary in the formulas they use.

The approaches considered relate to White-box attacks. It means that you know all information about the model. Such as architecture, weights, training strategies, etc. It helps us to find the function to minimize the loss of the model $J(\theta, X, y)$ where X-input data, y-output class, θ-hyperparameters of the model. All white-box attack methods aim to maximize the change in the model's loss function while maintaining a small perturbation of the input image. The higher the dimensionality of the input image space, the easier it is to create adversarial examples that are hardly to distinguish by human.

Some methods use gradient $\nabla xL(x, y)$ to find the way to change input data. Those methods linearize the losses around the input x to find the direction of ρ to which the model's predictions for classification result y are most sensitive (1). $x0$ is the generated (distorted) input, $y0$ is the classification result of the generated input, $L(x,y)$ is the loss function.

$$L(x + \rho, y) \approx L(x, y) + \rho T \nabla xL(x, y) \tag{1}$$

First one is FGSM attack. This attack is based on the way the model were trained. To train a neural network, a gradient is used. Gradients are used to define the way where quality of recognition improves [8]. Therefore, to create examples, we change the input data in the direction opposite to the gradient (2). Then the image will acquire those features that confuse the neural network and will be recognized with an error.

$$X\prime = X + \varepsilon * \text{sign}(\nabla_x J(X, ytrue)) \tag{2}$$

Second one is Jacobian-Based Saliency Map Attack. JSMA is a targeted attack that uses a gradient to estimate the significance of each pixel in the input image [9]. This significance metric characterizes the influence of the selected pixel on the change in the classification result of the neural network from the original class y to the target class $y0$. At each iteration, only the pixels with the highest significance score are selected and modified to minimize the distortion of the image. It can be seen that the principle of determining significant fragments for implementing the attack is close to the LIME significance map determination method described in the previous section. To calculate the significance of JSMA, the Jacobian matrix (3) is used, which describes the effect of changing each of the N input values on the M output values.

$$S_{mn} = \frac{\partial f(x)_m}{\partial x_n} \tag{3}$$

3.4 Using of Adversarial Attack for Augmentations

The adversarial images are to be added to a train dataset. Our approach shows how to augment them with better efficiency using the FGSM and JSMA methods. To make it more efficient we should use a combination (ensemble) of several neural networks [10–13]. Each of them will have its own gradient. And we can make different adversarial images from original picture for each of them. For example, we can train neural network models with different hyperparameters or changed (for example rotated) dataset. Using of several neural networks allows us to increase the quality of this process.

The general scheme for preparing a training dataset for a robust network includes the following stages:

- preparing an ensemble of neural networks with various parameters and training based on the original training sample;
- augmentation by adding malicious images obtained by FGSM and JSMA methods from the original training set for each model included in the ensemble;
- augmentation by analyzing significance maps and eliminating the influence of fragments located outside the ROI mask (related to the background, not the object);
- preparing a validation part of a samples from many modifications of each image of the original valid part of the dataset using JSMA and FGSM attacks;
- preparing test samples from many modifications of each image of the original test dataset using JSMA and FGSM attacks;
- model training;
- model testing;
- comparison of performance indicators with the original model.

4 The Effectiveness of the Proposed Methods

4.1 The Effectiveness of the Two-Stage Approach

To analyze the effectiveness of the first proposed method we compared one-stage and two-stage schemes. All the results are in the Table 1 below. We used 10 test collages with 10 tooth with 10..100% of original size on each.

Table 1. Results for collage test

Collage number	Two-stage network			One-stage networks
	Detected ROIs	Detected plombs by original scheme	Detected plombs by advanced scheme	
1	9	9	9	6
2	10	9	10	7
3	9	9	9	6
4	10	10	10	8
5	10	9	10	8
6	10	9	10	8
7	10	9	10	9
8	10	8	10	8
9	9	8	9	7
10	10	10	10	9

We can see the improvement in detection accuracy up to 97% for the advanced two-stage scheme compared with one-stage scheme.

As a result, we have prepared the test dataset of 1000 original images ("in the wild") from the annotated dataset and passed it as an input to the advanced two-stage scheme model. We achieved 97% accuracy (Table 2).

Table 2. Results for tests on the dataset

Approach	Precision, % (tooth only)	Precision, % (tooth&holes)	mAP, %	False positive, %
One-stage	76,44	64,19	71,31	7,58
Original two-stage	92,93	92,02	92,87	5,57
Advanced two-stage	97,60	97,12	97,20	0,56

4.2 The Effectiveness of Augmentation of the Dataset

To evaluate the effectiveness of the second approach we use a metric of uncertainty $U(x)$ (4). That value reflects the magnitude of the spread of predictions for images in the test sample [14].

$$U(x) = \left(\frac{1}{L} \sum_{i=1}^{L} \|F_r(x)\| \right) - \| \frac{1}{L} \sum_{i=1}^{L} F_r(x) \| \quad (4)$$

We performed a test for a neural network without (Fig. 5) and with (Fig. 6) added adversarial images. Left (blue) column shows results for original test images, and the right (red) one shows results for adversarial test images.

We can compare these diagrams and conclude that the number of test images for which the degree of uncertainty is zero or close to zero is increasing and the distribution $U(x)$ for malicious images is shifting to the right, which indicates an improvement in the stability (robustness) of the model to distortions.

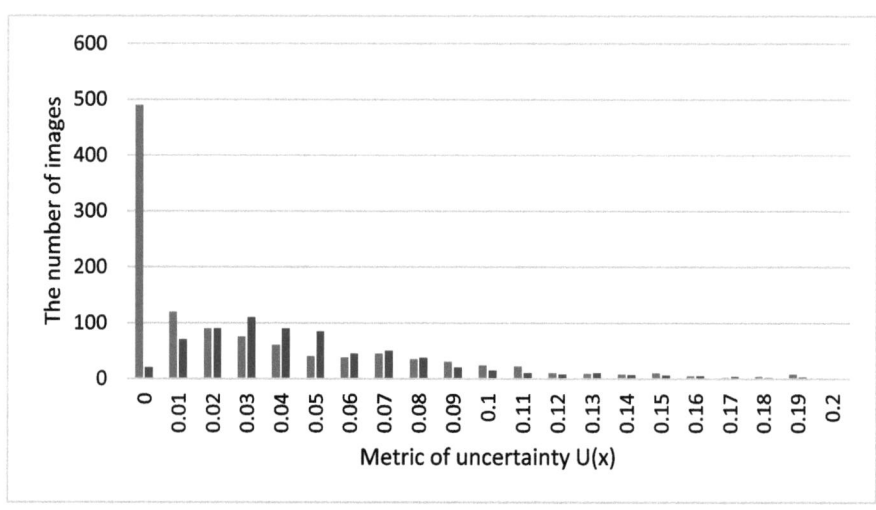

Fig. 5. Test results for original dataset

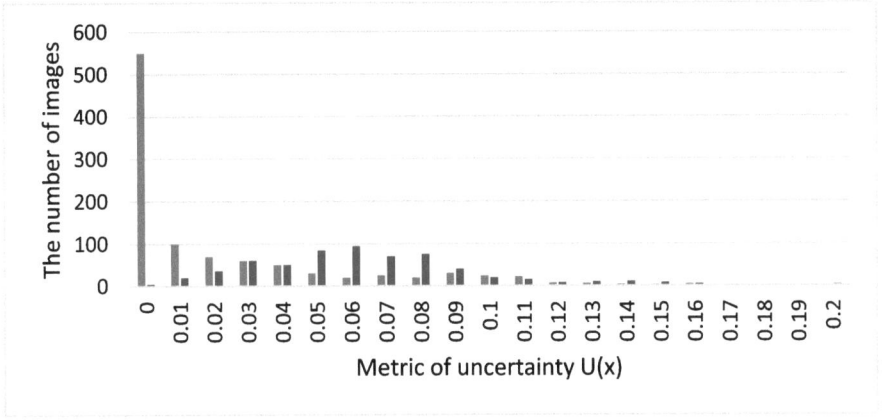

Fig. 6. Test results for dataset included adversarial images

5 Conclusion

The main subject of the work is robustness of neural networks. It is important feature because neural networks often work with corrupted data. Even the small number of changes can confuse neural network. And it is important to prepare them for possible violations. Those violations may be caused by nature: flare, blur, etc. Or can be a consequence of a human attack.

We propose two different ways to increase the quality of neural networks which works with corrupted data. First part of the article shows approach using an intermediate step before classification. This step helps us to make a detection in high quality only to a part of the image. The two-stage scheme is described and the way to upgrade it is shown. The modification of this scheme has lots of benefits in comparison with common one-stage scheme.

Second part of the article is about using different approaches to control and modify train dataset in order to prepare it to the problem data. The main task of images augmentation in a dataset is to exclude features that can lead to errors. The influence of those features should be decreased. Otherwise correct impact of correct features should be increased. Some approaches that allow to create this data are shown. Including those images in dataset can prepare neural network to corrupted data in the future detections.

References

1. Zhao, Z.Q., Zheng, P., Xu, S.T., Wu, X.: Object detection with deep learning: a review. IEEE Trans. Neural Netw. Learn. Syst. **30**(11), 3212–3232 (2019)
2. Chen, P.-Y., Hsieh, C.-J.: Adversarial Robustness for Machine Learning. Academic Press (2022)
3. A Practical Guide To Adversarial Robustness. https://towardsdatascience.com/a-practical-guide-to-adversarial-robustness-ef2087062bec
4. Redmon, J., Divvala, S., Girshick, R., Farhadi, A.: You only look once: unified, realtime object detection. In: Proceedings of the IEEE Conference on Computer Vision and Pattern Recognition, pp.779–788 (2016)

5. Gonzalez, J., Zaccaroa, C., Alvarez Garcia, J., Morilloa, L., Caparrinib, F.: Real-time gun detection in CCTV: an open problem. Neural Netw. **132**, 297–308 (2020)
6. Warr, K.: Strengthening Deep Neural Networks: Making AI Less Susceptible to Adversarial Trickery. Piter, St. Petersburg, 272 p., ill (2021)
7. Naveed, A., Ajmal, M.: Threat of adversarial attacks on deep learning in computer vision: a survey. IEEE Access **6**, 14410–14430 (2018)
8. Szegedy, C., et al.: Intriguing properties of neural networks. arXiv preprint arXiv:1312.6199 (2013)
9. Goodfellow, I.J., Shlens, J., Szegedy, C.: Explaining and Harnessing Adversarial Examples, arXiv e-prints (2014). https://doi.org/10.48550/arXiv.1412.6572
10. Вербицкий М.. Начальный курс топологии. Задачи и теоремы. — Litres, 2018-12-20. С. 163–164, 346 с
11. Кокоулин Р.А., Кокоулин А.Н. The hierarchical approach for image processing in objects recognition system. In: Proceedings of the 2022 IEEE Conference of Russian Young Researchers in Electrical and Electronic Engineering (ElConRus), St. Petersburg, Russia, с. 340–344, 25–28 January 2022
12. Тур А.И., Кокоулин А.Н.: Recyclable waste optical sorting system based on neural networks ensemble. In: Proceedings of the 2022 IEEE Conference of Russian Young Researchers in Electrical and Electronic Engineering (ElConRus), St. Petersburg, Russia, с. 345–348, 25–28 January 2022
13. Тур А.И., Ахметзянов К.Р., Южаков А.А., Кокоулин А.Н.: Вопросы применения иерархических систем распознавания в системах видеонаблюдения. Вестник Пермского национального исследовательского политехнического университета. Электротехника, информационные технологии, системы управления, № 34, с. 75–89 (2020)
14. Carlini, N., Wagner, D.: Towards evaluating the robustness of neural networks, arXiv e-prints (2016). https://doi.org/10.48550/arXiv.1608.04644

Reconstruction Derivatives from Values of Functions Belonging to Nikolskii-Besov Classes of Mixed Smoothness in Domains of a Certain Kind

S. N. Kudryavtsev[✉]

Federal Research Center Computer Science and Control of RAS, Vavilova Street 44-2,
Moscow 119333, Russia
kudrsn@yandex.ru
http://www.springer.com/gp/computer-science/lncs

Abstract. This paper investigates Nikolskii and Besov function spaces defined using L_p-averaged mixed moduli of continuity of appropriate orders, instead of relying on classical moduli of continuity tied to specific mixed derivatives. We provide upper and lower bounds for the optimal accuracy of derivative reconstruction based on function values at a fixed number of points, for functions belonging to these classes on domains of a certain type. These bounds are either comparable or tighter than previously established results by the author for such function classes on the unit cube I^d. This work also significantly extends the range of Nikol'skii and Besov spaces with mixed smoothness for which such estimates have been derived.

Keywords: Accuracy · Reconstruction · Derivative · Function values · Nikolskii-Besov classes of mixed smoothness

1 Introduction and Preliminary Definitions

This section introduces notations and definitions that are used throughout in this paper. Some auxiliary facts are also provided.

Let $d \in \mathbb{N}$. Define the set:

$$\mathbb{Z}_+^d = \left\{\lambda = (\lambda_1, ..., \lambda_d) \in \mathbb{Z}^d : \lambda_j \geq 0, \ j = 1, ..., d\right\}.$$

For $d \in \mathbb{N}$ and any $l \in \mathbb{Z}_+^d$, define the subset:

$$\mathbb{Z}_+^d(l) = \left\{\lambda \in \mathbb{Z}_+^d : \lambda_j \leq l_j, \ j = 1, ..., d\right\}.$$

Let $P_{d,l}$ denote the space of real-valued polynomials in $d \in \mathbb{N}$ variables of degree at most $l \in \mathbb{Z}_+^d(l)$, i.e., all functions $f : \mathbb{R}^d \mapsto \mathbb{R}$ of the form:

$$f(x) = \sum_{\lambda \in \mathbb{Z}_+^d(l)} a_\lambda x^\lambda, \quad x \in \mathbb{R}^d,$$

where $a_\lambda \in \mathbb{R}$, and $x^\lambda = x_1^{\lambda_1} \cdots x_d^{\lambda_d}$.

Given a topological space T, we denote by $C(T)$ the space of all continuous real-valued functions defined on T.

Suppose $D \subset \mathbb{R}^d$ is a Lebesgue measurable set and $1 \leq p \leq \infty$. Then $L_p(D)$ denotes the space of all real-valued measurable functions f on D with finite norm:

$$\|f\|_{L_p(D)} = \begin{cases} \left(\int_D |f(x)|^p dx\right)^{1/p}, & 1 \leq p < \infty, \\ \operatorname{supvrai}_{x \in D} |f(x)|, & p = \infty. \end{cases}$$

For all vectors $x, y \in \mathbb{R}^d$, define:

$$(x, y) = \sum_{j=1}^{d} x_j y_j, \quad xy = (x_1 y_1, ..., x_d y_d).$$

For $x \in \mathbb{R}^d$ and $A \subset \mathbb{R}^d$, define

$$xA = \{xy : y \in A\}.$$

If $x \in \mathbb{R}^d$ has non-zero components x_j, define $x^{-1} = (x_1^{-1}, ..., x_d^{-1})$, $j = 1, ..., d,$. For $x, y \in \mathbb{R}^d$ we write $x \leq y(x < y)$ if $x_j \leq y_j (x_j < y_j)$ for all $j = 1, \ldots, d,$.

Define $x_+ = ((x_1)_+, ..., (x_d)_+)$, where $t_+ = \frac{1}{2}(t + |t|)$ for $d \in \mathbb{N}$, $x \in \mathbb{R}^d$ and $t \in \mathbb{R}$.

Let $\mathbb{R}^d_+$ denote the set of all $x \in \mathbb{R}^d$ such that $x_j > 0$ for all j.

Define:

$$I^d = \{x \in \mathbb{R}^d : 0 < x_j < 1, j = 1, \ldots, d\},$$
$$\overline{I}^d = \{x \in \mathbb{R}^d : 0 \leq x_j \leq 1, j = 1, \ldots, d\},$$
$$B^d = \{x \in \mathbb{R}^d : -1 \leq x_j \leq 1, j = 1, \ldots, d\}.$$

Let $e = (1, ..., 1) \in \mathbb{R}^d$, and denote by e_j the standard basis vector in $\mathbb{R}^d$, with a 1 in the j-th position and zeros elsewhere.

For $\lambda \in \mathbb{Z}^d_+$, define the differential operator:

$$D^\lambda = \frac{D^{|\lambda|}}{Dx_1^{\lambda_1} \ldots Dx_d^{\lambda_d}}, \quad |\lambda| = \sum_{j=1}^{d} \lambda_j.$$

Let $s(x) = \{j = 1, \ldots, d : x_j \neq 0\}$, and for a subset $J \subset \{1, ..., d\}$, let $\chi_J \in \mathbb{R}^d$ be the vector with

$$(\chi_J)_j = \begin{cases} 1, & \text{when } j \in J; \\ 0, & \text{when } j \in (\{1, \ldots, d\} \setminus J). \end{cases}$$

For $d \in \mathbb{N}$, $x \in \mathbb{R}^d$ and $J = \{j_1, \ldots, j_k\} \subset \mathbb{N} : 1 \leq j_1 < j_2 < \ldots < j_k \leq d$, by x^J define the vector $x^J = (x_{j_1}, \ldots, x_{j_k}) \in \mathbb{R}^k$.

For a set $A \subset \mathbb{R}^d$ define $A^J = \{x^J : x \in A\}$.

Let $D \subset \mathbb{R}^d$ be an open set, and let $h \in \mathbb{R}^d$, $l \in \mathbb{Z}_+^d$. Define the set:

$$D_h^l = \{x \in D : x + tlh \in D \ \forall t \in \overline{I}^d\} =$$
$$\{x \in D : (x + \sum_{j \in s(l)} t_j l_j h_j e_j) \in D \ \forall t^{s(l)} \in (\overline{I}^d)^{s(l)}\}.$$

For any $d \in \mathbb{N}$ let D be open set in $\mathbb{R}^d$ and let $1 \leq p \leq \infty$. The mixed difference of order l corresponding to the vector h is defined for a function $f \in L_p(D)$ vectors $h \in \mathbb{R}^d, l \in \mathbb{Z}_+^d$ as:

$$(\Delta_h^l f)(x) = \left(\left(\prod_{j=1}^d \Delta_{h_j e_j}^{l_j}\right) f\right)(x) = \left(\left(\prod_{j \in s(l)} \Delta_{h_j e_j}^{l_j}\right) f\right)(x) =$$
$$\sum_{k \in \mathbb{Z}_+^d(l)} (-e)^{l-k} C_l^k f(x + kh), x \in D_h^l,$$

where $C_l^k = \prod_{j=1}^d C_{l_j}^{k_j}$.

Let D be a set in $\mathbb{R}^d$, and let $1 \leq p < \infty$. For $f \in L_p(D)$ and $l \in \mathbb{Z}_+^d$ define the "averaged" mixed modulus of continuity in $L_p(D)$ of order l of the function f as:

$$\Omega^{\prime l}(f, t^{s(l)})_{L_p(D)} = \begin{cases} \left((2t^{s(l)})^{-e^{s(l)}} \int_{t^{s(l)}(B^d)^{s(l)}} \|\Delta_\xi^l f\|_{L_p(D_\xi^l)}^p d\xi^{s(l)}\right)^{1/p} = \\ \left((2t^{s(l)})^{-e^{s(l)}} \int_{(tB^d)^{s(l)}} \int_{D_\xi^{l\chi_{s(l)}}} |\Delta_\xi^{l\chi_{s(l)}} f(x)|^p dx d\xi^{s(l)}\right)^{1/p}, \end{cases}$$
$$t^{s(l)} \in (\mathbb{R}_+^d)^{s(l)}.$$

Let $d \in \mathbb{N}, \alpha \in \mathbb{R}_+^d, 1 \leq p < \infty$ and D be a domain in $\mathbb{R}^d$. Then, we define the vector $l = l(\alpha) \in \mathbb{N}^d$ setting $l_j = \min\{m \in \mathbb{N} : \alpha_j < m\}, j = 1, \ldots, d$, and we denote by $(S_p^\alpha H)'(D)((S_p^\alpha \mathcal{H})'(D))$ the set of all functions $f \in L_p(D)$, having such property, that for any non-empty set $J \subset \{1, \ldots, d\}$ the following inequality holds:

$$\sup_{t^J \in (\mathbb{R}_+^d)^J} (t^J)^{-\alpha^J} \Omega^{\prime l\chi_J}(f, t^J)_{L_p(D)} = \sup_{t^J \in (\mathbb{R}_+^d)^J} (\prod_{j \in J} t_j^{-\alpha_j}) \Omega^{\prime l\chi_J}(f, t^{s(l\chi_J)})_{L_p(D)} < \infty (\leq 1).$$

Under the same conditions on the $\alpha, p, D, l = l(\alpha)$ and $\theta \in \mathbb{R} : 1 \leq \theta < \infty$ by the $(S_{p,\theta}^\alpha B)'(D)((S_{p,\theta}^\alpha \mathcal{B})'(D))$ we denote the set of all functions $f \in L_p(D)$, that for any non-empty set $J \subset \{1, \ldots, d\}$ satisfy the condition

$$\left(\int_{(\mathbb{R}_+^d)^J} (t^J)^{-e^J - \theta \alpha^J} (\Omega^{\prime l\chi_J}(f, t^J)_{L_p(D)})^\theta dt^J\right)^{1/\theta} =$$
$$\left(\int_{(\mathbb{R}_+^d)^J} (\prod_{j \in J} t_j^{-1-\theta \alpha_j})(\Omega^{\prime l\chi_J}(f, t^{s(l\chi_J)})_{L_p(D)})^\theta \prod_{j \in J} dt_j\right)^{1/\theta} < \infty (\leq 1).$$

When $\theta = \infty$ we put $(\mathcal{S}^\alpha_{p,\infty}\mathcal{B})'(D) = (\mathcal{S}^\alpha_p\mathcal{H})'(D), (\mathcal{S}^\alpha_{p,\infty}\mathcal{B})'(D) = (\mathcal{S}^\alpha_p\mathcal{H})'(D)$. The following inclusion holds:

$$(\mathcal{S}^\alpha_{p,\theta}\mathcal{B})'(D) \subset c_0(\alpha)(\mathcal{S}^\alpha_p\mathcal{H})'(D), \tag{1.1}$$

where $c_0(\alpha) = \prod_{j=1}^{d} 2^{2+\alpha_j}$.

For a Banach space X (over $\mathbb{R}$) we put $B(X) = \{x \in X : \|x\|_X \le 1\}$.

We present some information about multiple series, which shall used subsequently.

When $d \in \mathbb{N}$ for $y \in \mathbb{R}^d$ we put

$$\mathfrak{m}(y) = \min_{j=1,\dots,d} y_j$$

and for a Banach space X, a vector $x \in X$ and a family $\{x_\kappa \in X, \kappa \in \mathbb{Z}^d_+\}$ we write $x = \lim_{\mathfrak{m}(\kappa)\to\infty} x_\kappa$, if for any $\epsilon > 0$ exists $n_0 \in \mathbb{N}$, such that the inequality $\|x - x_\kappa\|_X < \epsilon$ is valid when $\kappa \in \mathbb{Z}^d_+$ and $\mathfrak{m}(\kappa) > n_0$.

Suppose that X is a Banach space, $d \in \mathbb{N}$ and $\{x_\kappa \in X, \kappa \in \mathbb{Z}^d_+\}$ is a family of vectors. Then, the sum of the series $\sum_{\kappa \in \mathbb{Z}^d_+} x_\kappa$ is defined as a vector $x \in X$, for which the equality $x = \lim_{\mathfrak{m}(k)\to\infty} \sum_{\kappa \in \mathbb{Z}^d_+(k)} x_\kappa$ holds.

For $d \in \mathbb{N}$ by Υ^d we denote the set

$$\Upsilon^d = \{\epsilon \in \mathbb{Z}^d : \epsilon_j \in \{0,1\}, j = 1, \dots, d\}.$$

The following lemma is correct.

Lemma 1.1. Suppose that X is a Banach space, and for a vector $x \in X$ and a family $\{x_\kappa \in X, \kappa \in \mathbb{Z}^d_+\}$ the equality $x = \lim_{\mathfrak{m}(\kappa)\to\infty} x_\kappa$ is valid. Then, for the family $\{\mathcal{X}_\kappa \in X, \kappa \in \mathbb{Z}^d_+\}$ whose elements are defined by the equality

$$\mathcal{X}_\kappa = \sum_{\epsilon \in \Upsilon^d : s(\epsilon) \subset s(\kappa)} (-\epsilon)^\epsilon x_{\kappa-\epsilon}, \kappa \in \mathbb{Z}^d_+,$$

the equality

$$x = \sum_{\kappa \in \mathbb{Z}^d_+} \mathcal{X}_\kappa$$

holds.

We should now introduce spaces of piecewise polynomial functions and define operators applicable to them that can be used to create approximatiions.

We first define a system of partitions of unity on open sets. It is used to construct piecewise polynomial functions from the spaces which are interest to us. Let $\psi^{1,0}$ be the characteristic function of an interval I, that is, let

$$\psi^{1,0}(x) = \begin{cases} 1, & \text{for } x \in I; \\ 0, & \text{for } x \in \mathbb{R} \setminus I. \end{cases}$$

When $m \in \mathbb{N}$ we set
$$\psi^{1,m}(x) = \int_I \psi^{1,m-1}(x-y)dy,$$

and when $d \in \mathbb{N}, m \in \mathbb{Z}_+^d$ we define
$$\psi^{d,m}(x) = \prod_{j=1}^d \psi^{1,m_j}(x_j), x = (x_1, \ldots, x_d) \in \mathbb{R}^d.$$

When $d \in \mathbb{N}, m, n \in \mathbb{Z}^d : m \leq n$, we denote
$$\mathcal{N}_{m,n}^d = \{\nu \in \mathbb{Z}^d : m \leq \nu \leq n\} = \prod_{j=1}^d \mathcal{N}_{m_j,n_j}^1.$$

If $m \in \mathbb{N}$ for all $x \in \mathbb{R}$ (if $m = 0$ for almost all $x \in \mathbb{R}$) the equality
$$\psi^{1,m}(x) = \sum_{\mu \in \mathcal{N}_{0,m+1}^1} a_\mu^m \psi^{1,m}(2x - \mu) \qquad (1.2)$$

is correct, where $a_\mu^m = 2^{-m} C_{m+1}^\mu$. When $d \in \mathbb{N}$ for $t \in \mathbb{R}^d$ by 2^t we denote the vector $2^t = (2^{t_1}, \ldots, 2^{t_d})$.

For $d \in \mathbb{N}, m, \kappa \in \mathbb{Z}_+^d, \nu \in \mathbb{Z}^d$ we denote
$$g_{\kappa,\nu}^{d,m}(x) = \psi^{d,m}(2^\kappa x - \nu) = \prod_{j=1}^d \psi^{1,m_j}(2^{\kappa_j} x_j - \nu_j), x \in \mathbb{R}^d.$$

For $d \in \mathbb{N}, m, \kappa \in \mathbb{Z}_+^d, \nu \in \mathbb{Z}^d$ the equality
$$\operatorname{supp} g_{\kappa,\nu}^{d,m} = 2^{-\kappa}\nu + 2^{-\kappa}(m + \mathbf{e})\overline{I}^d$$

is valid.

When $d \in \mathbb{N}, \kappa \in \mathbb{Z}_+^d, \nu \in \mathbb{Z}^d$ we denote
$$Q_{\kappa,\nu}^d = 2^{-\kappa}\nu + 2^{-\kappa} I^d.$$

When $d \in \mathbb{N}, m, \kappa \in \mathbb{Z}_+^d$ for any open set $U \subset \mathbb{R}^d$ for almost all $x \in U$ the equality
$$\sum_{\nu \in \mathbb{Z}^d : \operatorname{supp} g_{\kappa,\nu}^{d,m} \cap U \neq \emptyset} g_{\kappa,\nu}^{d,m}(x) =$$

is true.

Suppose that $d \in \mathbb{N}, l \in \mathbb{Z}_+^d, m \in \mathbb{N}^d, \kappa \in \mathbb{Z}_+^d$ and $U \subset \mathbb{R}^d$ is open set. Put
$$N_\kappa^{d,m,U} = \{\nu \in \mathbb{Z}^d : \operatorname{supp} g_{\kappa,\nu}^{d,m} \cap U \neq \emptyset\}, \qquad (1.3)$$

and write $\mathcal{P}_\kappa^{d,l,m,U}$ for the linear space of functions $f : \mathbb{R}^d \mapsto \mathbb{R}$, for each of them there exists a collection of polynomials $\{f_\nu \in \mathcal{P}^{d,l}, \nu \in N_\kappa^{d,m,U}\}$ such that, for all $x \in \mathbb{R}^d$ the equality

$$f(x) = \sum_{\nu \in N_\kappa^{d,m,U}} f_\nu(x) g_{\kappa,\nu}^{d,m}(x) \qquad (1.4)$$

holds.

Lemma 1.2. Suppose that $d \in \mathbb{N}, l \in \mathbb{Z}_+^d, m \in \mathbb{N}^d, \kappa \in \mathbb{Z}_+^d, U$ is an open bounded set in $\mathbb{R}^d$. When $j = 1,\ldots,d$ let $H_\kappa^{j,d,l,m,U} : \mathcal{P}_\kappa^{d,l,m,U} \mapsto \mathcal{P}_{\kappa+e_j}^{d,l,m,U}$, be the linear operator sending every function $f \in \mathcal{P}_\kappa^{d,l,m,U}$ written in the form (1.4) to the function

$$(H_\kappa^{j,d,l,m,U} f)(x) =$$

$$\sum_{\nu \in N_{\kappa+e_j}^{d,m,U}} \left(\sum_{\substack{\nu' \in N_\kappa^{d,m,U}, \mu_j \in \mathcal{N}_{0,m_j+1}^1 : \\ 2\nu'_j + \mu_j = \nu_j, \nu'_i = \nu_i, i=1,\ldots,d, i \neq j}} a_{\mu_j}^{m_j} f_{\nu'}(x) \right) g_{\kappa+e_j,\nu}^{d,m}(x), x \in \mathbb{R}^d. \qquad (1.5)$$

Then, for each $f \in \mathcal{P}_\kappa^{d,l,m,U}$ we have

$$(H_\kappa^{j,d,l,m,U} f)|_U = f|_U.$$

For $m \in \mathbb{N}^d, \epsilon \in \Upsilon^d, \nu \in \mathbb{Z}^d$ we denote by $\mathfrak{M}_\epsilon^m(\nu)$ the set of number collections

$$\mathfrak{M}_\epsilon^m(\nu) = \{\mathfrak{m}^\epsilon = \{\mathfrak{m}_j \in \mathcal{N}_{0,m_j+1}^1, j \in s(\epsilon)\} :$$
$$(\nu_j - \mathfrak{m}_j)/2 \in \mathbb{Z} \;\forall j \in s(\epsilon)\} =$$
$$\prod_{j \in s(\epsilon)} \{\mathfrak{m}_j \in \mathcal{N}_{0,m_j+1}^1 : (\nu_j - \mathfrak{m}_j)/2 \in \mathbb{Z}\} = \prod_{j \in s(\epsilon)} \mathfrak{M}_1^{m_j}(\nu_j),$$

and associate each pair $\nu \in \mathbb{Z}^d, \mathfrak{m}^\epsilon \in \mathfrak{M}_\epsilon^m(\nu)$ with an element $\mathfrak{n}_\epsilon(\nu, \mathfrak{m}^\epsilon) \in \mathbb{Z}^d$, by putting

$$(\mathfrak{n}_\epsilon(\nu, \mathfrak{m}^\epsilon))_j = \begin{cases} (\nu_j - \mathfrak{m}_j)/2, j \in s(\epsilon); \\ \nu_j, j \in \{1,\ldots,d\} \setminus s(\epsilon). \end{cases}$$

To state the Lemma 1.3 we introduce the following notation. When $d \in \mathbb{N}$ for $j \in \{1,\ldots,d\}$ we denote by $\eta^j : \mathbb{R}^d \times \mathbb{R}^d \mapsto \mathbb{R}^d$ the mapping which is defined by relation

$$(\eta^j(\xi,x))_i = \begin{cases} \xi_i, i = 1,\ldots,j; \\ x_i, i = j+1,\ldots,d, \end{cases} \quad \xi, x \in \mathbb{R}^d.$$

Lemma 1.3. Supose that $d \in \mathbb{N}, l \in \mathbb{Z}_+^d, m \in \mathbb{N}^d, \kappa \in \mathbb{Z}_+^d, \epsilon \in \Upsilon^d : s(\epsilon) \subset s(\kappa), U$ is a bounded open set in $\mathbb{R}^d$. Then, the linear operator $H_{\kappa,\kappa-\epsilon}^{d,l,m,U} : \mathcal{P}_{\kappa-\epsilon}^{d,l,m,U} \mapsto \mathcal{P}_\kappa^{d,l,m,U}$, whose value on a function $f \in \mathcal{P}_{\kappa-\epsilon}^{d,l,m,U}$ is defined by the equality

$$H_{\kappa,\kappa-\epsilon}^{d,l,m,U} f = \begin{cases} f, \text{ when } \epsilon = 0; \\ (\prod_{j \in s(\epsilon)} H_{\eta^j(\kappa-\epsilon,\kappa)}^{j,d,l,m,U}) f, \text{ when } \epsilon \neq 0, (\text{ see } (1.5)), \end{cases} \qquad (1.6)$$

has the following properties:
1) for $f \in \mathcal{P}_{\kappa-\epsilon}^{d,l,m,U}$ the following equality

$$(H_{\kappa,\kappa-\epsilon}^{d,l,m,U} f)\,|_U = f\,|_U$$

is true;
2) for any $f \in \mathcal{P}_{\kappa-\epsilon}^{d,l,m,U}$ written in the form

$$f = \sum_{\nu' \in N_{\kappa-\epsilon}^{d,m,U}} f_{\kappa-\epsilon,\nu'} g_{\kappa-\epsilon,\nu'}^{d,m}, \{f_{\kappa-\epsilon,\nu'} \in \mathcal{P}^{d,l}, \nu' \in N_{\kappa-\epsilon}^{d,m,U}\},$$

the representation

$$H_{\kappa,\kappa-\epsilon}^{d,l,m,U} f = \sum_{\nu \in N_\kappa^{d,m,U}} f_{\kappa,\nu} g_{\kappa,\nu}^{d,m}$$

where

$$f_{\kappa,\nu} = \sum_{\mathfrak{m}^\epsilon \in \mathfrak{M}_\epsilon^m(\nu)} A_{\mathfrak{m}^\epsilon}^m f_{\kappa-\epsilon, \mathfrak{n}_\epsilon(\nu,\mathfrak{m}^\epsilon)},$$

and

$$A_{\mathfrak{m}^\epsilon}^m = \prod_{i \in s(\epsilon)} a_{\mathfrak{m}_i}^{m_i}, (\text{ see } (1.2)), \mathfrak{m}^\epsilon \in \mathfrak{M}_\epsilon^m(\nu), \nu \in N_\kappa^{d,m,U},$$

is valid.

Further, when $l \in \mathbb{Z}_+$ in the interval I we fix the system of $(l+1)$ different points $\{\xi_{1,0}^{1,l,\lambda} \in I, \lambda = 0,\ldots,l\}$ and construct the system of polynomials $\{\pi_{1,0}^{1,l,\lambda} \in \mathcal{P}^{1,l}, \lambda = 0,\ldots,l\}$, having such property that, when $\lambda, \mu = 0,\ldots,l$ these equalities

$$\pi_{1,0}^{1,l,\lambda}(\xi_{1,0}^{1,l,\mu}) = \begin{cases} 1, & \text{when } \lambda = \mu; \\ 0, & \text{when } \lambda \neq \mu, \end{cases}$$

are satisfied.

When $l \in \mathbb{Z}_+, \delta \in \mathbb{R}_+, x^0 \in \mathbb{R}$ we define the system of points $\{\xi_{\delta,x^0}^{1,l,\lambda} \in x^0 + \delta I, \lambda = 0,\ldots,l\}$ and the system of polynomials $\{\pi_{\delta,x^0}^{1,l,\lambda} \in \mathcal{P}^{1,l}, \lambda = 0,\ldots,l\}$, by putting

$$\xi_{\delta,x^0}^{1,l,\lambda} = x^0 + \delta \xi_{1,0}^{1,l,\lambda}, \pi_{\delta,x^0}^{1,l,\lambda}(x) = \pi_{1,0}^{1,l,\lambda}(\delta^{-1}(x - x^0)).$$

When $d \in \mathbb{N}, l \in \mathbb{Z}_+^d, \delta \in \mathbb{R}_+^d, x^0 \in \mathbb{R}^d$ we construct the system of points $\{\xi_{\delta,x^0}^{d,l,\lambda} \in x^0 + \delta I^d, \lambda \in \mathbb{Z}_+^d(l)\}$ and the system of polynomials $\{\pi_{\delta,x^0}^{d,l,\lambda} \in \mathcal{P}^{d,l}, \lambda \in \mathbb{Z}_+^d(l)\}$, by settinh

$$(\xi_{\delta,x^0}^{d,l,\lambda})_j = \xi_{\delta_j,x_j^0}^{1,l_j,\lambda_j}, j = 1,\ldots,d, \pi_{\delta,x^0}^{d,l,\lambda}(x) = \prod_{j=1}^d \pi_{\delta_j,x_j^0}^{1,l_j,\lambda_j}(x_j).$$

It's clear that, when $\lambda, \mu \in \mathbb{Z}_+^d(l)$ the following equalities

$$\pi_{\delta,x^0}^{d,l,\lambda}(\xi_{\delta,x^0}^{d,l,\mu}) = \begin{cases} 1, & \text{when } \lambda = \mu; \\ 0, & \text{when } \lambda \neq \mu, \end{cases}$$

are correct.

When $d \in \mathbb{N}, l \in \mathbb{Z}_+^d, \delta \in \mathbb{R}_+^d, x^0 \in \mathbb{R}^d$ we denote by $A_{\delta,x^0}^{d,l} : \mathbb{R}^{(l+e)^e} \mapsto \mathcal{P}^{d,l}$ the linear operator sending each collection of numbers $t = \{t_\lambda \in \mathbb{R}, \lambda \in \mathbb{Z}_+^d(l)\}$ to the polynomial

$$A_{\delta,x^0}^{d,l} t = \sum_{\lambda \in \mathbb{Z}_+^d(l)} t_\lambda \pi_{\delta,x^0}^{d,l,\lambda} \in \mathcal{P}^{d,l},$$

and we denote by $\phi_{\delta,x^0}^{d,l}$ the linear mapping of the space $C(x^0 + \delta I^d)$ to the space $\mathbb{R}^{(l+e)^e}$, which every function $f \in C(x^0 + \delta I^d)$ send to the collection of its values $\{f(\xi_{\delta,x^0}^{d,l,\lambda}), \lambda \in \mathbb{Z}_+^d(l)\}$, and define the linear operator $\mathcal{P}_{\delta,x^0}^{d,l} : C(x^0 + \delta I^d) \mapsto \mathcal{P}^{d,l}$ by the equality

$$\mathcal{P}_{\delta,x^0}^{d,l} = A_{\delta,x^0}^{d,l} \circ \phi_{\delta,x^0}^{d,l} = \sum_{\lambda \in \mathbb{Z}_+^d(l)} f(\xi_{\delta,x^0}^{d,l,\lambda}) \pi_{\delta,x^0}^{d,l,\lambda}.$$

As it may be seen from definitions, when $\lambda \in \mathbb{Z}_+^d(l)$ the following equality

$$(\mathcal{P}_{\delta,x^0}^{d,l} f)(\xi_{\delta,x^0}^{d,l,\lambda}) = f(\xi_{\delta,x^0}^{d,l,\lambda})$$

holds.

For $d \in \mathbb{N}, l, \kappa \in \mathbb{Z}_+^d, \nu \in \mathbb{Z}^d$ we define the linear operator $R_{\kappa,\nu}^{d,l} : C(Q_{\kappa,\nu}^d) \mapsto \mathcal{P}^{d,l}$ by the equality

$$R_{\kappa,\nu}^{d,l} f = \mathcal{P}_{\delta,x^0}^{d,l} f, f \in C(x^0 + \delta I^d),$$

when $\delta = 2^{-\kappa}, x^0 = 2^{-\kappa}\nu$.

When $d \in \mathbb{N}$ suppose that, a bounded domain $D \subset \mathbb{R}^d$, its open subset $U \subset D$ and $\kappa \in \mathbb{Z}_+^d$ are such that, the inequality

$$\{\nu' \in \mathbb{Z}^d : Q_{\kappa,\nu'}^d \subset D\} \neq \emptyset \tag{1.7}$$

is valid, and $m \in \mathbb{N}^d$. We fix some mapping

$$\nu_\kappa = \nu_\kappa^{d,m,D,U} : N_\kappa^{d,m,U} \ni \nu \mapsto \nu_\kappa^{d,m,D,U}(\nu) \in \{\nu' \in \mathbb{Z}^d : Q_{\kappa,\nu'}^d \subset D\} \text{ (see. (1.3))}, \tag{1.8}$$

and when $l \in \mathbb{Z}_+^d$ we define the linear operator

$$R_\kappa^{d,l,m,D,U,\nu_\kappa} : C(D) \mapsto \mathcal{P}_\kappa^{d,l,m,U}$$

by the equality

$$R_\kappa^{d,l,m,D,U,\nu_\kappa} f = \sum_{\nu \in N_\kappa^{d,m,U}} (R_{\kappa,\nu_\kappa^{d,m,D,U}(\nu)}^{d,l} f) g_{\kappa,\nu}^{d,m}. \tag{1.9}$$

Also, we introduce the folloeing notation.

Suppose that, $d \in \mathbb{N}$, a bounded domain $D \subset \mathbb{R}^d$ and $\kappa^0 \in \mathbb{Z}_+^d$ are such that, the relation (1.7) with κ^0, instead of κ, is true, and $U \subset D$ is an open subset of D. When $m \in \mathbb{N}^d$ we consider some family of mappings $\mathcal{N} = \{\nu_{\kappa^0+\kappa}^{d,m,D,U}, \kappa \in \mathbb{Z}_+^d\}$

of the form (1.8) with $\kappa^0 + \kappa$, instead of κ, and when $\kappa, l \in \mathbb{Z}_+^d$, based on the (1.9) and (1.6), we define the linear operator $\mathcal{R}_{\kappa^0,\kappa}^{d,l,m,D,U,\mathcal{N}} : C(D) \mapsto \mathcal{P}_{\kappa^0+\kappa}^{d,l,m,U}$, by setting

$$\mathcal{R}_{\kappa^0,\kappa}^{d,l,m,D,U,\mathcal{N}} = \sum_{\epsilon \in \Upsilon^d : s(\epsilon) \subset s(\kappa)} (-\mathfrak{e})^\epsilon H_{\kappa^0+\kappa,\kappa^0+\kappa-\epsilon}^{d,l,m,U} R_{\kappa^0+\kappa-\epsilon}^{d,l,m,D,U,\nu_{\kappa^0+\kappa-\epsilon}}. \quad (1.10)$$

2 Disclosure of Interests

At this subsection we first present an auxiliary assertions, on which relies the proof of the upper bound for the quantity that is studied here.

Proposition 2.1. Suppose that $d \in \mathbb{N}, \alpha \in \mathbb{R}_+^d, 1 \leq p < \infty$ satisfy the condition

$$\alpha - p^{-1}\mathfrak{e} > 0. \quad (2.1)$$

Then, for any domain $D \subset \mathbb{R}^d$ for any function $f \in (S_p^\alpha H)'(D)$ there exists the function $F \in C(D)$, for which the equality $F(x) = f(x)$ is valid for almost all $x \in D$.

Lemma 2.2. When $d \in \mathbb{N}$, let a domain $D \subset \mathbb{R}^d$ and its open subset $U \subset D$ are such that, there exists a $\delta \in \mathbb{R}_+^d$, for which the inclusion $(U + \delta I^d) \subset D$ holds. Then, for any $m \in \mathbb{N}^d$ there exist constants $\mathcal{K}^0 = \mathcal{K}^0(d, m, D, U) \in \mathbb{Z}_+^d, \Gamma^0 = \Gamma^0(d, m, D, U) \in \mathbb{R}_+^d$, for which there exists families of mappings

$$\mathcal{N} = \mathcal{N}^{d,m,D,U} = \{\nu_{\mathcal{K}^0+\kappa}^{d,m,D,U} : N_{\mathcal{K}^0+\kappa}^{d,m,U} \mapsto \mathbb{Z}^d, \kappa \in \mathbb{Z}_+^d\}, \{n_{\mathcal{K}^0+\kappa}^{d,m,D,U} : N_{\mathcal{K}^0+\kappa}^{d,m,U} \mapsto \mathbb{Z}^d, \kappa \in \mathbb{Z}_+^d\},$$

having the following properties:
1) when $\kappa \in \mathbb{Z}_+^d$ for each $\nu \in N_{\mathcal{K}^0+\kappa}^{d,m,U}$ this inclusion

$$(Q_{\mathcal{K}^0+\kappa,\nu_{\mathcal{K}^0+\kappa}^{d,m,D,U}}^d(\nu) \cup Q_{\mathcal{K}^0+\kappa,n_{\mathcal{K}^0+\kappa}^{d,m,D,U}}^d(\nu)) \subset D \cap (2^{-\mathcal{K}^0-\kappa}\nu + \Gamma^0 2^{-\mathcal{K}^0-\kappa}B^d)$$

is true;
2) when $\kappa \in \mathbb{Z}_+^d, \nu \in N_{\mathcal{K}^0+\kappa}^{d,m,U}, \epsilon \in \Upsilon^d : s(\epsilon) \subset s(\kappa), \mathfrak{m}^\epsilon \in \mathfrak{M}_\epsilon^m(\nu)$ for the objects which are defined by relations

$$\mathcal{D}_{\mathcal{K}^0+\kappa,\nu,\epsilon,\mathfrak{m}^\epsilon}^{d,m,D,U} = \chi_{\mathcal{K}^0+\kappa,\nu,\epsilon,\mathfrak{m}^\epsilon}^{d,m,D,U} + \delta_{\mathcal{K}^0+\kappa,\nu,\epsilon,\mathfrak{m}^\epsilon}^{d,m,D,U} I^d,$$

where the point $\chi_{\mathcal{K}^0+\kappa,\nu,\epsilon,\mathfrak{m}^\epsilon}^{d,m,D,U} \in \mathbb{R}^d$ and the vector $\delta_{\mathcal{K}^0+\kappa,\nu,\epsilon,\mathfrak{m}^\epsilon}^{d,m,D,U} \in \mathbb{R}_+^d$ are defined by equalities

$$(\chi_{\mathcal{K}^0+\kappa,\nu,\epsilon,\mathfrak{m}^\epsilon}^{d,m,D,U})_j =$$

$$\min(2^{-\mathcal{K}_j^0-\kappa_j}(n_{\mathcal{K}^0+\kappa}^{d,m,D,U}(\nu))_j, 2^{-\mathcal{K}_j^0-\kappa_j+\epsilon_j}(\nu_{\mathcal{K}^0+\kappa-\epsilon}^{d,m,D,U}(\mathfrak{n}_\epsilon(\nu,\mathfrak{m}^\epsilon)))_j), j \in \mathcal{N}_{1,d}^1;$$

$$(\delta_{\mathcal{K}^0+\kappa,\nu,\epsilon,\mathfrak{m}^\epsilon}^{d,m,D,U})_j = \max(2^{-\mathcal{K}_j^0-\kappa_j}(n_{\mathcal{K}^0+\kappa}^{d,m,D,U}(\nu))_j + 2^{-\mathcal{K}_j^0-\kappa_j},$$

$$2^{-\mathcal{K}_j^0-\kappa_j+\epsilon_j}(\nu_{\mathcal{K}^0+\kappa-\epsilon}^{d,m,D,U}(\mathfrak{n}_\epsilon(\nu,\mathfrak{m}^\epsilon)))_j + 2^{-\mathcal{K}_j^0-\kappa_j+\epsilon_j}) - (\chi_{\mathcal{K}^0+\kappa,\nu,\epsilon,\mathfrak{m}^\epsilon}^{d,m,D,U})_j, j \in \mathcal{N}_{1,d}^1,$$

and

$$\mathcal{D}_{\mathcal{K}^0+\kappa,\nu}^{d,m,D,U} = \{x \in \mathbb{R}^d : \min_{\epsilon \in \Upsilon^d : s(\epsilon) \subset s(\kappa), \mathfrak{m}^\epsilon \in \mathfrak{M}_\epsilon^m(\nu)}(\chi_{\mathcal{K}^0+\kappa,\nu,\epsilon,\mathfrak{m}^\epsilon}^{d,m,D,U})_j < x_j <$$

$$\max_{\epsilon \in \Upsilon^d : s(\epsilon) \subset s(\kappa), \mathfrak{m}^\epsilon \in \mathfrak{M}_\epsilon^m(\nu)}(\chi_{\mathcal{K}^0+\kappa,\nu,\epsilon,\mathfrak{m}^\epsilon}^{d,m,D,U})_j + (\delta_{\mathcal{K}^0+\kappa,\nu,\epsilon,\mathfrak{m}^\epsilon}^{d,m,D,U})_j, j = 1,\ldots,d\},$$

the inclision

$$\mathcal{D}_{\mathcal{K}^0+\kappa,\nu,\epsilon,\mathfrak{m}^\epsilon}^{d,m,D,U} \subset \mathcal{D}_{\mathcal{K}^0+\kappa,\nu}^{d,m,D,U} \subset D$$

is valid;

3) when $\kappa \in \mathbb{Z}_+^d, \nu \in N_{\mathcal{K}^0+\kappa}^{d,m,U}$ for all $\epsilon \in \Upsilon^d : s(\epsilon) \subset s(\kappa)$, and $\mathfrak{m}^\epsilon \in \mathfrak{M}_\epsilon^m(\nu)$ if $j \in \mathcal{N}_{1,d}^1 \setminus s(\epsilon)$ the equality

$$(\nu_{\mathcal{K}^0+\kappa-\epsilon}^{d,m,D,U}(\mathfrak{n}_\epsilon(\nu,\mathfrak{m}^\epsilon)))_j = (\nu_{\mathcal{K}^0+\kappa}^{d,m,D,U}(\nu))_j$$

is correct.

Lemma 2.3. Suppose that $d \in \mathbb{N}$, a bounded domain $D \subset \mathbb{R}^d$ and its open subset $U \subset D$ satisfy the conditions of Lemma 2.2. Also suppose that for $\alpha \in \mathbb{R}_+^d, 1 \leq p < \infty, 1 \leq q \leq \infty, \lambda \in \mathbb{Z}_+^d$ the inequality (2.1) holds, and also $m \in \mathbb{N}^d, \lambda \in \mathbb{Z}_+^d(m)$. Then, there exist the constants $c_1(d,\alpha,p,q,\lambda,m,D,U) > 0, c_2(d,m,D,U) > 0, c_3(d,\alpha,m,D,U) > 0$ and $\mathfrak{e} \in \mathbb{R}_+^d$ such that for any function $f \in (S_p^\alpha H)'(D)$ when $l = l(\alpha), \kappa \in \mathbb{Z}_+^d \setminus \{0\}$ the following inequality is valid:

$$\|D^\lambda \mathcal{R}_{\mathcal{K}^0,\kappa}^{d,l-\epsilon,m,D,U,\mathcal{N}} f\|_{L_q(\mathbb{R}^d)} \leq c_1 2^{(\kappa,\lambda+(p^{-1}-q^{-1})_+\mathfrak{e})} \left(\Omega^{l\chi_{s(\kappa)}}(f,(c_2 2^{-\kappa})^{s(\kappa)})_{L_p(D)} \right.$$

$$+ \sum_{J \subset \mathcal{N}_{1,d}^1 : J \neq \emptyset} \left(\int_{(c_3 I^d)^J} (\prod_{j \in J} u_j^{-p(\alpha_j-\epsilon_j)-1}) (\Omega^{l\chi_{J \cup s(\kappa)}}(f,(u\chi_{J \setminus s(\kappa)}))\right.$$

$$\left.\left. +2^{-\kappa} u\chi_{s(\kappa) \cap J} + c_2 2^{-\kappa} \chi_{s(\kappa) \setminus J})_{L_p(D)}^{J \cup s(\kappa)} du^J \right)^{1/p} \right)$$

(see . (1.10) with $\mathcal{K}^0, \mathcal{N}$ from the Lemma 2.2). (2.2)

Proposition 2.4. Suppose that the coditions of the Lemma 2.3 hold and the inequality

$$\alpha - \lambda - (p^{-1}-q^{-1})_+ \mathfrak{e} > 0 \qquad (2.3)$$

is satisfied. Then, for any function $f \in (S_p^\alpha H)'(D)$ if $l = l(\alpha)$

the following equality is true in $L_q(U)$:

$$\mathcal{D}^\lambda(f\mid_U) = \sum_{\kappa\in\mathbb{Z}_+^d}(\mathcal{D}^\lambda(\mathcal{R}_{\mathcal{K}^0,\kappa}^{d,l-\mathfrak{e},m,D,U,\mathcal{N}}f))\mid_U. \qquad (2.4)$$

Now we recall the problem statement of the reconstruction derivatives from the values of functions at a given number of points.

Let T be a topological space, and let X be a Banach space over $\mathbb{R}$, let $\mathcal{V}:D(\mathcal{V})\mapsto X$ be a linear operator defined on $D(\mathcal{V})\subset C(T)$, whose values belong X. Also let $K\subset D(\mathcal{V})$ be a certain class of functions. For every $n\in\mathbb{N}$ by $\Phi_n(C(T))$ we denote the set of all mappings $\phi:C(T)\mapsto\mathbb{R}^n$, for each of which there exists a collection of points $\{t^j\in T,\ j=1,\ldots,n\}$ such that the following equality is correct: $\phi(f)=(f(t^1),\ldots,f(t^n))$, $f\in C(T)$; also by $\mathcal{A}^n(X)(\overline{\mathcal{A}}^n(X))$ we denote the set of all mappings (of all linear mappings) $A:\mathbb{R}^n\mapsto X$. Then, for $n\in\mathbb{N}$ we put

$$\sigma_n(\mathcal{V},K,X) = \inf_{A\in\mathcal{A}^n(X),\phi\in\Phi_n(C(T))}\sup_{f\in K}\|\mathcal{V}f - A\circ\phi(f)\|_X,$$

and

$$\overline{\sigma}_n(\mathcal{V},K,X) = \inf_{A\in\overline{\mathcal{A}}^n(X),\phi\in\Phi_n(C(T))}\sup_{f\in K}\|\mathcal{V}f - A\circ\phi(f)\|_X.$$

Proposition 2.5. Suppose that the conditions of the Proposition 2.4 (except the conditions $m\in\mathbb{N}^d, \lambda\in\mathbb{Z}_+^d(m)$) are satisfied and $1\le\theta\le\infty$. Then, there exist the constants $c_4(d,\alpha,p,\theta,q,\lambda,D,U)>0$ and $n_0(d,\alpha,p,\theta,q,\lambda,D,U)\in\mathbb{N}$ such that for $n\ge n_0$ one can construct the mappings $\phi\in\Phi_n(C(D))$ and $A\in\overline{\mathcal{A}}^n(L_q(D))$ such that for any function $f\in(S_{p,\theta}^\alpha B)'(D)$ the following inequality holds:

$$\|\mathcal{D}^\lambda(f\mid_U) - (A\circ\phi(f))\mid_U\|_{L_q(U)} \le c_4 n^{-\mathfrak{m}}(\log n)^{(\mathfrak{m}+1-1/\max(p,\theta))(\mathfrak{c}-1)}, \qquad (2.5)$$

where

$$\mathfrak{m} = \mathfrak{m}(\alpha-\lambda-(p^{-1}-q^{-1})_++\mathfrak{e}),\ \mathfrak{c} = \mathfrak{c}(\alpha-\lambda-(p^{-1}-q^{-1})_++\mathfrak{e}) = \mathfrak{c}(\alpha-\lambda).$$

To prove the proposition, under its conditions, putting $\mathcal{J}=\{j\in\mathcal{N}_{1,d}^1:\alpha_j-\lambda_j-(p^{-1}-q^{-1})_+=\mathfrak{m}\}$, we fix a vector $\beta\in\mathbb{R}_+^d$, having the following properties: $\beta_j=1, j\in\mathcal{J}, \beta_j>1$ and $\beta_j^{-1}(\alpha_j-\lambda_j-(p^{-1}-q^{-1})_+)>\mathfrak{m}, j\in\mathcal{N}_{1,d}^1\setminus\mathcal{J}$.

Further, fixing $m\in\mathbb{N}^d$ such that $\lambda\in\mathbb{Z}_+^d(m)$, we take $\mathcal{K}^0=\mathcal{K}^0(d,m,D,U)\in\mathbb{Z}_+^d, \mathcal{N}=\{\nu_{\mathcal{K}^0+\kappa}^{d,m,D,U},\kappa\in\mathbb{Z}_+^d\}$ from the Lemma 2.2, and for $l=l(\alpha)$ we define the family of points

$$x_{\mathcal{K}^0+\kappa,\nu}^{d,l-\mathfrak{e},m,D,U} = \xi_{\delta,x^0}^{d,l-\mathfrak{e},\rho} = x^0+\delta\xi_{\mathfrak{e},0}^{d,l-\mathfrak{e},\rho} \text{ (see subsection 1.), when } x^0=2^{-\mathcal{K}^0-\kappa}\nu, \delta$$
$$= 2^{-\mathcal{K}^0-\kappa}, \rho\in\mathbb{Z}_+^d(l-\mathfrak{e}), \nu\in\mathbb{Z}^d: Q_{\mathcal{K}^0+\kappa,\nu}^d\subset D, \kappa\in\mathbb{Z}_+^d,$$

and for $r\in\mathbb{N}$ we consider the set of points

$$\{x_{\mathcal{K}^0+\kappa,\nu}^{d,l-\mathfrak{e},\rho,m,D,U}\mid\rho\in\mathbb{Z}_+^d(l-\mathfrak{e}),\nu\in\mathbb{Z}^d:Q_{\mathcal{K}^0+\kappa,\nu}^d\subset D,\kappa\in\mathbb{Z}_+^d:(\kappa,\beta)\le r\}.$$

The number of these points satisfy the following inequality:

$$\text{card}\{x^{d,l-\mathfrak{e},\rho,m,D,U}_{\mathcal{K}^0+\kappa,\nu} \mid \rho \in \mathbb{Z}_+^d(l-\mathfrak{e}), \nu \in \mathbb{Z}^d : Q^d_{\mathcal{K}^0+\kappa,\nu} \subset D, \kappa \in \mathbb{Z}_+^d : (\kappa,\beta) \leq r\} \leq$$

$$\sum_{\kappa \in \mathbb{Z}_+^d:(\kappa,\beta)\leq r} l^{\mathfrak{e}} \,\text{card}\{\nu \in \mathbb{Z}^d : Q^d_{\mathcal{K}^0+\kappa,\nu} \subset D\} \leq$$

$$c_5 2^r r^{\mathfrak{e}-1}.$$

For $n \geq n_0 = 2c_5$ we choose $r \in \mathbb{N}$ such that the relation

$$c_5 2^r r^{\mathfrak{e}-1} \leq n < c_5 2^{r+1}(r+1)^{\mathfrak{e}-1} \qquad (2.6)$$

is true, and construct the system of points

$$\{x^{d,l-\mathfrak{e},\rho,m,D,U}_{\mathcal{K}^0+\kappa,\nu} \mid \rho \in \mathbb{Z}_+^d(l-\mathfrak{e}), \nu \in \mathbb{Z}^d : Q^d_{\mathcal{K}^0+\kappa,\nu} \subset D, \kappa \in \mathbb{Z}_+^d : (\kappa,\beta) \leq r\}$$

corresponding to this $r \in \mathbb{N}$. Then, under the conditions of the proposition (also see (1.1) and the Proposition 2.1) one can easily see that there exist the mappings $\phi \in \Phi_n(C(D))$ and $A \in \overline{\mathcal{A}}^n(L_q(D))$ such that for any $f \in (\mathcal{S}^\alpha_{p,\theta}\mathcal{B})'(D)$ the representation

$$A \circ \phi(f) = \sum_{\kappa \in \mathbb{Z}_+^d:(\kappa,\beta)\leq r} (\mathcal{D}^\lambda(\mathcal{R}^{d,l-\mathfrak{e},m,D,U,\mathcal{N}}_{\mathcal{K}^0,\kappa} f))\mid_D$$

is valid. At that, for any $f \in (\mathcal{S}^\alpha_{p,\theta}\mathcal{B})'(D)$ the inequality (2.5) can be deduced using (2.4), (2.2) taking into account (1.1), (2.6).

When $d \in \mathbb{N}$ we denote the set

$$\Sigma^d = \{\sigma \in \mathbb{Z}^d : \sigma_j \in \{-1,1\}, j = 1,\ldots,d\}.$$

Theorem 2.6. When $d \in \mathbb{N}$, suppose that D is a bounded domain in $\mathbb{R}^d$, for which there exists a system of open subsets $\{U_i \subset D, i = 1,\ldots,\mathcal{I}\}$ such that when $i = 1,\ldots,\mathcal{I}$ there exist $\delta^i \in \mathbb{R}_+^d$ and $\sigma^i \in \Sigma^d$, for which the inclusion $(U_i + \sigma^i \delta^i I^d) \subset D$ holds and $D = \cup_{i=1}^{\mathcal{I}} U_i$. Also, suppose that $\alpha \in \mathbb{R}_+^d, 1 \leq p < \infty, 1 \leq \theta, q \leq \infty, \lambda \in \mathbb{Z}_+^d$ and the inequalities (2.1) and (2.3) are satisfied. Then, there exist the constants $c_6(d,\alpha,p,\theta,q,\lambda,D) > 0$ and $N_0 \in \mathbb{N}$ such that for $\mathcal{V} = \mathcal{D}^\lambda, D(\mathcal{V}) = \{f \in C(D) : \mathcal{D}^\lambda f \in L_q(D)\}, X = L_q(D), K = (\mathcal{S}^\alpha_{p,\theta}\mathcal{B})'(D)$ if $N \geq N_0$ the following inequality holds:

$$\overline{\sigma}_N(\mathcal{V},K,X) \leq c_6 N^{-\mathfrak{m}}(\log N)^{(\mathfrak{m}+1-1/\max(p,\theta))(\mathfrak{e}-1)}.$$

Thus the upper estimate of the quantity under consideration is obtained.

For lower estimate it was possible to establish the following assertion.

Theorem 2.7. When $d \in \mathbb{N}$, suppose that $D \subset \mathbb{R}^d$ is bounded domain, and for $\alpha \in \mathbb{R}_+^d, 1 \leq p < \infty, 1 < q < \infty, \lambda \in \mathbb{Z}_+^d$ the relations (2.3), (2.1) are valid and $1 \leq \theta \leq \infty$. Also suppose that $\mathcal{V} = \mathcal{D}^\lambda, D(\mathcal{V}) = \{f \in C(D) :$

$\mathcal{D}^\lambda f \in L_q(D)\}, X = L_q(D), K = B((S_{p,\theta}^\alpha B)'(D))$. Then, there exists a constant $c_7(\mathcal{V}, K, X) > 0$ such that if $n \in \mathbb{N}$ the following inequality is true:

$$\sigma_n(\mathcal{V}, K, X) \geq$$
$$c_7 \begin{cases} n^{-(\mu-(1/p-1/q)_+)} \times \\ \quad (\log n)^{(\gamma-1)(\mu-(1/p-1/q)-(1/\theta-1/p)_+)-(1/q-1/\mathfrak{Q})_+}, & \text{when } q \leq p \text{ or } 2 \leq p < q, \\ n^{-(\mu-(1/p-1/q)_+)}(\log n)^{(\gamma-1)(1/\mathfrak{Q}-1/\theta)_+}, & \text{when } p < \min(2,q), \end{cases}$$

where $\mu = \mathfrak{m}(\alpha - \lambda), \gamma = \mathfrak{c}(\alpha - \lambda), \mathfrak{Q} = \max(2, q)$.

References

1. Nikolskii, S.M.: Approximation of Functions of Several Variables and Imbedding Theorems. Springer-Verlag, Berlin (1975)
2. Besov, O.V., Il'in, V.P., Nikolskii, S.M.: Integral Representations of Functions and Imbedding Theorems, Wiley (1978)
3. Kudryavtsev, S.N.: Reconstruction of derivatives from values of functions in Nikolskii-Besov spaces. Proc. Steklov Inst. Math. (2020)
4. Triebel, H.: Theory of Function Spaces II. Birkhäuser Verlag, Basel (1992)

A Neural Network Approach to Longitudinal Vehicle Acceleration Control

Ivan Gromov

Federal Research Center "Computer Science and Control" of the Russian Academy of Sciences, 44/2 Vavilova Street, 119333 Moscow, Russia
8357743@gmail.com

Abstract. In this paper, a self-driving vehicle longitudinal acceleration controller based on a neural network inverse model of the control object is being developed. The advantages of the neural network approach to the inverse model in comparison with classical control theory methods are discussed. Moreover, the paper describes an approach to data collection and processing for training an artificial neural network car longitudinal acceleration controller. A metric for evaluating the quality of the acceleration controller is proposed. The dependencies between the parameters of an artificial neural network controller and the quality of control are revealed.

Keywords: Longitudinal control · Inverse vehicle model · Acceleration tracking · Self-driving vehicle · Artificial neural network

1 Introduction

In recent years, thanks to the intensive self-driving vehicles development by large global companies such as Waymo, Tesla, Huawei, Yandex and others, the quality of popular open source solutions in the field of self-driving vehicles control has been growing. This phenomenon leads to an increase in the use of open source solutions [1,2] by startups, universities and enthusiastic engineers. Unfortunately, trying to apply a ready-made solution to driving a real car, researchers inevitably face a number of difficulties. The great part of the problems is caused by connecting software development, necessary to convert the upper-level self-driving system control commands to the format, expected by the drive-by-wire car application programming interface.

The task of control action conversion is solved in most other works [3–7], dedicated to the longitudinal control of autonomous vehicles, using an inverse model of the control object. An inverse model, as described in [9], is artificial dynamic system, which is used to compensate the physical system dynamics. Thus, with respect to vehicle dynamics, inverse model usually converts the expected output of direct vehicle model to its input, need to be applied to reach expected output.

In [3], the authors emphasize the nonlinear nature of self-driving car model and the presence of a delay in it, which complicates development of such a model. Therefore, the authors of this paper propose to define the inverse model as a "black box" and afterwards solve the problem of its identification. In [4], the same researchers, based on this approach, presented their longitudinal control action transformation algorithm, based on the nonparametric identification of the correspondence map between the vehicle speed, the control action magnitude and the traction force. Similar identification methods have formed the basis of other works [5–7] devoted to the self-driving vehicle control algorithms, currently widely used in industry.

In [8–11], researchers rely on analytical inverse models in the form of differential equation systems, the parameters of which are either determined by the identification procedure or specified according to the available documentation data about the control object. The key disadvantage of this approach is the high complexity of the inverse vehicle model, and, consequently, the high labor costs for its development and identification.

In addition, a number of researchers in their works [12–15] solved the problem of control action conversion without an inverse model by building a feedback acceleration tracking system based on classical control methods. Thus, there are solutions using fuzzy logic controllers to control longitudinal acceleration [12], nonlinear model-predictive control [13], analytical methods of classical nonlinear control theory [14], PID regulation [15], etc. However, such methods do not take into account the change in the dynamic properties of the car while it aims at determination of robust solution that does not take into account the nonlinear nature of the control object model.

Algorithms for the longitudinal acceleration of a car tracking based on an inverse vehicle model have gained more popularity than approaches based on classical control methods, but most of these methods are labor-intensive. In this paper, an alternative method of longitudinal control action conversion, based on a multi-layer artificial neural network, using for an vehicle inverse model obtaining, is proposed.

2 Problem Statement

The application software interface of self-driving vehicle usually involves controlling the engine and brake separately from each other using different control signals. At the same time, existing high-level software products for self-driving cars development usually do not focus on the application programming interface of a particular car at all and operate with reference longitudinal acceleration, as an output. Thus, self-driving cars developers are often faced with the need to develop a binding software that converts the output control action of a high-level control system into actions, expected by the application software interface of the car.

As part of the current work, a solution to the problem of a longitudinal control action conversion, based on a multi-layer artificial neural network is presented.

A real car with an application software interface is considered as control object, waiting for an input engine control signal in percentages from 0 to 100 and a brake control signal in m/ss from −7 to 0, which implies the presence of a deceleration regulator in the car, implemented in the electronic stability control unit in the car hardware system, which in practice occurs quite often.

Thus, in the case of acceleration of the car, the problem posed is reduced to converting the reference acceleration in m/ss into a engine torque request in percents, and, in the case of deceleration of the car, to end-to-end transmission of the control action, as shown in Fig. 1. To solve the data conversion problem under consideration, various information about the condition of the car can be used, for example, the current longitudinal acceleration, current speed, etc.

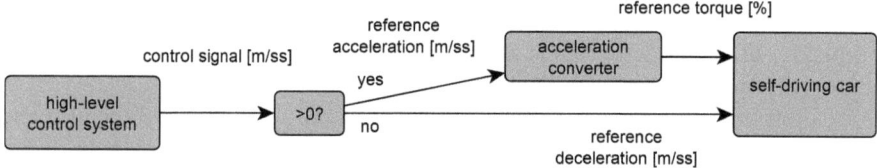

Fig. 1. Problem statement

3 PID-Based Approach

It should be remembered that problem solution based on a multi-layer artificial neural network is only one of many possible solutions. Nevertheless, it is the proposed approach that allows, unlike others, to obtain a solution of satisfactory quality in a short time. So, for comparison with the proposed approach, let's consider one of the most obvious solutions by development a car longitudinal acceleration feedback control system based on a PID controller (see Fig. 2). Figure 3 shows acceleration changes of a car during PID controller based solution test, containing graphs of the dependence on the time of the reference longitudinal acceleration of the car and the current longitudinal acceleration of the car. As

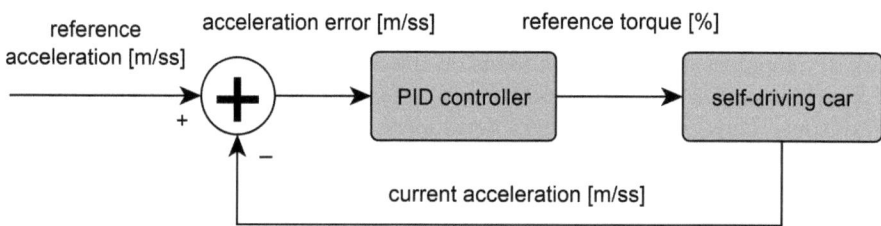

Fig. 2. Acceleration PID-based solution

you can see, the error between the reference acceleration and the current acceleration the end of test is significantly less than in other parts of experiment. This is due to the fact that the car, as an control object, has different dynamic properties at different transmission gears and, therefore, requires individual adjustment of the PID controller for each of the gears. In the example under consideration, the PID controller was adjusted for the second gear? on which it works in the end of experiment. Individual adjustment of the PID controller for each of the 7 transmissions of the car's transmission is labor-intensive work. In addition, such a solution has an additional disadvantage: self-oscillations arising from the use of a regulator with variable coefficients (Fig. 4). Thus, the experiment showed that, despite its apparent simplicity, longitudinal vehicle acceleration tracking approaches based on classical methods require high labor costs because of difficult controller parameters tuning by an expert and often do not allow obtaining a robust solution with satisfactory quality for the entire range of conditions of the control object. It should be noted that other classical methods, for example, fuzzy controllers, assume no less labor-intensive adjustment of internal parameters than in the presented experiment and are equally inclined to provide sufficient control quality only in a very limited range of states of the control object. Attempts to make the controller adaptive further complicate the development process, requiring tuning more internal parameters and, sometimes, leading to specific behavior of a nonlinear control object, for example, to self-oscillation, as was shown in the example with an adaptive PID controller.

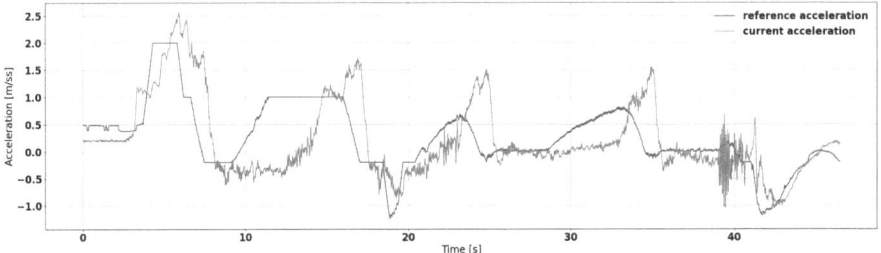

Fig. 3. The result of testing a solution based on a PID acceleration controller

4 Data Collection and Preprocessing

To train a neural network model for acceleration conversion into the required engine torque, a data set was collected on a 2 km long road section. It should be noted that the route on which the data set was collected has a rectilinear shape. This is due to the fact that the desired law of acceleration conversion at the moment of the engine does not significantly depend on whether the car is turning: at the current time or not.

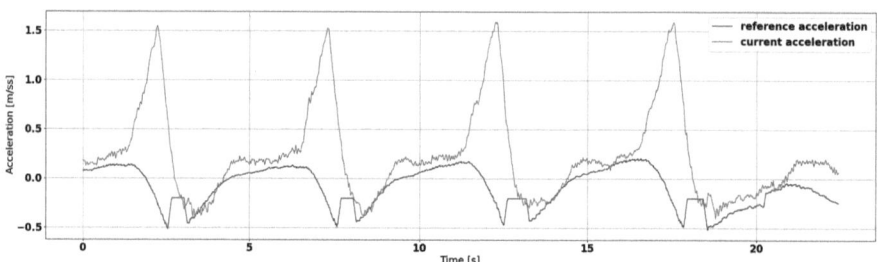

Fig. 4. Self-oscillation of acceleration when using a PID controller with variable coefficients

A slightly more serious problem is the lack of ascents and descents on the track, since the acceleration of the car is measured using a MEMS accelerometer, and the tilt of the car introduces a static error in the measured acceleration value. This problem will be solved at the data preprocessing stage by creating an acceleration observer that combines data obtaining from the accelerometer and wheel speed sensors.

The training data collecting process consisted of several experiments, during each of which the car was given some constant input engine torque request and the state of the car was measured. Graphs of car state changes during the experiment are shown in Fig. 5.

They vividly illustrate the dependence of the car longitudinal acceleration on the current transmission gear. As you can see, while the car is moving in first gear, there is an obvious dependence between longitudinal acceleration and the reference torque of engine that is close to proportional and only slightly dependence between longitudinal acceleration and current speed of the car. In should be noted, that during the transmission gear shifting, the proportionality coefficient between the reference torque and the acceleration of the car changes dramatically. Thus, the acceleration of car depends on the current gear, the current speed and the reference torque of the engine. However, these three values have a greater impact not on the current acceleration of the car, but on the acceleration of the car at some future point in time, the time interval to which is determined by the delay in the control system. The control delay for the car given in the experiment was estimated by the results of the experiment at 0.8 s.

As a result, a dataset consisting of 5 columns was formed based on the recorded log file: current transmission gear, current speed, engine reference torque, future longitudinal acceleration, current longitudinal acceleration (since it obviously also affects the value of future longitudinal acceleration). Then, a Kalman filter was applied to the columns containing acceleration to eliminate the random noise of the MEMS accelerometer and eliminate its error from the tilt of the car by fusing the accelerometer measured data with the first derivative of the current car speed, measured due to wheel speed sensors. Eventually, after clearing the data from emissions and areas where the car was manually controlled, a dataset was formed to train the inverse model of the car.

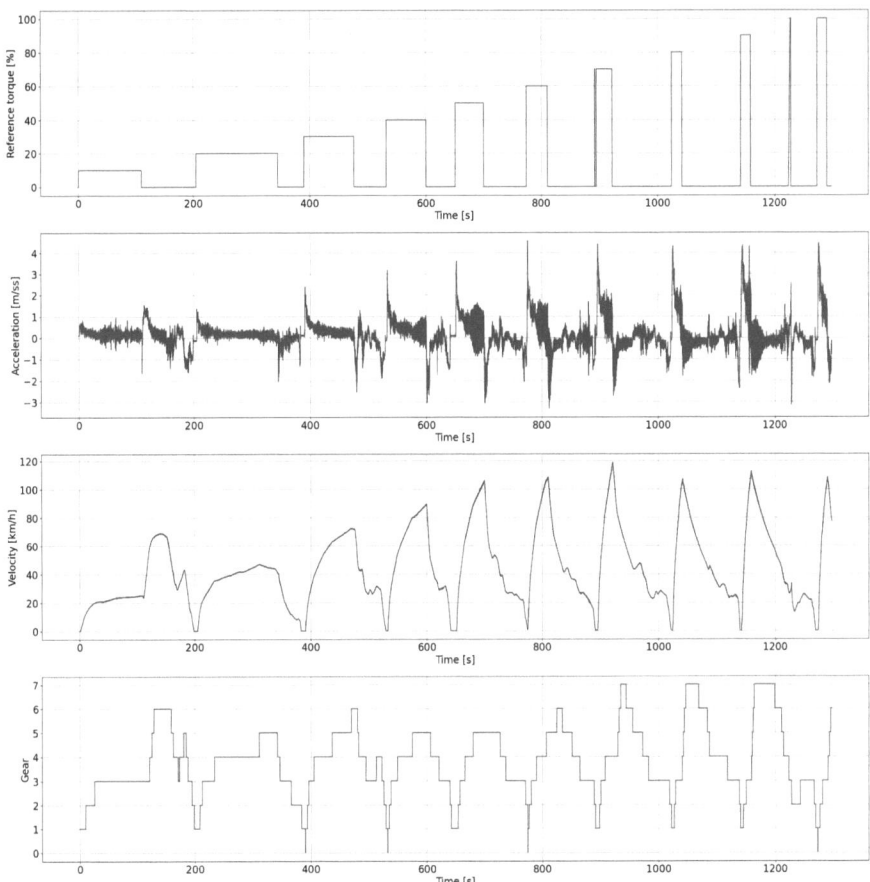

Fig. 5. The collected data set

5 Multi-layer Neural Network Approach

As was shown in the previous chapter, there is a dependence between the future longitudinal acceleration of the car and the following values: the current longitudinal acceleration, the current speed, the current transmission gear and the current reference torque of the engine. Equation (1), reflecting this dependence, is a model of the longitudinal motion of the car. However, as part of the vehicle reference acceleration conversion to the required engine torque task, an inverse dependence is required (2). Equation (2) is called the inverse longitudinal motion model and allows you to determine the reference torque of the engine, necessary to achieve a given future acceleration at certain current speed, longitudinal acceleration and transmission gear. In approach considered in this paper, dependence (2) is proposed to be approximated using an artificial neural network.

$$a_{long}(t+\delta) = f(a_{long}(t), v_{long}(t), gear(t), T_{ref}(t)) \tag{1}$$

$$T_{ref}(t) = g(a_{long}^{ref}(t), a_{long}(t), v_{long}(t), gear(t)) \qquad (2)$$

It is proposed to use a multi-layer artificial neural network with 4 inputs and 1 output as am inverse model (Fig. 6). In addition to an artificial neural network, a frequency filter is included in the proposed solution in Fig. 6. Its necessity is due to random fluctuations in the output of the neural network, which in the presented solution can cause spontaneous longitudinal oscillations of the car, therefore negatively affecting the comfort of passengers. The choice of the limitation value was based on the fact that the first derivative of the reference moment is approximately proportional to the first derivative of the longitudinal acceleration (i.e., the longitudinal jerk of the car). The proportionality coefficient between the derivative of the reference torque and the longitudinal jerk is assumed to be equal to the proportionality coefficient between the reference torque and the longitudinal acceleration for each of the transmission gears. Thus, the estimation of the limitation of the first derivative of the neural network output T'_{imax} is carried out according to the following formula (3), the parameters of which were taken from a previously collected dataset.

$$T'_{imax} = j_i^{max} \frac{T_i}{a_{mid}} \qquad (3)$$

where a_{mid} is the average longitudinal acceleration in i-th gear at a reference torque of T_i percents, j_i^{max} is the maximum longitudinal jerk taken for reasons of passenger comfort [16].

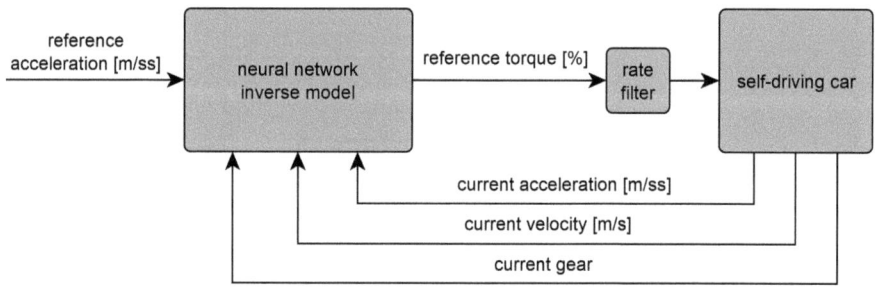

Fig. 6. Controller based on artificial neural network inverse model approach

6 Experimental Results

During the experiment, 10 neural network models were trained on previously obtained and preprocessed data, which differ from each other in the number of layers, the number of neurons in each layer, the size of the control delay taken into account at the data preprocessing stage and the use of a dropout. The PyTorch framework was used to train neural networks, after which the neural network was implemented into self-driving vehicle control system that sends a control action to the car with a frequency of 100 Hz. After tests on the validation part of the training sample, the LeakyReLU function with parameter 0.15 was selected as the activation function of the neural network. The AdaDelta method was used to optimize the weights of the neural network. Each neural network was trained under the same conditions with the same number of epochs and the same batch size. Each of the trained neural networks was tested on the same straight track, performing the task of holding several consecutive previously specified values of speed. Table 1 presents the test results of control action conversion algorithms based on various neural network models. Two metrics were used to assess the quality of acceleration tracking. The first one (4) is the square root of car longitudinal acceleration error square integral, related to the length of the route. The second metric (5) is equal to the first metric multiplied by the time, car passes the test route and is a dimensionless value. Based on the results of the experiments, it can be concluded that dropout and a decrease in the control delay taken into account lead to an improvement in the quality of the solution (at the same time, delay taken into account should not be reduced to a value significantly lower than the actual delay). Figure 7 shows the result of the acceleration converter in one of the best experiments. The blue graph shows the reference acceleration, the orange graph shows the current acceleration.

$$J_1 = \frac{\sqrt{\int_0^{t_{total}} (a_{long}^{ref}(t) - a_{long}(t))^2 \, dt}}{s_{total}} \tag{4}$$

$$J_2 = J_1 t_{total} \tag{5}$$

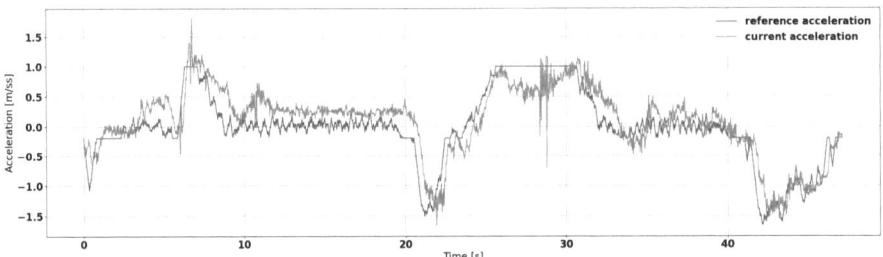

Fig. 7. The result of testing a solution based on a neural network inverse model

Table 1. Artificial neural network inverse model parameters affecting on solution performance

№	Hidden layers	Neurons per hidden layer	Delay [s]	Dropout	Metric (4)	Metric (5)
1	3	6	1.0	–	0.04	4.079
2	6	20	1.0	–	0.034	1.706
3	3	6	0.8	–	0.037	1.716
4	6	20	0.8	–	0.036	1.838
5	3	6	0.6	–	0.036	1.835
6	6	20	0.6	–	0.025	1.3
7	3	6	0.8	+	0.028	1.418
8	3	6	0.6	+	0.026	0.85
9	3	6	1.0	+	0.027	1.221
10	1	5	0.8	+	0.029	1.27

7 Conclusion

Thus, in this paper, a conversion self-driving vehicle longitudinal acceleration problem solution was presented. The proposed approach may be useful for researchers in the field of self-driving cars to combine open source software solutions and self-driving hardware platforms. The method based on the inverse neural network model differs from other methods in its simplicity and speed of deployment on board the car computer. The disadvantages of the presented solution include the complexity of interpreting the results of the neural network, which is an important aspect in the context of vehicle control tasks. It is planned to solve this problem in future works.

References

1. Kato, S., et al.: Autoware on board: enabling autonomous vehicles with embedded systems. In: 2018 ACM/IEEE 9th International Conference on Cyber-Physical Systems (ICCPS), IEEE (2018)
2. Fan, H., et al.: Baidu apollo em motion planner. arXiv preprint arXiv:1807.08048 (2018)
3. Wang, Y., Bin, Y., Li, K.: Longitudinal acceleration tracking control of low speed heavy-duty vehicles. Tsinghua Sci. Technol. **13**(5), 636–643 (2008)
4. Feng, G., JianQiang, W., Keqiang, L.: Hierarchical switching control of longitudinal acceleration with large uncertainties. In: 2006 IEEE International Conference on Vehicular Electronics and Safety, IEEE (2006)
5. Zhu, Q., et al.: An adaptive longitudinal control method for autonomous follow driving based on neural dynamic programming and internal model structure. Int. J. Adv. Robot. Syst. **14**(6), 1729881417740711 (2017)
6. Wang, J., et al.: Adaptive speed tracking control for autonomous land vehicles in all-terrain navigation: an experimental study. J. Field Robot. **30**(1), 102–128 (2013)

7. Sharma, S., Tewolde, G., Kwon, J.: Lateral and longitudinal motion control of autonomous vehicles using deep learning. In: 2019 IEEE International Conference on Electro Information Technology (EIT), IEEE (2019)
8. Wang, S., et al.: Neural network sliding mode control of intelligent vehicle longitudinal dynamics. IEEE Access **7**, 162333–162342 (2019)
9. Dias, J.E.A., Pereira, G.A.S., Palhares, R.M.: Longitudinal model identification and velocity control of an autonomous car. IEEE Trans. Intell. Transp. Syst. **16**(2), 776–786 (2014)
10. Rajamani, R.: Vehicle Dynamics and Control. Springer (2011)
11. Bünte, T., et al.: Inverse model based torque vectoring control for a rear wheel driven battery electric vehicle. IFAC Proc. **47**(3), 12016–12022 (2014)
12. Qu, T., et al.: Multi-mode switching-based model predictive control approach for longitudinal autonomous driving with acceleration estimation. IET Intell. Transp. Syst. **14**(14), 2102–2112 (2020)
13. Caporale, D., et al.: Towards the design of robotic drivers for full-scale self-driving racing cars. In: 2019 International Conference on Robotics and Automation (ICRA), IEEE (2019)
14. Pedone, S., Fagiolini, A.: Robust longitudinal control of self-driving racecar models. In" 2022 European Control Conference (ECC), IEEE (2022)
15. Marcano, M., et al.: Low speed longitudinal control algorithms for automated vehicles in simulation and real platforms. Complexity **1**(2018), 7615123 (2018)
16. Bae, I., Moon, J., Seo, J.: Toward a comfortable driving experience for a self-driving shuttle bus. Electronics **8**(9), 943 (2019)

Algorithms of ECG Time Series Processing in EDF-Format

Sinan V. Kurbanov[1], Denis A. Andrikov[1(✉)], and Aleksandr E. Khramov[2]

[1] Peoples' Friendship University of Russia named after Patrice Lumumba, 6 Miklukho-Maklaya Street, Moscow 117198, Russia
andrikovdenis@mail.ru
[2] Baltic Centre for Neurotechnology and Artificial Intelligence, 14A Nevskogo Street, Kaliningrad 236016, Russia

Abstract. The Python script for reading digital cardiac signal in EDF format and its spectral analysis was developed. The discrete Fourier series decomposition of one period of the cardiac signal was performed. The deviation of the Fourier image from the original cardiac signal was investigated. This deviation was found to be minimal if the number of harmonics is two times less than the number of cardiac signal readings within the duration of the R-R interval. The adequacy of the author's software for spectral research was confirmed by synthesizing the signal from the Fourier image and comparing the original and synthetic signals. A practically functional dependence of the spectrum on the type of cardiac signal envelope was revealed. The sensitivity of the spectral analysis compared with visual identification of the cardiac waveform is evaluated. The applicability of the Fourier image of the cardiac signal for diagnosing temporary distortions of the heart rhythm, as well as the use of this image as a vector of the state of the heart, has been established. The prospects for developing this research are noted, for example, using methods of mathematical statistics and practical testing of the results with the participation of cardiologists. The theoretical significance of the proposed study consists in specifying the methodology of the Fourier transform application for the analysis of cardiac signals for diagnostic purposes using computer technologies. The results obtained in the course of the study are of practical value for the development of medical devices and software.

Keywords: Cardiac Signal · EDF-file · Fourier Transformation · Cardiac Signal Spectra

1 Introduction

Computer modeling of the human cardiovascular system is an urgent task in modern science. The construction of a maximally complete model would allow intensifying scientific research, for example, because of the possibility of conducting virtual experiments. Currently, there are many approaches to modeling the cardiovascular system. Attempts have been made to develop such a model using various methods. Among them there are geometric and physical models [1], two-chamber and four-chamber models, kinetic and three-dimensional geometric models. There are models based on the decomposition of a complex biological system of the heart, with the allocation of subsystems amenable to adequate mathematical description in accordance with their functional role. These are the so-called modular models [2]. An approach is also used, the essence of which is the replacement of a complex biological object by some technical construction. Within the framework of such an approach [3], the heart is represented by a four-chamber pump, and the circulatory system by a rather complex pipeline. The model uses hydrodynamic equations to describe structural units and their interactions. All these models are rather complex and, as a rule, cannot represent the heart in all its diversity. Therefore, the proposed method proposes a hypothesis about the possibilities of assessing the state of the heart as a dynamic system based on the spectrum of an electrocardiographic signal limited to one R-R interval. The mathematical model describing the state of cardiac activity is a vector of several first amplitudes of decomposition of one period of heartbeats into Fourier series, and, therefore, the graphical representation of this vector (spectrogram) provides an assessment of the state of the heart. The algorithm is loading, preprocessing, Fourier transform, and visualization of what is happening. The algorithm is implemented in Python and uploaded on GitHub. The results of the study may have practical value as a tool for applied use in cardiology—both for medicine and for use in medical technology.

2 Problem Statement

The modern trend in functional diagnostics is to obtain maximum information with minimal impact on the patient's body. Such a method of non-invasive research is the method of registration electrocardiograms, as the most common in clinical practice at present. Spectral diagnostic methods [4–7] based on Fourier transform and wavelet transform are being developed. As stated in [8], the assessment of heart rate variability, also called heart rate variability (HRV) analysis, as a clinical practice has been developed since the early 1960s. This process was facilitated by the application of mathematical statistics methods, algorithms of biological signal processing, and the development of physiological interpretation of the obtained data. In the future, HRV will be separated into an independent, non-invasive method in cardiology. The method is actively developing at present [9–12]. The subject of HRV analysis is mainly the so-called sinus arrhythmia, reflecting the complex processes of interaction of various circuits of heart rhythm regulation. The HRV method is based on the recognition and measurement of time intervals between R-beats of the electrocardiogram (ECG), construction of dynamic series of cardiac intervals, and subsequent analysis of the obtained numerical series by various mathematical methods. When analyzing, a distinction is made between short-term (up to units of hours) and long-term (lasting a day or more) recordings. Methods of cardiac interval analysis are divided into visual and mathematical methods [9]. Mathematical methods fall into three broad classes:

1. The study of overall variability (statistical methods or temporal analysis);
2. Study of periodic components of HRV (frequency analysis);
3. Study of the internal organization of the dynamic series of cardiointervals (autocorrelation analysis).

For the purposes of identification of heart states as a system, we detect cardiac rhythm disorders by studying spectral densities of short-term ECG recordings. Spectral analysis is a sensitive research tool based on the Fourier transform. This transformation, based on the time function $f(t)$ known for some signal, allows the construction of a frequency function $F(\omega)$ describing the same signal. The Fourier transform has some mathematical limitations for the original time function, but it can be performed on any physical signal [13], in particular on the cardiac signal [7,14].

As it is known, an electrocardiographic signal is periodic with period T, so this signal can be decomposed into a Fourier series of the following form (1):

$$f(t) = a_0 + \sum_{n=1}^{\infty}(a_n * cos\omega_n t + b_n * sin\omega_n t); \omega_n = \frac{2\pi}{T} * n; \qquad (1)$$

where

$$a_0 = \frac{1}{T}\int_0^T f(t)\,dt; a_n = \frac{2}{T}\int_0^T f(t)cos\omega_n t\,dt; b_n = \frac{2}{T}\int_0^T f(t)sin\omega_n t\,dt \qquad (2)$$

Due to the replacement of analog devices by digital ones, the actual standard for recording electrocardiographic signals is the EDF-format, where the signal $f(t)$ is represented by the grid function $f(t_k)$ – a set of voltage values at some fixed moments of time t_k. The ECG signal recording is presented in EDF-file format. The European Data Format (EDF) was introduced in 1992 as the standard for EEG and PSG (sleep) recordings [15]. A Python script for reading ECG data and processing it is available in the public domain.

The improved EDF+ format allows multiple non-contiguous recordings to be stored in a single file. This is the only incompatibility with EDF. Using EDF+, all signals, annotations, and events recorded in a single session using a single recording system can be safely stored together in a single file. EDF+ can also store only events and annotations without any signals. This flexibility allows you to choose the optimal combination.

A Python script reads a header and an arbitrary record from a file according to a standard information structure. The header contains general information (patient ID, start of record, end of record, etc.). The data of all signals is returned in a data matrix; each column of this matrix is a discrete set of all values of one of the signals.

The first 256 bytes of the EDF file contain general information about the format itself, patient data, and data about the signal recordings, including the number of signals (ns). This data is supplemented by blocks of 256 bytes for each signal, indicating the type of signal by the nature of the information contained (e.g., body temperature, cardiogram, etc.), the amplitude analog and digital, the duration of the signal in seconds, and the number of discrete values. Thus, the header block contains 256 + (ns * 256) bytes.

The header block is followed by an array of discrete values of signals; each value is allocated 2 bytes; the value is discrete, represented by an integer with a sign. The position of the individual entries in the header block is as follows (each byte represents an ASCII character):

8 bytes: data format version (default 0);
80 bytes: patient identification data;
80 bytes: record identification data;
8 bytes: record start date (dd.mm.yy);
8 bytes: record start time (hh.mm.ss);
8 bytes: number of bytes in the header block;
44 bytes reserved;
8 bytes: number of data records (–1 if unknown, each record may contain multiple signals);
8 bytes: duration of the signal record in seconds;
4 bytes: number of signals in the record (ns);
further, the multiplier (ns ×) means that the parameter is recorded for each signal:
ns × 16 bytes: ns of signal labels (meaning signal names, e.g. 'ECGV2Ref' or 'Body temp');
ns × 80 bytes: type of converter (for example, Ag/AgCl electrode);
ns × 8 bytes: unit of measurement of the physical quantity (mV - millivolts, degree - degrees Celsius, etc.);
ns × 8 bytes: physical minimum of the signal;
ns × 8 bytes: physical maximum of the signal;
ns × 8 bytes: digital minimum of the signal;
ns × 8 bytes: digital maximum signal;
ns × 80 bytes: filtering parameters when recording the signal (e.g., filter bandwidth);
ns × 8 bytes: number of digital values (nr) in the signal recording (product of the recording time by the sampling rate; e.g., a 3330 s recording with a sampling rate of 200 Hz would record 666000 here);
ns × 32 bytes reserved.

The array of discrete signal values is an nr × ns matrix (nr rows and ns columns), each column is a signal.

The Python script processes the first signal from the EDF file, by the specified number, which can be any within the ns parameter.

3 Research Method

3.1 Visual Analysis of Characteristic Areas of the Cardiogram and Allocation of the Heartbeat Cycle

To develop algorithms for processing ECG, an open database of critical conditions of the RSHMANE Institute is used: http://rohmine.org/baza-dannykh-rokhmine/testovaya-baza-kriticheskikh-sostoyaniy-2019-g/.

Figure 1 shows the waveform of the signal recovered from the EDF-file 01_GUSA.edf.

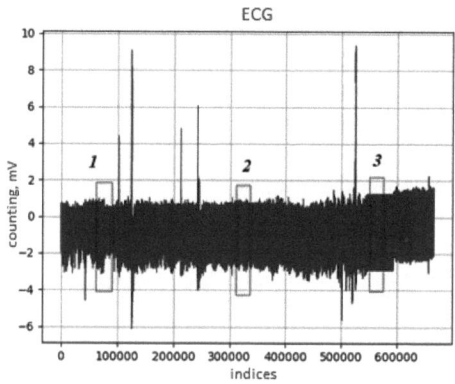

Fig. 1. Full cardiogram from EDF-file

The cardiogram is shown to scale, where the entire recorded signal is displayed over a long recording period. In such a form, no analysis can be made. Nevertheless, even in this form, at least three different sections can be distinguished on the cardiogram, labeled in Fig. 1 as 1, 2, 3. You can view the cardiogram in more detail by changing the time scale. For example, by selecting 400 points in Sect. 1, the part of the signal shown in Fig. 2 is obtained.

Here we can already observe a characteristic view of the cardiogram, which allows us to highlight one period of the heart rhythm. This period is presented in Fig. 3. The period is limited by points with indices 80232 and 80350.

Similarly, cardiograms were obtained in Sects. 2 (Fig. 4, the period is limited to points with indices 320095 and 320179) and 3 (Fig. 5, the period is limited to points with indices 550010 and 550063). Cardiograms were recorded with a sampling frequency of 200 Hz (found as the ratio of hdr.samples/hdr.duration); the number of values for the period and the value of the period in seconds for the Sections were 118 and 0.590 s; 84 and 0.420 s; 53 and 0.265 s, respectively.

Comparing the cardiograms in Sects. 1, 2, 3, we can conclude that the cardiogram in Sect. 1 is close to normal, in Sect. 2 there is a slight rhythm disturbance, and in Sect. 3, a pacemaker should probably be used. The first visual stage of analysis requires the participation of an expert cardiologist or the use of an advanced expert system.

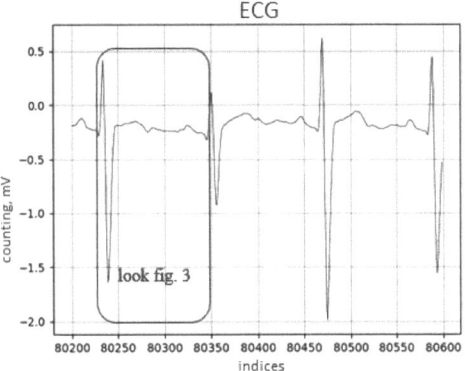

Fig. 2. Part of the cardiogram in Sect. 1

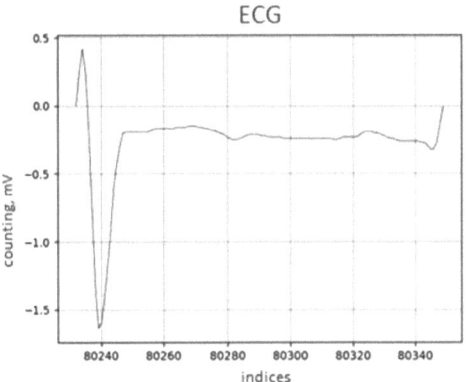

Fig. 3. The period of the cardiogram in Sect. 1

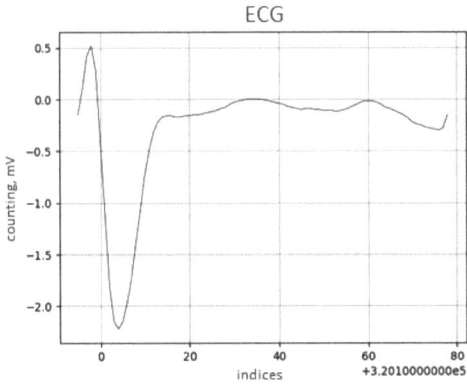

Fig. 4. The period of the cardiogram in Sect. 2

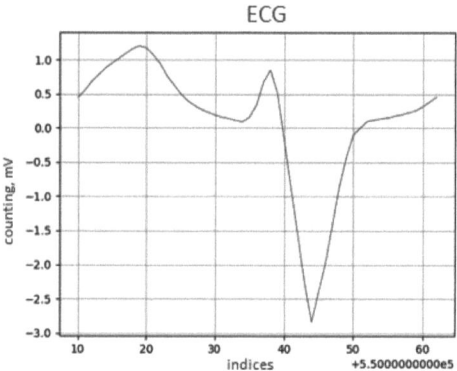

Fig. 5. The period of the cardiogram in Sect. 3

3.2 Spectrum Construction and Fourier Transform of Cardiac Signal

To expand the analysis capabilities, we will perform a Fourier transform to study the features of the ECG-signal spectrum from the presented EDF files. There are several algorithms for Fourier series decomposition [16]. These algorithms usually require a fixed and multiple of 2^N number of points per period; if this and a number of other conditions are met, such algorithms provide a gain in computational speed. In the case of cardiac signals, there is considerable variation in the number of points per period due to heart rate variability. Discretizing a cardiac signal with 2^N number of points per period in this case would require approximating the signal with polynomials and computing the values of the approximating function on a grid of 2^N values of the argument. Such transformations distort the original signal and level out the gain in speed. In the light of the above, due to the small number of points per period and the presumably small number of calculated harmonics, preference was given to the direct calculation of integrals by formula (2) using the trapezoidal method, which is a compromise between other methods in terms of accuracy and volume of calculations. The calculation formulas take the form:

$$\omega_n = \frac{2\pi}{T} * n; \quad a_0 = \frac{1}{2T} \sum_{k=1}^{m-1} (f_k + f_{k+1}) * (t_{k+1} - t_k); \tag{3}$$

$$a_n = \frac{1}{T} \sum_{k=1}^{m-1} (f_k * cos\omega_n t_k + f_{k+1} * cos\omega_n t_{k+1}) * (t_{k+1} - t_k); \tag{4}$$

$$b_n = \frac{1}{T} \sum_{k=1}^{m-1} (f_k * sin\omega_n t_k + f_{k+1} * sin\omega_n t_{k+1}) * (t_{k+1} - t_k); \tag{5}$$

where m is the number of cardiac signal samples within the R-R interval duration. The adequacy of the transformation was confirmed by synthesizing the signal based on part of the Fourier transform (1), and comparing the original and synthetic signals.

4 Results and Discussions

4.1 Analysis of Cardiac Signal Spectra - Numerical Results

The results of the experimental determination of the optimal number of harmonics are shown below. At $n = 5$ harmonics on the 3rd section according to Fig. 1 are shown in Figs. 6 and 7.

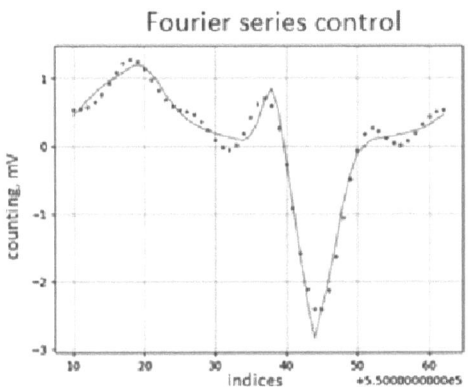

Fig. 6. Initial cardiogram (blue line) and partial sum of Fourier series (red points) at $n = 5$ harmonics (Color figure online)

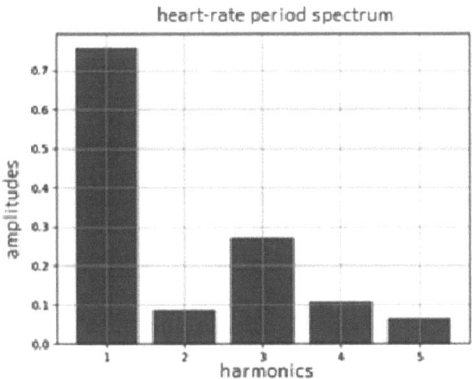

Fig. 7. Spectrum of the cardiogram $A(\omega)$ at $n = 5$ harmonics

As can be seen from Fig. 6, five harmonics are clearly not enough for a good approximation of the cardiac signal. For $n = 10$, the figure is Fig. 8, for $n = 20$ – Fig. 9, for $n = 40$ – Fig. 10.

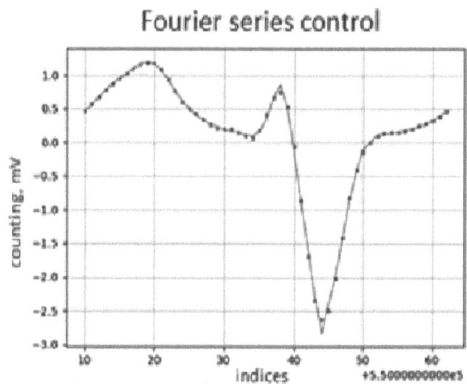

Fig. 8. Initial cardiogram (blue line) and partial sum of Fourier series (red points) at $n = 10$ harmonics (Color figure online)

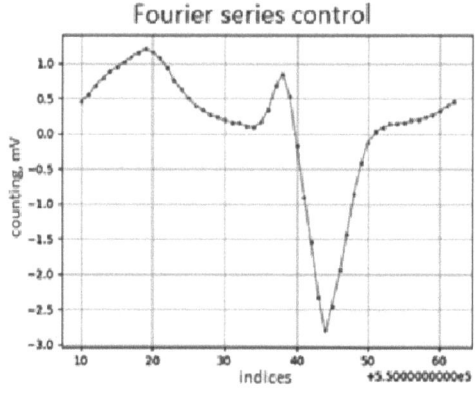

Fig. 9. Initial cardiogram (blue line) and partial sum of Fourier series (red points) at $n = 20$ harmonics (Color figure online)

Comparison of the figures shows that the deviation is minimal if the number of harmonics is twice less than the number of cardiac signal samples within the duration of the R-R interval; in the considered case it corresponds to $n = 26$. Experimental confirmation of this hypothesis is presented in Fig. 11. In this case, the coincidence of the partial sum of the Fourier series and the cardiac signal values is observed. A much smaller number of harmonics, as well as a much

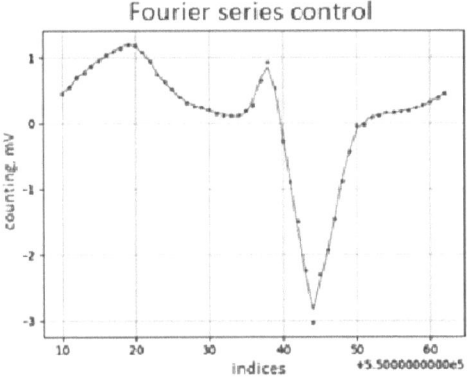

Fig. 10. Initial cardiogram (blue line) and partial sum of Fourier series (red points) at $n = 40$ harmonics (Color figure online)

larger number of harmonics, causes distortion of the shape of the reconstructed cardiac signal. Thus, the number of harmonics in Fourier series decomposition should be equal to half the number of cardiac signal values per period.

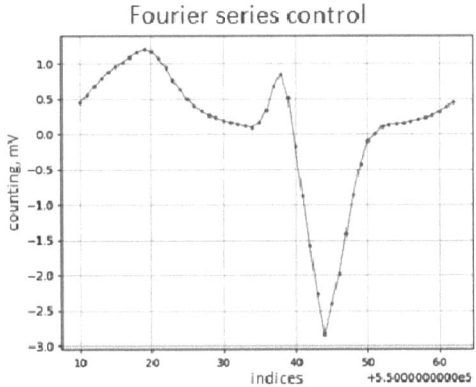

Fig. 11. Initial cardiogram (blue line) and partial sum of Fourier series (red points) at $n = 26$ harmonics (Color figure online)

The next step is comparison of the spectra in different sections of the cardiogram. The cardiograms used in this work were provided by the Russian Society of Holter Monitoring and Noninvasive Electrophysiology (RSHMANE). The spectra are shown in Figs. 12, 13 and 14.

Visual comparative analysis of cardiogram spectra allows us to assume that in normal cardiac signal shape there is a significant number of higher harmonics in the spectrum, their amplitude being approximately equal to the amplitude of the

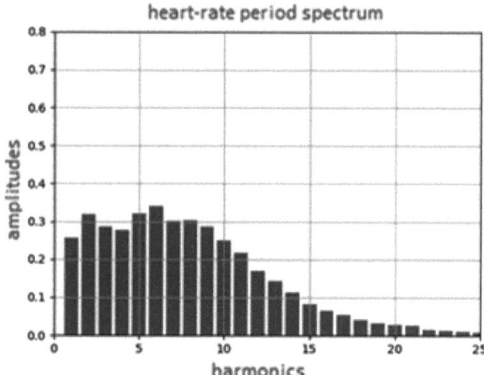

Fig. 12. Spectrum of the cardiogram in the first section

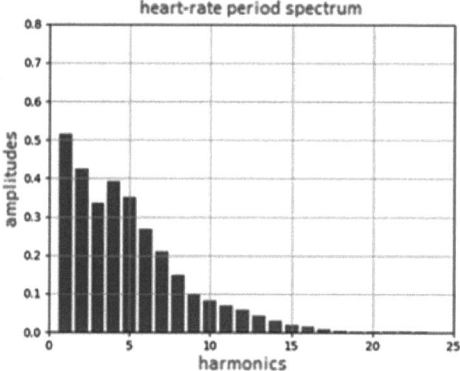

Fig. 13. Spectrum of the cardiogram in the second section

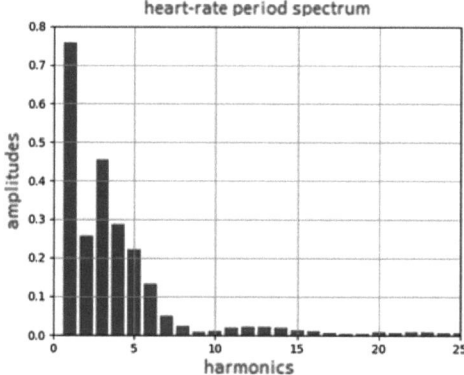

Fig. 14. Spectrum of the cardiogram in the third section

first harmonic and even exceeding it. As the arrhythmia develops, the amplitude of higher harmonics decreases compared to the first harmonic, and the number of harmonics of significant magnitude decreases. A visual comparison of the spectra suggests that the most pronounced changes in the spectrum are characteristic of the first 10...12 harmonics. The observed spectrum transformations allow us to conclude about the applicability of the Fourier image of a cardiac signal for diagnosing temporary heart rate distortions; at the same time, spectral analysis allows to determine the features of human states hidden in the time domain. To identify the patterns, a large volume of experiments and interpretation of the results by specialized specialists in the field of medicine are required.

5 Conclusion

In the process, the specification of digital cardiograms in EDF-format is reviewed, and a Python script is developed to extract this information. The decomposition into the classical Fourier series for one period of the cardiac signal was performed. It was found that the deviation of the Fourier image from the original cardiac signal is minimal if the number of harmonics is twice less than the number of cardiac signal samples within the duration of the R-R interval. The adequacy of the author's software for spectral study was confirmed by synthesizing the signal using the Fourier image and comparing the original and synthetic signals. By comparative analysis of the sections of the cardiogram, the clear dependence of the spectrum on the cardiac signal shape was confirmed, which allows us to conclude that the Fourier image of the cardiac signal is applicable for diagnosing temporary distortions of the heart rhythm. The hypothesis that characteristic changes in the spectrum can be detected earlier than they appear visually on ECG is formulated. At the same time, the identification of regularities requires a large volume of experiments and interpretation of their results by specialized specialists for diagnostic use. Nevertheless, the part of the cardiac signal spectrum within the duration of one R-R interval can be considered as a vector of the heart state and this limited spectrum can be used for diagnostics and construction of heart rhythm regulators as part of a pacemaker. To develop work on spectral research and design an interface that is understandable to the end user, the developed scripts are provided in the public domain at: https://github.com/TAUforPython/BioMedAI.git and the design of an interface is understandable to the end user. Without developing the software functionality, studying the dynamics of changes in the ECG spectrum seems difficult.

References

1. Drozd, D.D.: Basics of application of mathematical models in cardiology. Bull. Med. Internet Conf. **5**(9), 1140–1142 (2015). (in Russian)
2. Kiselev, I.N., et al.: Modular modeling of the human cardiovascular system. Math. Biol. Bioinf. **7**(2), 703–736 (2012). (in Russian)

3. Lebedenko, I.S., et al.: The mathematical model of the heart. Biotechnosphere **3**, 24–31 (2009). (in Russian)
4. Zakharov, S.M., Znaiko, G.G.: Spectral analysis of electrocardiosignals. Voprosy radioelektroniki **3**, 110–115 (2017). (in Russian)
5. Sergeychik, O.I.: Models and algorithms for spectral analysis of processing cardiac time series: abstract of the dissertation for the degree of Candidate of Technical Sciences. Institute of the Problems of Northern Development SB RAS; Tyumen Cardiology Center, Tyumen (2007). (in Russian)
6. Genlang, C., Zhiqing, H., Yongjuan, G., Chaoyi, P.: A cascaded classifier for multi-lead ECG based on feature fusion. Comput. Methods Programs Biomed. **178**, 135–143 (2019)
7. Sun, J.: Automatic cardiac arrhythmias classification using CNN and attention-based RNN network. Healthcare Technol. Lett. **10** (2023). https://doi.org/10.1049/htl2.12045, https://www.researchgate.net/publication/370168517
8. Snezhitsky, V.A., et al.: Heart rate variability: applications in cardiology. GrSMU, Grodno (2010). (in Russian)
9. Baevsky, R.M, Chernikova, A.G.: Heart rate variability analysis: physiologic foundations and main methods. Cardiometry 66—76 (2017). (in Russian)
10. Latfullin, I.A., in Russian, et al.: High-resolution ECG: from present to future. Russian J. Cardiol. **2**(82), 29–34 (2010)
11. Grinevich, A.A., Chemeris, N.K.: Spectral analysis of heart rate variability based on the Hilbert-Huang method. Proc. Russian Acad. Sci. Life Sci. **511**(1), 395—398 (2023). (in Russian)
12. Moskalenko, A.V., Makhortykh, S.A.: On spectral analysis of the regulation of the main cardiac rhythm. In: Proceedings of the International Conference "Mathematical Biology and Bioinformatics, vol. 9, Article (textnumero) e41. IMPB RAS, Pushchino (2022). (in Russian)
13. Rauscher, C.: Fundamentals of Spectrum Analysis. Rohde & Schwarz, Munich (2001)
14. Sawant, N., Patidar, S.: Diagnosis of cardiac abnormalities applying scattering transform and fourier-bessel expansion on ECG signals. 2021 Comput. Cardiol. (CinC) **48**, 1–4 (2021)
15. Alvarez-Estevez, D.: European data format, https://www.edfplus.info, Accessed 15 Dec 2023
16. Kandidov, V.P., Chesnokov, S.S., Shlenov, S.A.: Discrete Fourier transform. Faculty of Physics of Moscow State University, Moscow (2019). (in Russian)

On Homogeneous Observers for Linear Systems

V. Zhdanov(✉), D. Galkina, D. Konovalov, A. Kremlev, and K. Zimenko

ITMO University, 49 Kronverkskiy av., 197101 Saint Petersburg, Russia
viktor_zhdanov@itmo.ru

Abstract. The paper addresses the finite-time homogeneous observers design problem for linear systems. The stability conditions are formulated in the form of linear matrix inequalities being less conservative in comparison with other works. The proposed stability analysis relies on the homogeneity property and the implicit Lyapunov function method. The proposed approach can be further developed for a number of other homogeneous applications with similar sector nonlinearities.

Keywords: finite-time observer · homogeneity · implicit Lyapunov function method

1 Introduction

The challenge of designing nonlinear observers remains a dynamic area of research. Time-constrained state estimation, such as finite-/fixed-/prescribed-time observers (see, e.g., [1–11]), is particularly essential for fast control systems, time-critical control processes, and fault detection. Homogeneous observers (e.g., [1,6,10]) provide not only fast finite-/fixed-time convergence but also a range of robust properties: Input-to-State Stability (ISS) with respect to additive disturbances and measurement noise, and robustness with respect to delays (see, e.g., results [12–14] on robust properties of homogeneous systems).

In [1,10], homogeneous observers are proposed for linear systems with stability analysis (parameters tuning) based on Linear Matrix Inequalities (LMIs). However, these results have significant limitations (e.g., on the order of the observer, degree of homogeneity, settling time estimates, etc.) due to the conservative nature of the LMIs used for stability analysis. For instance, such constraints result in LMIs being solvable in cases where the homogeneous observer approximates the linear observer (with a degree of homogeneity close to 0).

In the paper [11], for homogeneous observers as in [3], stability conditions and parameter tuning are formulated to be less conservative based on the use of new Implicit Lyapunov Function (ILF) modification and analysis of system nonlinearities. This paper is an extension of [11]. The paper presents a new analysis of nonlinearities used in a homogeneous observer, which significantly relaxes stability conditions represented with LMIs for a wide range of homogeneity degree. The proposed approach can be further developed for a number of other homogeneous applications with similar sector nonlinearities, such as distributed observers (e.g., as in [9]), consensus protocols for multi-agent systems (e.g., as in [15]), and others.

The paper is organized as follows. Section 2 presents the mathematical notation and fundamental definitions used in this work. Section 3 provides the necessary theoretical background and preliminary results. The main contributions regarding stability analysis of homogeneous finite-time observers are developed in Sect. 4. Finally, Sect. 5 concludes the paper with a summary of key findings and potential research directions.

2 Notation

Through the paper the following notation will be used:

- $\mathbb{R}$ ($\mathbb{R}_+$) denotes the set of real (positive real) numbers;
- $\mathbb{R}^n$ and $\mathbb{R}^{n \times n}$ represents the n-dimensional Euclidean space equipped with the vector norm $|\cdot|$
- $\mathbb{D}_+^n$ represents the set of $n \times n$ nonnegative diagonal matrices;
- $\overline{1,m}$ indicates the integer sequence $1, 2, \ldots, m$;
- For a square matrix $P \in \mathbb{R}^{n \times n}$, $\lambda_i(P)$ refers to its i-th eigenvalue, with $\lambda_{\max}(P)$ and $\lambda_{\min}(P)$ denoting the respective maximum and minimum eigenvalues;
- The left eigenvector of a matrix $A \in \mathbb{R}^{n \times n}$ corresponding to the i-th eigenvalue is denoted by $\mathbf{u_l}(A)$, satisfying the relation $\mathbf{u_l}(A)A = \lambda_i(A)\mathbf{u_l}(A)$;
- diag$\{\lambda_i\}i = 1^n$ constructs an $n \times n$ diagonal matrix with entries $\lambda_1, \lambda_2, \ldots, \lambda_n$ along its main diagonal;
- The $m \times n$ zero matrix is denoted by $O_{m \times n}$;
- The vector $e_s(i) \in \mathbb{R}^{s \times 1}$ corresponds to the i-th canonical basis vector of $\mathbb{R}^s$, with its i-th component equal to 1 and all other components 0.

3 Preliminaries

3.1 Finite-Time Stability

Consider the system

$$\dot{x}(t) = f(x(t)), \quad x(0) = x_0, \quad t \geq 0, \tag{1}$$

The system's state is represented by the vector $x(t) \in \mathbb{R}^n$, evolving according to a nonlinear continuous vector field $f : \mathbb{R}^n \to \mathbb{R}^n$ that satisfies the equilibrium condition $f(0) = 0$.

Definition 1 [16, 17]. *The origin of (1) is said to be globally finite-time stable if it is globally uniformly asymptotically stable and there exists a locally bounded function $T : \mathbb{R}^n \to [0, +\infty)$ such that any solution $x(t, x_0)$ of the system (1) for all $x_0 \in \mathbb{R}^n$ satisfies $x(t, x_0) = 0, \forall t \geq T(x_0)$. The function T is referred to as a settling-time estimate.*

The following theorem presents the ILF method [18, 19] for finite-time stability analysis.

Theorem 1 [20]. *If there exists a continuous function*

$$Q: \mathbb{R}_+ \times \mathbb{R}^n \to \mathbb{R}$$
$$(V,x) \mapsto Q(V,x)$$

such that

C1) $Q(V,x)$ is continuously differentiable for $\forall x \in \mathbb{R}^n \setminus \{0\}$ and $\forall V \in \mathbb{R}_+$;
C2) for any $x \in \mathbb{R}^n \setminus \{0\}$ there exist $V^- \in \mathbb{R}_+$ and $V^+ \in \mathbb{R}_+$:

$$Q(V^-,x) < 0 < Q(V^+,x); \tag{2}$$

C3) for $\Omega = \{(V,x) \in \mathbb{R}^{n+1} : Q(V,x) = 0\}$

$$\lim_{\substack{x \to 0 \\ (V,x) \in \Omega}} V = 0^+, \quad \lim_{\substack{V \to 0^+ \\ (V,x) \in \Omega}} \|x\| = 0, \quad \lim_{\substack{\|x\| \to \infty \\ (V,x) \in \Omega}} V = +\infty;$$

C4) the inequality

$$-\infty < \frac{\partial Q(V,x)}{\partial V} < 0$$

holds for $\forall V \in \mathbb{R}_+$ and $\forall x \in \mathbb{R}^n \setminus \{0\}$;
C5) the inequality

$$\frac{\partial Q(V,x)}{\partial x} f(x) \leq \sigma V^{1-\mu} \frac{\partial Q(V,x)}{\partial V}$$

holds $\forall (V,x) \in \Omega$, where $0 < \mu \leq 1$ and $\sigma \in \mathbb{R}_+$ are some constants.

Therefore, the origin of system (1) is globally finite-time stable, with its settling time bounded by the following estimate:

$$T(x_0) \leq \frac{V_0^\mu}{\sigma \mu},$$

where $V_0 \in \mathbb{R}_+ : Q(V_0, x_0) = 0$.

3.2 Homogeneity

The concept of homogeneity represents a fundamental symmetry property with respect to dilation transformations, as established in the literature [21–23]. Formally, a linear dilation is characterized by a matrix-valued function $\mathbf{d}: \mathbb{R} \to \mathbb{R}^{n \times n}$ defined through the exponential mapping $\mathbf{d}(s) = \exp(G_{\mathbf{d}} s)$, where the matrix $G_{\mathbf{d}}$ (the generator of dilation) satisfies the anti-Hurwitz condition.

Definition 2 [23]. *A vector field $f: \mathbb{R}^n \to \mathbb{R}^n$ (a function $g: \mathbb{R}^n \to \mathbb{R}$) is said to be $\mathbf{d}$-homogeneous of degree $\nu \in \mathbb{R}$ iff*

$$f(\mathbf{d}(s)x) = e^{\nu s}\mathbf{d}(s)f(x), \quad \forall x \in \mathbb{R}^n \setminus \{0\}, \quad \forall s \in \mathbb{R}.$$
$$(\text{resp. } g(\mathbf{d}(s)x) = e^{\nu s}g(x), \quad \forall x \in \mathbb{R}^n \setminus \{0\}, \quad \forall s \in \mathbb{R}.) \tag{3}$$

The system $\dot{x} = f(x)$ is called $\mathbf{d}$-homogeneous if $f(x)$ is $\mathbf{d}$-homogeneous.

The case where the dilation generator matrix $G_{\mathbf{d}}$ takes the form of an identity matrix ($G_{\mathbf{d}} = I_n$) defines the special class of weighted homogeneity (standard homogeneity).

It is important to note that the homogeneity property implies robustness against external perturbations, measurement noise, and time delays (see, for example, [12–14]).

3.3 Stability Analysis of Homogeneous Systems with Sector Nonlinearities

Consider the following system:

$$\dot{x}(t) = A_0 x(t) + A_1 f(Ux(t)), \quad t \geq 0, \tag{4}$$

where $x(t) \in \mathbb{R}^n$ is the state vector, $f(Ux) = [f_1(e_n^T(1)Ux)\ldots f_n(e_n^T(n)Ux)]^T$ is a continuous function forward-time existence of solutions for system (4) in forward time, $f(0) = 0$, U, A_0 and A_1 are the matrices with dimensions $\mathbb{R}^{n \times n}$.

he analysis of system (4) proceeds under the following technical assumptions:

- **Assumption 1.** For any $i = \overline{1,n}$:

$$sf_i(s) > 0 \quad \forall s \in \mathbb{R} \setminus \{0\}. \tag{5}$$

- **Assumption 2.** The system (4) is **d**-homogeneous of degree $\mu < 0$ with the generator $G_\mathbf{d}$, i.e.

$$\mathbf{d}(s)(A_0 x + A_1 f(Ux)) = e^{-s\mu}(A_0 \mathbf{d}(s)x + A_1 f(U\mathbf{d}(s)x)), \forall s \in \mathbb{R}, \forall x \in \mathbb{R}^n \tag{6}$$

and $f(Ux) = e^{-s\mu} \mathbf{d}(s)^{-1} f(U\mathbf{d}(s)x)$ is homogeneous of degree μ. We observe that the negative homogeneity degree μ implies the existence of positive constants $\psi_i \in \mathbb{R}_+$ for each $i = \overline{1,n}$.

$$sf_i(s) > \psi_i s^2 \quad \forall s \in [\underline{s}_i, \overline{s}_i] \setminus \{0\} \tag{7}$$

for some $-\infty < \underline{s}_i < 0 < \overline{s}_i < +\infty$.

- **Assumption 3.** Each row of the matrix U is a left eigenvector of $G_\mathbf{d}$, meaning that for all $i \in \overline{1,n}$, there exists $j \in \overline{1,k}$ such that $e_n^T(i)U = \mathbf{u}j(G\mathbf{d})$, where k represents the number of disjoint real eigenvalues of $G_\mathbf{d}$.

Systems of the form (4), which satisfy Assumptions 1–3, are a particular case of the systems considered in [24] and are characteristic of a range of applications in finite-time control synthesis (see, for example, [15, 25–27]) and observer design (for example, [3,9,28]). Based on [24][Corollary 1], let us present a result for analyzing the finite-time stability of system (4).

Consider the following ILF modification:

$$Q(V,x) = x^T \mathbf{d}^T(-\ln V) P \mathbf{d}(-\ln V)x + 2\sum_{i=1}^n \Lambda_i \int_0^{\mathbf{u}_i \mathbf{d}(-\ln V)x} f_i(s)ds - \chi, \tag{8}$$

where $\Lambda = \text{diag}\{\Lambda_i\}_{i=1}^n \in \mathbb{D}_+^n$, $\chi \in \mathbb{R}_+$, and $P \in \mathbb{R}^{n \times n}$ is a symmetric positive semi-definite matrix.

Define $\Psi = \text{diag}\{\psi_i\}_{i=1}^n \in \mathbb{D}_+^n$. For $i = \overline{1,n}$ denote $\mathbf{u}_i = e_n^T(i)U$, and in accordance with Assumption 3 define the corresponding eigenvalue of the generator $G_\mathbf{d}$ as $\lambda_{\mathbf{u}_i}(G_\mathbf{d})$, i.e., $\mathbf{u}_i G_\mathbf{d} = \lambda_{\mathbf{u}_i}(G_\mathbf{d})\mathbf{u}_i$. Define

$$\mathscr{I} \in \mathbb{R}^{n \times n} : \mathscr{I}_{i,j} = \begin{cases} \text{sign}(\mathbf{u}_i \mathbf{u}_j^T), & \text{if } \mathbf{u}_j, \mathbf{u}_i \text{ are parallel} \\ 0, & \text{otherwise} \end{cases}$$

for $i, j = \overline{1,n}$.

Lemma 1 [24]. *Let Assumptions 1, 2, and 3 hold. Consider parameters $\alpha, \iota_1, \iota_2 \in \mathbb{R}_+$ selected to satisfy the following LMI system*

$$P \geq 0; \quad \Gamma > 0; \quad \iota_1 P + U^T \Lambda U > 0; \tag{9a}$$

$$PG_\mathbf{d} + G_\mathbf{d}^T P \geq 0; \tag{9b}$$

$$U^T \Lambda U + \iota_2 (PG_\mathbf{d} + G_\mathbf{d}^T P) > 0; \tag{9c}$$

$$Q_1 + Q_3 + \alpha Q_2 \leq 0, \tag{9d}$$

is feasible for some $P = P^T \in \mathbb{R}^{n \times n}$; $\Lambda \in \mathbb{D}_+^n$, $0 \leq \Xi = \Xi^T \in \mathbb{R}^{2n \times 2n}$, where for $X = P + U^T \Lambda \Psi U$ and $\tilde{A}_0 = A_0 + A_1 \Psi U$

$$Q_1 = \begin{bmatrix} X\tilde{A}_0 + \tilde{A}_0^T X & XA_1 + \tilde{A}_0^T U^T \Lambda \\ \Lambda U \tilde{A}_0 + A_1^T X & \Lambda U A_1 + A_1^T U^T \Lambda \end{bmatrix};$$

$$Q_2 = \begin{bmatrix} XG_\mathbf{d} + G_\mathbf{d}^T X & U^T \Lambda^1 H \\ H\Lambda^1 U & 0_{n \times n} \end{bmatrix};$$

$$Q_3 = \Xi + \begin{bmatrix} 0_{n \times n} & U^T \Upsilon_{0,1} \\ \Upsilon_{0,1} U & \Upsilon_{1,1} \end{bmatrix},$$

$\Upsilon_{0,1} \in \mathbb{D}_+^n$, $\Upsilon_{1,1} = \Upsilon_{1,1}^T \in \mathbb{R}^{n \times n}$: $e_n^T(\eta)\Upsilon_{1,1}e_n(l) \begin{cases} \geq 0, & \text{if } e_n^T(\eta)\mathscr{I}e_n(l) = 1 \\ \leq 0, & \text{if } e_n^T(\eta)\mathscr{I}e_n(l) = -1 \\ = 0, & \text{otherwise} \end{cases}$ *for $\eta, l = \overline{1,n}$, and $H = \text{diag}\{\lambda_{\mathbf{u}_i}(G_\mathbf{d})\}_{i=1}^l$.*

Then the origin is globally finite-time stable and

$$T(x_0) \leq -\frac{V_0^{-\mu}}{\alpha \mu},$$

where $Q(V_0, x_0) = 0$ for $\chi < n \min_{i \in \overline{1,n}} \{\underline{s}_i^2, \overline{s}_i^2\} \lambda_{\min}(X) \|U\|^{-2}$.

Remark 1 [24]. In order to obtain less conservative LMIs the following approach can be used. In (9) some elements $e_n^T(\eta)\Upsilon_{i,1}e_n(l)$, $\eta, l \in \overline{1,n}$ of the matrices $\Upsilon_{i,1}$, $i \in \overline{0,1}$ that correspond to the case $e_n^T(\eta)\mathscr{I}e_n(l) = 1$ can be selected to be negative if there are positive elements of $\Upsilon_{i,1}$, $l \in \overline{0,1}$ that correspond to terms with identical multipliers but possessing faster or equivalent dynamics near the origin; and where the sum of these elements dominates (i.e., is greater than or equal to) the sum of absolute values of their negative counterparts associated with terms exhibiting slower dynamics. An analogous statement with reversed sign conditions applies when $e_n^T(\eta)\mathscr{I}e_n(l) = -1$.

4 Main Result

Consider the linear system in the following form

$$\begin{aligned} \dot{x}(t) &= Ax(t) + Bu(t), \\ y &= Cx(t), \end{aligned} \tag{10}$$

where $x(t) \in \mathbb{R}^n$ is the state vector, $y(t) \in \mathbb{R}$ represents the measured output, $u : \mathbb{R} \to \mathbb{R}^m$ denotes the control input function, $B \in \mathbb{R}^{n \times m}$ is the input matrix, and

$$A = \begin{bmatrix} 0 & 1 & 0 & \cdots & 0 \\ 0 & 0 & 1 & \cdots & 0 \\ \vdots & \vdots & \vdots & \ddots & \vdots \\ 0 & 0 & 0 & \cdots & 1 \\ 0 & 0 & 0 & \cdots & 0 \end{bmatrix} \in \mathbb{R}^{n \times n},$$

$$C = \begin{bmatrix} 1 & 0 & \cdots & 0 \end{bmatrix} \in \mathbb{R}^{1 \times n},$$

In [3] the following finite-time homogeneous observer is proposed

$$\dot{\hat{x}}(t) = A\hat{x}(t) + Bu(t) + \operatorname{diag}\{l_i\}_{i=1}^n \begin{bmatrix} |\hat{x}_1(t) - y(t)|^{1+\mu} \operatorname{sign}(\hat{x}_1(t) - y(t)) \\ |\hat{x}_1(t) - y(t)|^{1+2\mu} \operatorname{sign}(\hat{x}_1(t) - y(t)) \\ \vdots \\ |\hat{x}_1(t) - y(t)|^{1+n\mu} \operatorname{sign}(\hat{x}_1(t) - y(t)) \end{bmatrix}, \quad (11)$$

where $\mu \in (-1/n, 0)$, and the gains $l_i \in \mathbb{R}$ for $i = \overline{1,n}$ remain to be designed. Considering the estimation error $e(t) = \hat{x}(t) - x(t)$, the error dynamics are governed by

$$\dot{e}(t) = Ae(t) + \operatorname{diag}\{l_i\}_{i=1}^n \begin{bmatrix} |e_1(t)|^{1+\mu} \operatorname{sign}(e_1(t)) \\ |e_1(t)|^{1+2\mu} \operatorname{sign}(e_1(t)) \\ \vdots \\ |e_1(t)|^{1+n\mu} \operatorname{sign}(e_1(t)) \end{bmatrix}, \quad (12)$$

which corresponds to (4) with $A_0 = A$, $A_1 = \operatorname{diag}\{l_i\}_{i=1}^n$,

$$U = \begin{bmatrix} 1 & 0 & \cdots & 0 \\ 1 & 0 & \cdots & 0 \\ \cdots & \cdots & \cdots & \cdots \\ 1 & 0 & \cdots & 0 \end{bmatrix}, \quad f(Ue(t)) = \begin{bmatrix} |e_1(t)|^{1+\mu} \operatorname{sign}(e_1) \\ |e_1(t)|^{1+2\mu} \operatorname{sign}(e_1) \\ \vdots \\ |e_1(t)|^{1+n\mu} \operatorname{sign}(e_1) \end{bmatrix}.$$

Note that Assumptions 1-3 are satisfied for the system (12). Indeed, since $f_i(s) = |s|^{1+i\mu} \operatorname{sign}(s)$, $i = \overline{1,n}$, Assumption 1 holds. It can be straightforwardly verified that the system (12) is homogeneous with negative degree $\mu < 0$, where $G_\mathbf{d} = \operatorname{diag}\{1 + (i-1)\mu\}_{i=1}^n$, and $e_n^T(i)U = \begin{bmatrix} 1 & 0 & \cdots & 0 \end{bmatrix} = \mathbf{u}_1(G_\mathbf{d})$ for $i = \overline{1,n}$, i.e., Assumptions 2 and 3 hold. Note also that $\mathbf{u}_i G_\mathbf{d} = \mathbf{u}_i$, and $H := \operatorname{diag}\{\lambda_{\mathbf{u}_i}(G_\mathbf{d})\}_{i=1}^l = I$.

The homogeneity property ensures that the observer (11) is input-to-state stable (ISS) with respect to exogenous disturbances and measurement noise. For analytical clarity, this work focuses on the disturbance-free case.

In the paper [11], for homogeneous observers (11) stability conditions and parameter tuning are formulated to be less conservative based on the use of the ILF modification (8). Let us present an extension of the work [11], in which the use of new analysis of nonlinearities in system (12) allowed for significantly less conservative LMIs for the case where μ is sufficiently close to $-1/n$.

Theorem 3 Let the gains $l_i \in \mathbb{R}$ for $i = \overline{1,n}$ be chosen such that $A_0 + A_1 U$ is Hurwitz. Let $\alpha_1, \alpha_2, \iota, \in \mathbb{R}_+$ and $\beta_i \geq 0$, $i \in \overline{1,n}$ to satisfy the following LMI system

$$P \geq \tilde{P} \geq 0; \tag{13a}$$

$$\iota P + U^T \Lambda U > 0; \tag{13b}$$

$$P G_{\mathbf{d}} + G_{\mathbf{d}}^T P + 2 U^T \Lambda U > 0; \tag{13c}$$

$$Q_1 + \alpha_1 Q_2 + Q_3 + \alpha_2 Q_4 \leq 0; \tag{13d}$$

$$0 \leq \beta \leq \tilde{P}_1 + 2 \sum_{i=1}^{n} \frac{\Lambda_i}{2 + i\mu} \tag{13e}$$

is feasible for $\tilde{P}, \Lambda \in \mathbb{D}_+^n$, $\tilde{P}_1 = \lambda_1(\tilde{P})$, $P, \Xi \geq 0$, $R = \begin{bmatrix} 0 & 0 & \cdots & 0 \\ 0 & 1 & \cdots & 0 \\ \vdots & \vdots & \ddots & \vdots \\ 0 & 0 & \cdots & 1 \end{bmatrix} \in \mathbb{D}_+^n$, where for $\tilde{A}_0 = A_0 + A_1 U$, $A_1 = \mathrm{diag}\{l_i\}_{i=1}^n$ and $X = P + U^T \Lambda U$

$$Q_1 = \begin{bmatrix} X\tilde{A}_0 + \tilde{A}_0^T X & X A_1 + \tilde{A}_0^T U^T \Lambda \\ \Lambda U \tilde{A}_0 + A_1^T X & \Lambda U A_1 + A_1^T U^T \Lambda \end{bmatrix};$$

$$Q_2 = \begin{bmatrix} X G_{\mathbf{d}} + G_{\mathbf{d}}^T X & U^T \Lambda \\ \Lambda U & O_{n \times n} \end{bmatrix};$$

$$Q_3 = \Xi + \begin{bmatrix} O_{n \times n} & U^T \Upsilon_{0,1} \\ \Upsilon_{0,1} U & \Upsilon_{1,1} \end{bmatrix};$$

$$Q_4 = \begin{bmatrix} P + 2 U^T \tilde{\Lambda} U - \beta e_n(1) e_n^T(1) & U^T \tilde{\Lambda} - \beta e_n(1) e_n^T(n) \\ \tilde{\Lambda} U - \beta e_n(n) e_n^T(1) & -\beta e_n(n) e_n^T(n) \end{bmatrix},$$

and $\Upsilon_{0,1} \in \mathbb{D}_+^n$, $\Upsilon_{1,1} = \Upsilon_{1,1}^T \in \mathbb{R}^{n \times n}$ satisfy the conditions of Remark 1. Under these design constraints, the observer (11) guarantees finite-time convergence of the observation error to zero, while

$$T(e_0) \leq -\frac{V_0^{-\mu}}{\alpha_1 \mu},$$

where $Q(V_0, e_0) = 0$ with $\chi = \tilde{P}_1 + 2 \sum_{i=1}^{n} \frac{\Lambda_i}{2+i\mu}$.

The method for constructing the matrices $\Upsilon_{0,1}, \Upsilon_{1,1}$ in Q_3 is presented in [11]:

Remark 2 [11]. According to Remark 1 the matrices $\Upsilon_{i,j}, i,j \in \overline{0,1}$ ($\Upsilon_{0,1} \in \mathbb{D}^n$, $\Upsilon_{1,1} \in \mathbb{R}^{n \times n}$) can be constructed as follows: some elements $e_n^T(\eta_1) \Upsilon_{i,j} e_n(\eta_2)$, $\eta_1, \eta_2 \in \overline{1,n}$ can be selected to be negative if there are positive elements of $\Upsilon_{l,q}, l, q \in \overline{0,1}$ that correspond to multiplication of terms $\hat{f}_{\eta_3}(U\mathbf{d}(-\ln V)e)$ and $\hat{f}_{\eta_4}(U\mathbf{d}(-\ln V)e)$, $\eta_3, \eta_4 \in \overline{1,n}$ with the same multipliers, where

$$\hat{f}_{\eta_3}(U\mathbf{d}(-\ln V)e)\hat{f}_{\eta_4}(U\mathbf{d}(-\ln V)e) \geq \hat{f}_{\eta_1}(U\mathbf{d}(-\ln V)e)\hat{f}_{\eta_2}(U\mathbf{d}(-\ln V)e);$$

for $V^{-1}e_1 \leq 1$, and the sum of these elements is greater or equal than the sum of absolute values of negative counterparts that correspond to terms with slower dynamics. For example, for $n = 3$ the matrix Q_3 can take the form

$$Q_3 = \Xi + \begin{bmatrix} 0 & 0 & 0 & \xi_1 & \xi_2 & \xi_3 \\ 0 & 0 & 0 & 0 & 0 & 0 \\ 0 & 0 & 0 & 0 & 0 & 0 \\ \xi_1 & 0 & 0 & -2(\xi_2+\xi_3+\xi_5+\xi_6) & \xi_5 & \xi_6 \\ \xi_2 & 0 & 0 & \xi_5 & -2\xi_4 & \xi_4 \\ \xi_3 & 0 & 0 & \xi_6 & \xi_4 & 0 \end{bmatrix},$$

$$\xi_i \geq 0, \ i = \overline{1,6}$$

where $\Upsilon_{0,1} = \mathrm{diag}\{\xi_i\}_{i=1}^3$,

$$\Upsilon_{1,1} = \begin{bmatrix} -2(\xi_2+\xi_3+\xi_5+\xi_6) & \xi_5 & \xi_6 \\ \xi_5 & -2\xi_4 & \xi_4 \\ \xi_6 & \xi_4 & 0 \end{bmatrix},$$

and the following inequalities $|\hat{f}_i(U\mathbf{d}(-\ln V)e)| \leq |\hat{f}_{i+1}(U\mathbf{d}(-\ln V)e)|$ for $i = \overline{1,2}$ and $e_1 \hat{f}_j(U\mathbf{d}(-\ln V)e) \geq [\hat{f}_1(U\mathbf{d}(-\ln V)e)]^2$ for $j = \overline{2,3}$ are considered.

The main difference between Theorem 3 and the result from [11, Theorem 3] lies in the matrix Q_4. Since the matrix UA_1 is not diagonally stable, the main idea is to construct such a matrix Q_4 satisfying $y^T Q_4 y \geq 0$ that imply $Q_1 + Q_3 + \alpha_2 Q_4 < 0$. The approach proposed in [11] became less conservative as $|\mu|$ decreases (note that the homogeneous observer (11) is close to the linear one as $|\mu| \to 0$). In the proposed result the construction of Q_4 is mainly based on the upper bound $|\hat{f}_n(s)| \leq 1 - s$ for $s \in [0,1]$ that is more accurate as $\mu \to -1/n$ (see Fig. 3). So, the proposed methodology yields less conservative stability conditions for the strongly nonlinear observer (11), formulated as linear matrix inequalities (LMIs), particularly when the homogeneity degree μ approaches its theoretical lower bound of $-1/n$ (Fig. 1).

Example. For comparison purposes, let us consider the same example as in [11], i.e., we consider the 3rd order system (10) and the observer (11) with $\mu = -0.3$ and

$$A_1 = \begin{bmatrix} -0.5 & 0 & 0 \\ 0 & -0.3 & 0 \\ 0 & 0 & -0.025 \end{bmatrix}.$$

The system (13) is feasible with $\alpha_1 = 0.05$, $\alpha_2 = 0.1$ and

$$Q_3 = \begin{bmatrix} 0 & 0 & 0 & 7.85 & 0.48 & 10.34 \\ 0 & 0 & 0 & 0 & 0 & 0 \\ 0 & 0 & 0 & 0 & 0 & 0 \\ 7.85 & 0 & 0 & -100.69 & 22.29 & 17.24 \\ 0.48 & 0 & 0 & 22.29 & -34.71 & 17.35 \\ 10.34 & 0 & 0 & 17.24 & 17.35 & 0.01 \end{bmatrix},$$

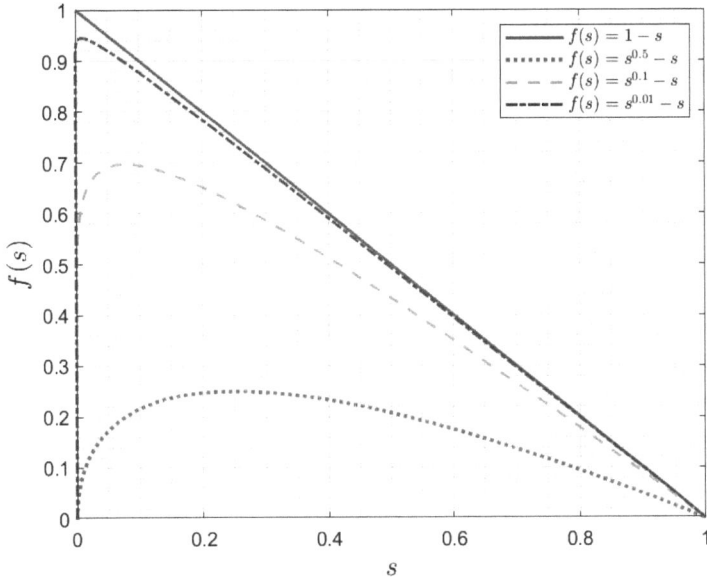

Fig. 1. Functions $1-s$ and $|\hat{f}_n(s)|$ for different $\mu \in (0,1)$ and $s \in [0,1]$

$$Q_4 = \begin{bmatrix} 56.9 & -102.9 & -128.5 & 0 & 29 & -76 \\ -102.9 & 353.1 & -358.6 & 0 & 0 & 0 \\ -128.5 & -358.6 & 2386.6 & 0 & 0 & 0 \\ 0 & 0 & 0 & 0 & 0 & 0 \\ 29 & 0 & 0 & 0 & 0 & 0 \\ -76 & 0 & 0 & 0 & 0 & -93.9 \end{bmatrix}.$$

Simulation results are presented in Fig. 2, while Fig. 3 illustrates the logarithmic-scale evolution of the error vector norm $|e|$, demonstrating the finite-time convergence property.

Note that for the same parameters the result from [11] is applicable only for $\alpha_1 = 0.005$, which indicates more conservative stability conditions. Thus, the approach proposed in Theorem 3 allows to relax the conditions on the choice of the gain matrix A_1 and obtain a less conservative estimates of the settling time (in the given example the settling time estimate differs by 10 times from the result of [11]).

In addition, Theorem 3 requires selection of only three (i.e., $\iota, \alpha_1, \alpha_2 \in \mathbb{R}_+$) parameters to solve LMIs (9), while the result [11] requires $n+3$ parameters to be selected by user.

Remark 3. As part of the proof of Theorem 3, in order to guarantee $\frac{\partial Q}{\partial V} < 0$, the inequality $PG_\mathbf{d} + G_\mathbf{d}^T P \geq 0$ is not required (only (13c) is required), which is a relaxation of LMIs compared to [11]. The results of the papers [11, 15] can be slightly improved with a similar relaxation of LMI-based conditions.

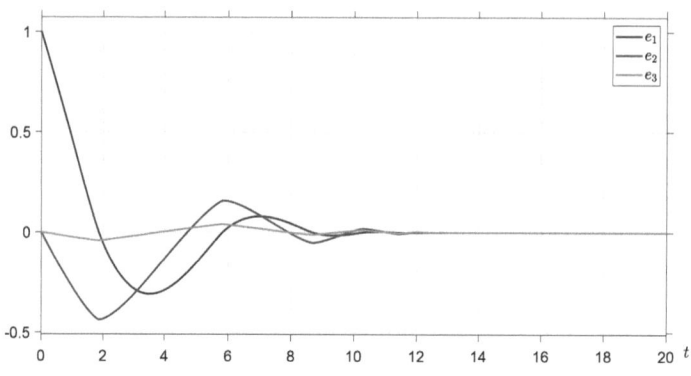

Fig. 2. Simulation plot of e

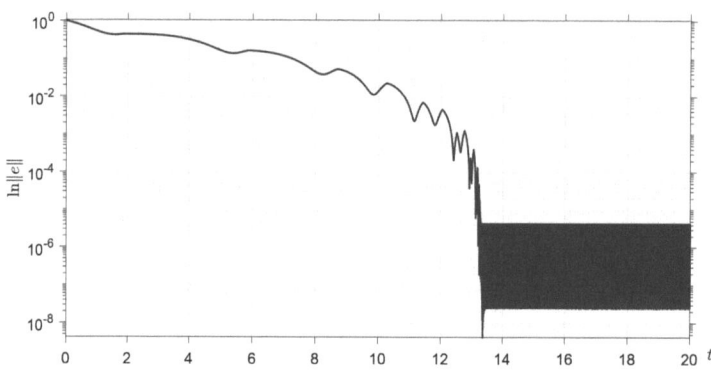

Fig. 3. Simulation plot of $\ln \|e\|$

5 Conclusions

This research focuses on stability analysis and parameters tuning for the finite-time observer (11). The stability conditions and parameter tuning methods presented here are less conservative than in [11], especially for the highly nonlinear case $\mu \to -1/n$. The methods used in the paper can also be applied to other applications (for example, consensus control protocol as in [15] and distributed observers as in [9]), where stability analysis is necessary for systems in the form (4). Extension of the proposed results for the observer design problem for MIMO systems is a direction for further research.

Acknowledgments. This work is supported by RSF under grant 24-19-00454 in ITMO University.

Disclosure of Interests. The authors have no competing interests.

References

1. Lopez-Ramirez, F., Polyakov, A., Efimov, D., Perruquetti, W.: Finite-time and fixed-time observer design: implicit Lyapunov function approach. Automatica **87**, 52–60 (2018)
2. Angulo, M., Moreno, J., Fridman, L.: Robust exact uniformly convergent arbitrary order differentiator. Automatica **49**(8), 2489–2495 (2013)
3. Perruquetti, W., Floquet, T., Moulay, E.: Finite-time observers: application to secure communication. IEEE Trans. Autom. Control **53**(1), 356–360 (2008)
4. Holloway, J., Krstić, M.: Prescribed-time observers for linear systems in observer canonical form. IEEE Trans. Autom. Control **64**, 3905–3912 (2019)
5. Levant, A.: Robust exact differentiation via sliding mode technique. Automatica **34**, 379–384 (1998)
6. Levant, A.: Homogeneity approach to high-order sliding mode design. Automatica **41**(5), 823–830 (2005)
7. Li, Y., Sanfelice, R.G.: A finite-time convergent observer with robustness to piecewise-constant measurement noise. Automatica **57**, 222–230 (2015)
8. Engel, R., Kreisselmeier, G.: A continuous-time observer which converges in finite time. IEEE Trans. Autom. Control **47**, 1202–1204 (2002)
9. Silm, H., Efimov, D., Michiels, W., Ushirobira, R., Richard, J.-P.: A simple finite-time distributed observer design for linear time-invariant systems. Syst. Control Lett. **141**, 104707 (2020)
10. Zimenko, K., Polyakov, A., Efimov, D., Kremlev, A.: Homogeneity based finite/fixed-time observers for linear MIMO systems. Int. J. Robust Nonlinear Control **33**(15), 8870–8889 (2023)
11. Zimenko, K., Efimov, D., Polyakov, A., Kremlev, A.: On finite-time observers for linear systems. In: 22nd IFAC World Congress, vol. 56, no. 2, pp. 1667–1671. IFAC-PapersOnLine (2023)
12. Bernuau, E., Polyakov, A., Efimov, D., P.W: Verification of ISS, iISS and IOSS properties applying weighted homogeneity. Syst. Control Lett. **62**(12), 1159–1167 (2013)
13. Efimov, D., Polyakov, A., Perruquetti, W., Richard, J.-P.: Weighted homogeneity for time-delay systems: finite-time and independent of delay stability. IEEE Trans. Automatic Control **61**(1), 210–215 (2016)
14. Zimenko, K., Efimov, D., Polyakov, A., Perruquetti, W.: A note on delay robustness for homogeneous systems with negative degree. Automatica **79**, 178–184 (2017)
15. Zimenko, K., Efimov, D.V., Polyakov, A., Ping, X.: Finite-time control protocol for uniform allocation of second-order agents. In: 62nd IEEE Conference on Decision and Control. CDC 2023, pp. 1427–1431. IEEE, Singapore (2023)
16. Bhat, S.P., Bernstein, D.S.: Finite-time stability of continuous autonomous systems. SIAM J. Control. Optim. **38**(3), 751–766 (2000)
17. Orlov, Y.: Finite time stability and robust control synthesis of uncertain switched systems. SIAM J. Control. Optim. **43**(4), 1253–1271 (2004)
18. Korobov, V.I.: A solution of the problem of synthesis using a controllability function. In: Doklady Akademii Nauk. **248**(5), 1051–1055. Russian Academy of Sciences, Russia (1979)
19. Adamy, J., Flemming, A.: Soft variable-structure controls: a survey. Automatica **40**(11), 1821–1844 (2004)
20. Polyakov, A., Efimov, D., Perruquetti, W.: Finite-time and fixed-time stabilization: Implicit lyapunov function approach. Automatica **51**, 332–340 (2015)
21. Zubov, V.I.: Systems of ordinary differential equations with generalized-homogeneous right-hand sides. Izvestiya Vysshikh Uchebnykh Zavedenii. Matematika **1**, 80–88 (1958)
22. Polyakov, A.: Generalized homogeneity in systems and control. Springer (2020)

23. Kawski, M.: Homogeneous feedback stabilization. New Trends in Systems Theory. Progress in Systems and Control Theory **7** (1991)
24. Zimenko, K., Efimov, D., Polyakov, A., Kremlev, A.: Finite-time stability analysis of homogeneous systems with sector nonlinearities. Automatica **170**, 111872 (2024)
25. Sanchez, T., Moreno, J.A.: On a sign controller for the triple integrator. In: 52nd IEEE Conference on Decision and Control, pp. 3566–3571 (2013)
26. Parsegov, S., Polyakov, A., Shcherbakov, P.: Nonlinear fixed-time control protocol for uniform allocation of agents on a segment. Dokl. Math. **87**, 133–136 (2012)
27. Parsegov, S., Polyakov, A., Shcherbakov, P.: Fixed-time consensus algorithm for multi-agent systems with integrator dynamics. IFAC Proc. Volumes **46**, 110–115 (2013)
28. Davila, J., Fridman, L., Levant, A.: Second-order sliding-mode observer for mechanical systems. IEEE Trans. Autom. Control **50**(11), 1785–1789 (2005)

Intelligent Systems for Urban Planning: Well-Being and Residents' Perception

Nailia Gabdrakhmanova[1] and Maria Pilgun[2]

[1] Peoples Friendship University of Russia, 117198 Moscow, Russia
[2] Department of General and Comparative-Historical Linguistics, Lomonosov Moscow State University, 119991 Moscow, Russia
pilgunm@yandex.ru

Abstract. This article presents a model-based approach that integrates mathematical methods and neural network semantic algorithms to analyze well-being, identify the presence or absence of social tension, and provide predictive analytics for the development of events related to a major urban planning project. The database included relevant materials, comprising a total of 33,700,700 tokens. Data collection was conducted from January 1, 2021, at 23:59:59 to December 19, 2021, at 23:59:59. All algorithms produced identical results, indicating a positive attitude of residents towards the project's implementation and the absence of social tension. The analysis and interpretation of the data enabled the construction of a forecast for the conflict-free development of the situation surrounding the project, which was subsequently confirmed by the actual course of events.

Keywords: Predictive Analytics · Clustering Time Series · neural networks · Data-Driven Reasoning · Well-Being · Social Tension

1 Introduction

1.1 The Development of Intelligent Systems

Urban planning and the development of transportation systems aim to enhance residents' well-being. The creation of intelligent systems capable of assessing citizens' perceptions in real time has become particularly relevant, as these systems enable the incorporation of residents' opinions, the adjustment of decision-making processes in urban planning, and prompt responses to feedback during the implementation of specific urban projects. It is well established that urban planning, the development of a metropolis's road transport system, and environmental factors directly impact the well-being and health of urban populations [1,2].

1.2 Background and Related Work

Social well-being is examined in this article through the lens of the WHO-5 Wellness Index, developed by the Psychiatric Research Unit at the WHO Collaborating Center for Mental Health, Frederiksborg General Hospital, DK-3400

Hillerød. A common methodology for modeling various sources of social tension is analyzed [3].

The study of residents' perceptions during the planning and implementation of urban projects is crucial for maintaining and preventing social tension [4]. The analysis of content generated by social media actors allows for real-time exploration of citizens' opinions and evaluations, providing predictive analytics. The behavior of online community users is discussed in [5–7], while the features of information dissemination in social networks are described in [8].

Of particular interest to researchers is the analysis of influence maximization, which aims to identify k nodes that can influence the largest number of users in a social network [9]. Influence maximization in social networks, which models innovation diffusion in the presence of multiple competing innovations (e.g., when different companies promote competing products through viral marketing), is explored in [10]. Additionally, a general structural scheme for optimal multiprocessor scheduling has been identified and analyzed, demonstrating how to derive new dominance rules from optimal solution characteristics using an approach similar to reverse optimization [11].

To foster favorable conditions for the development of urban projects, it is crucial not only to identify and neutralize conflict zones in a timely manner but also to prevent their emergence. Therefore, real-time predictive analytics is becoming a key element in urban planning.

It is worth noting that predictive analytics is in high demand across various fields under modern conditions. Predictive analytics models are actively employed to address different social issues. For instance, researchers have developed a comprehensive econometric framework based on gravity equations to forecast migrant flows between countries [12]. Socioeconomic status (SES) indices have been developed as predictors of well-being in targeted programs using principal component analysis (PCA) and partial least squares (PLS) [13]. A study examined concurrent and longitudinal predictive effects of depression and positive youth development (PYD) qualities on nonsuicidal self-injury (NSSI) and suicidal self-injury (SSI) among adolescents [14].

Data analysis and algorithmic tools are also successfully used in crime mapping. Statistical models that predict the most likely times and locations for potential crimes are well-known and have been effectively applied in practical police prevention activities. Predictive models in the medical field have become one of the most important directions for industry development [15]. The expansion of judgment-based predictive analytics has been noted, with researchers analyzing behavioral biases that affect the accuracy of evaluative forecasting [16]. Predictive models are also applied in microbiology [17], mechanical engineering [18], and other fields.

Predictive analytics using large data sets is of particular importance. Data mining and Big Data are major trends in the development of the modern scientific paradigm [19]. Researchers are actively advancing various relevant fields, such as crypto-based privacy and security in data mining [20], advanced data mining applications [21], descriptive data mining [22], and foundational studies in data mining, data warehousing, machine learning, artificial intelligence, databases, statistics, knowledge engineering, and big data technologies [23], among others.

However, social media data mining and data-driven reasoning for analyzing socially significant issues related to urban development projects have not yet been comprehensively described.

The aim of this study is to develop and test an interdisciplinary approach that integrates mathematical models and neural network semantic algorithms to analyze situations, detect the presence or absence of social tension, and provide predictive analytics for events surrounding a specific urban development project.

This study was conducted using social media data related to the implementation of a major transportation infrastructure project in a metropolis.

1.3 Method

For data analysis, a model leveraging the integration potential of mathematical and neural network semantic algorithms was developed. The Kribrum software and hardware complex was utilized for monitoring and data collection.

1.3.1 Neural Network Semantic Analysis

For the neural network analysis of user-generated content and digital traces, the TextAnalyst technology

(analyst.ru/index.php?lang=eng&dir=content/products/&id=ta)

was employed.

Network analysis was conducted using ORA-LITE

(http://www.casos.cs.cmu.edu/projects/ora/), while content analysis utilized the AutoMap text mining tool

(http://www.casos.cs.cmu.edu/projects/automap/). For semantic data interpretation, GPT-4o, GPT-4o mini, and GPT-4

(https://chatgpt.com) were also used. For visual analytics, the Tableau platform (https://www.tableau.com) was employed.

The design of the neural network semantic study is presented in Fig. 1.

The author's method for calculating social well-being and stress indices is described in [24].

1.3.2 Mathematical Models

Through neural network text analysis, the initial data were transformed into a multidimensional numerical time series. Dynamic models were constructed from these time series to study society's negative and positive attitudes towards the projects. The algorithm for building the dynamic model includes solving the task of segmenting the time series and constructing autoregressive models [25] for each segment of the time series. Clustering time series is a significant task in the field of event stream data analysis and data mining [26,27]. The study developed a time series clustering algorithm based on the combination of methods for calculating

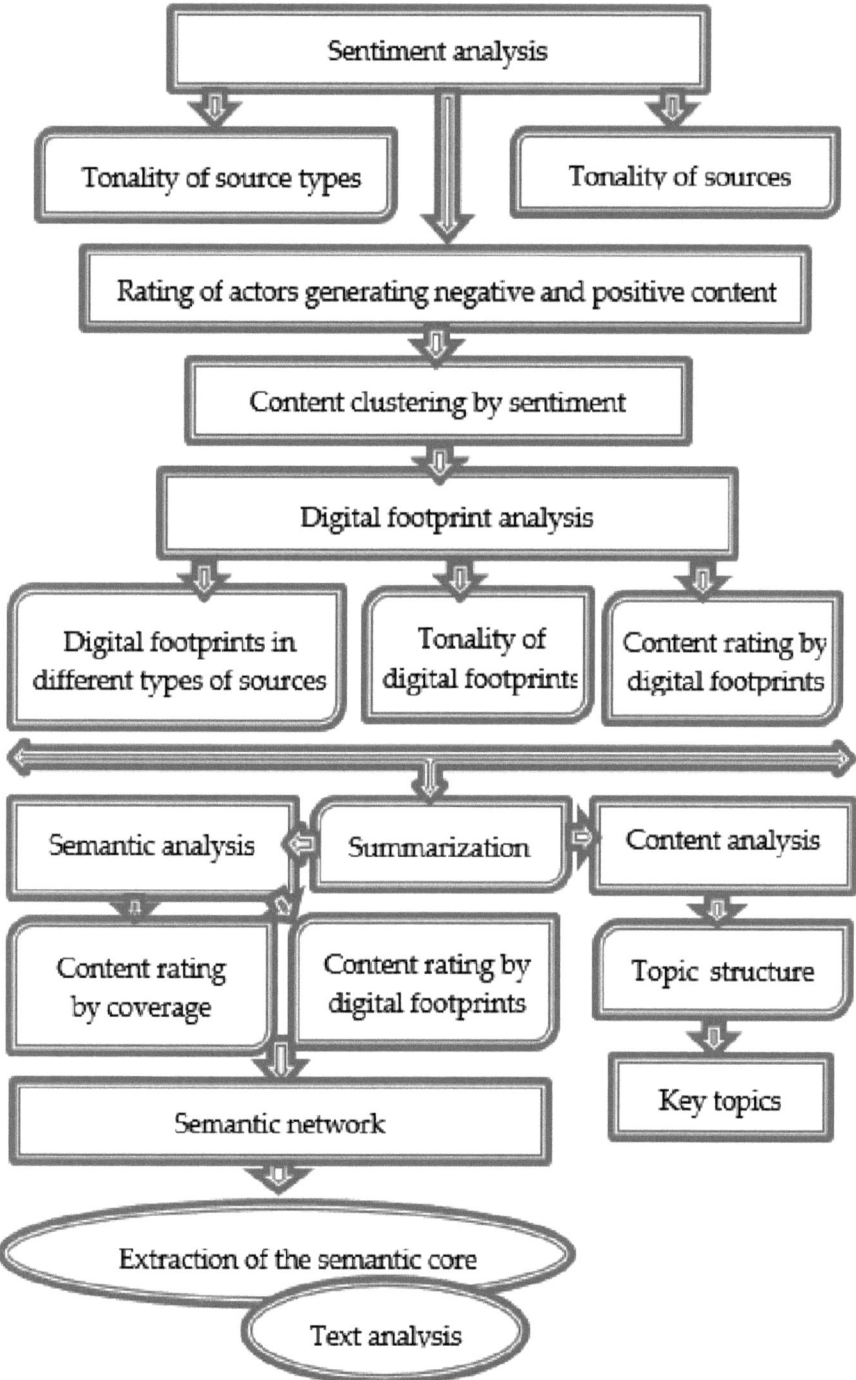

Fig. 1. Design of neural network semantic research.

distances between time series (DWT) [28] and topological data analysis (TDA). The use of TDA is justified by its robustness to noise, independence from metric choice, and ability to identify structures in time series data. A critical feature of TDA persistence diagrams is their resilience to noise, as introduced by H. Edelsbrunner [29]. Computational topology methods have further evolved in the works of G. Carlsson, A. Zomorodian [30,31], and others.

1.4 Data

The material for the study included social networks, online media, messengers, thematic portals, video hosting platforms, microblogs, blogs, forums, review sites, mapping services, and other sources related to the implementation of the urban development project. Data collection was conducted from January 1, 2021, at 23:59:59 to December 19, 2021, at 23:59:59. The quantitative characteristics of the database are presented in Table 1.

Table 1. Quantitative characteristics of the database

Description	Indicators
Number of tokens	33 700 700
Maximum messages per day	779
Number of active actors	1221
Activity (posts per actor)	11,36
Number of sources	322

The dynamics of the total number of messages and unique messages are presented in Figs. 2 and 3.

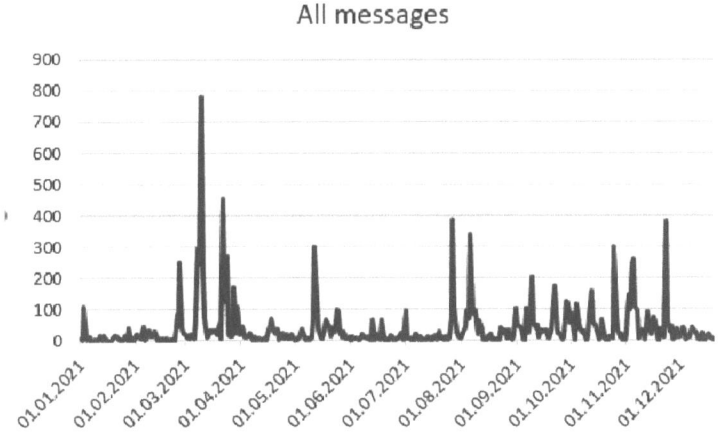

Fig. 2. Dynamics of all messages.

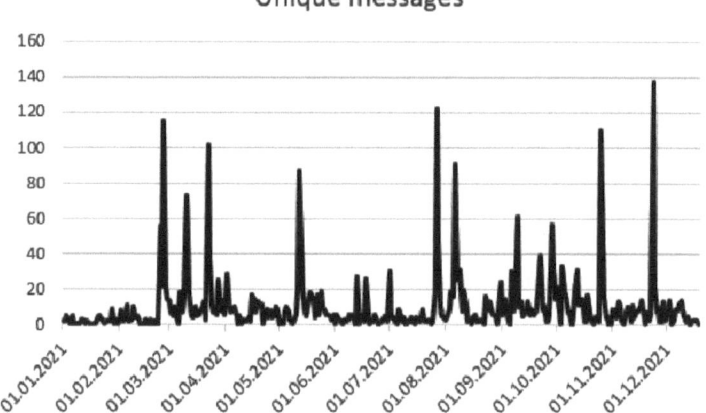

Fig. 3. Dynamics of unique messages.

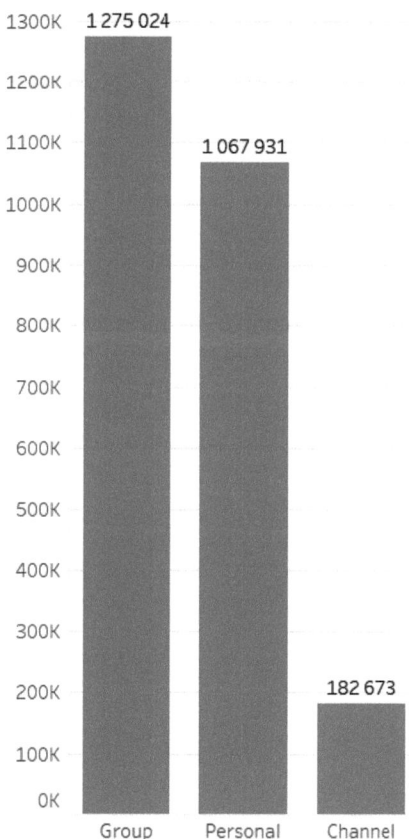

Fig. 4. Types of Actors.

Among the actors generating content related to the project, communities predominate, and there is also a significant number of personal accounts (Fig. 4).

When discussing issues related to the project's implementation, actors preferred to use social media (Fig. 5).

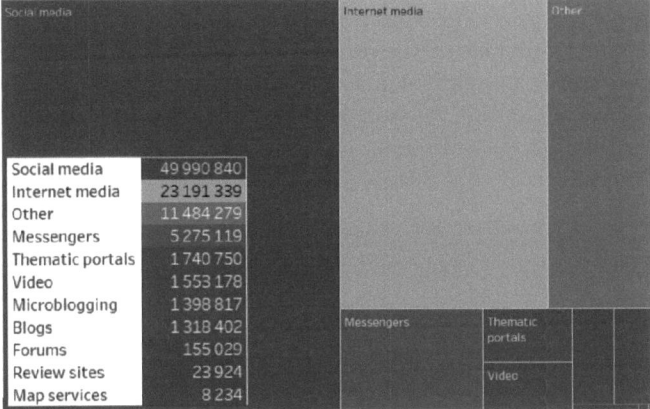

Fig. 5. Types of sources.

Meanwhile, microblogs received the greatest coverage among types of sources (Fig. 6).

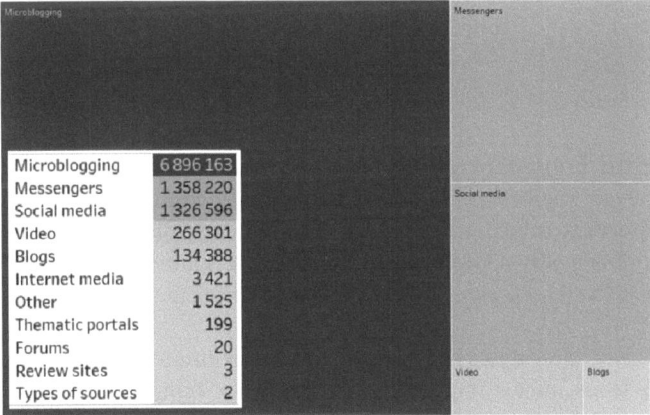

Fig. 6. Source types coverage.

2 Results and Disscution

2.1 Neural Network Semantic Algorithm

2.1.1 Sentiment Analysis

The formation and analysis of ratings of actors generating positive and negative content, as well as sentiment analysis of a consolidated database of content from various digital platforms and source types, allow for preliminary conclusions regarding residents' opinions on the project's implementation (Fig. 7). Sentiment analysis demonstrated a predominance of the positive cluster, which may indicate an absence of social tension surrounding the project's implementation and a positive perception by the residents.

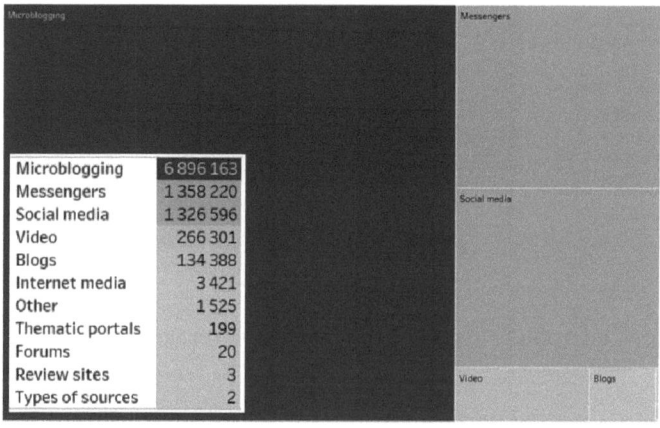

Fig. 7. Tonality of source types.

2.1.2 Digital Footprint Analysis

Analysis of users' digital traces allows us to study communicative behavior and provides important information about the opinions and assessments of actors. The predominance of positive tonality in comments, likes, reposts and duplicates suggests a positive assessment by the actors of the situation around construction (Fig. 8).

The study of messages that received the maximum number of positive and negative reactions makes it possible to detail the opinions and assessments of actors.

2.1.3 Semantic and Content Analysis

Before conducting the semantic analysis, the most significant texts were identified. In particular, audience reach and digital traces significantly characterize the attention of actors to specific semantic nuances articulated in particular

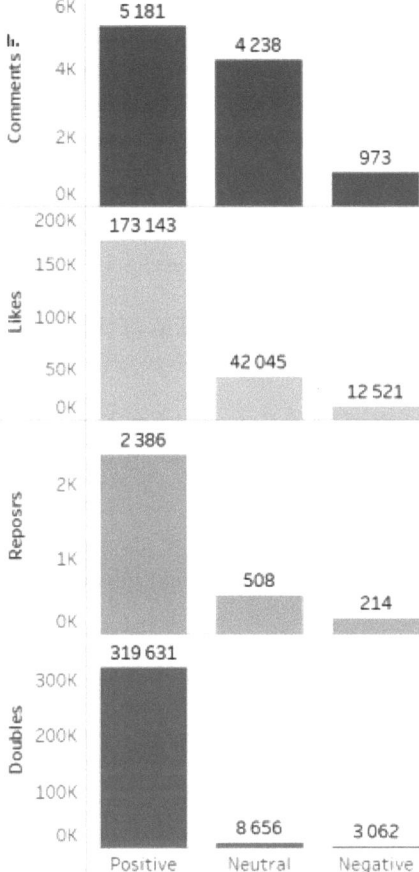

Fig. 8. Tonality of digital footprints.

messages. Based on audience reach data, the number of comments, likes, and duplicates received by the messages, corresponding rankings of user-generated messages were compiled.

As a result of summarization, content analysis, thematic structure extraction, semantic network construction, and the formation and analysis of the semantic core, the main themes and semantic accents that were most important for actors in the positive and negative clusters were identified. In the positive cluster, the key themes that generated positive user perception were:

- The Southeast Highway (SE WX) is expected to improve transportation links for 22 districts of Moscow, home to over 2 million people.
- New highway connections, including the Northwest Chord, Northeast Chord, Southern Rocade, and Southeast Chord, will alleviate traffic on central city streets and improve road conditions.

- The new transport framework will simplify routes, reduce vehicle detours, and positively impact the city's ecological situation.
- Seven overpasses will be constructed at the intersection of the Southern Rocade and the Southeast Highway in Moscow.
- Completion of tunnel drilling between three new metro stations in Moscow.
- Over 70 km of public roads were built in the capital in 2021.
- From 2011 to 2020, 1033.3 km of roads, 297 artificial structures, and 254 pedestrian underpasses were constructed in Moscow.
- Construction of overpasses, tunnels, interchanges, and the relocation of engineering communications.
- Construction and reconstruction of 9.3 km of roads, including five overpasses (2.3 km) and three underground pedestrian crossings near buildings on Kantemirovskaya Street.
- The section of the Southern Rocade from Maryinsky Park Street to the Moscow Ring Road (MKAD) is more than a third complete.
- By the end of 2023, the Southern Rocade will connect to the MKAD via Verkhniye Polya Street.
- More than 50
- In the negative cluster, the key themes that generated negative reactions from residents were:
- Lack of a full-fledged noise protection structure.
- Elimination of all exits to houses on the even side of Kantemirovskaya Street.
- Conversion of courtyard areas and fire access roads into secondary roads.
- No provision for compensatory landscaping.
- Elimination of parking spaces.
- Road closures and bus route changes due to construction.
- The Southeast Chord and the Southern Rocade are intended not to "improve district connectivity" but to reduce traffic on the Moscow Ring Road and the Third Transport Ring by diverting traffic through residential areas.
- Increase in the number of cars in the capital.
- Expansion of paid parking zones.

2.2 Mathematical Models

Problem Statement As a result of the neural network text analysis, a multidimensional time series [s(i), p(i), ng(i), nl(i)], i=1,2,...,n, was obtained, where n is the length of the time series. The components of the series are: s(i) – the number of messages, p(i) – the number of positive messages, ng(i) – the number of negative messages, and nl(i) – the number of neutral messages at time i. The task is to build a mathematical model to predict the values of the series several step.

A. Clustering Time Series Using DTW

Clustering multidimensional time series is a useful tool for detecting recurring patterns in temporal data. Once patterns are detected, the complex behavior of time series can be interpreted as a temporal sequence consisting of a small number of states or clusters. Time series clustering is a method used to group similar time series data based on their patterns and characteristics.

Clustering is an unsupervised learning task, meaning that we measure features but do not measure the outcome. Clustering algorithms can be organized in different ways. The most well-known methods are the K-means method and Dynamic Time Warping (DTW). DTW is an algorithm that performs clustering based on calculating the distance between time series. DTW allows for the "elastic" transformation of time series to detect similar patterns with different phases [31].

Let us consider time series of size n: X=[x(1),...,x(n)] and Y=[y(1),...,y(n)]. Let dv(X,Y) denote a function that calculates the distance between time series X and Y. In DTW algorithms, various metrics can be used depending on the task. In our tasks, Euclidean distance was used. The DTW method was used to compute distances for each month. The sample size is n=360, which corresponds to almost a whole year. A Python program was developed to compute the distances between time series. Figure 9 shows the dynamics of dv(,) changes. On the x-axis is the month number, and on the y-axis is the distance between the time series for that month. From Fig. 9, it can be seen that the series with negative and neutral sentiment are very close, while the series with positive and neutral sentiment and the series with positive and negative sentiment are far from each other.

Conclusion: Time series clustering can be achieved by introducing threshold values for the distance between time series with the number of positive and negative messages (Fig. 9).

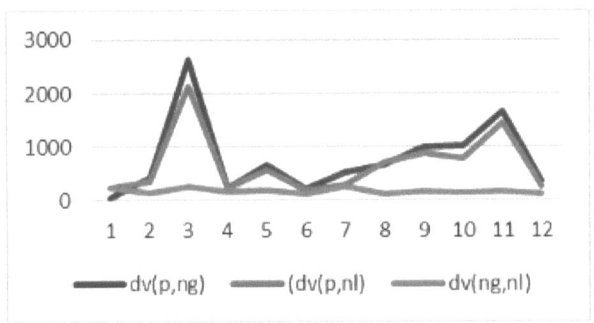

Fig. 9. Dynamic Time Warping (DTW).

B. Clustering Time Series Using TDA

Homology theory associates to any topological space X a sequence of abelian groups $H_k(X), k = 0, 1, 2, ...$, which are homotopy invariants of the space. In recent years, reliable and efficient data structures and algorithms for Topological Data Analysis (TDA) have been developed and are now implemented in libraries such as Python. In our work, Cech complexes are used in TDA. Let X be a finite point cloud and $\epsilon > 0$. The Cech complex is the nerve of the set of ϵ-balls centered at X.

To solve the problem of time series clustering using topological methods, the data were normalized. The normalization formulas used were:

$$pn(i) = p(i)/s(i); ngn(i) = ng(i)/s(i); nln(i) = nl(i)/s(i), i = 1, \ldots, n$$

These normalization formulas were used because the ratio of the number of negative to positive opinions is important, not their absolute values. Using Python programs developed for this purpose, persistence diagrams were constructed for each month for the point cloud in the two-dimensional plane $(ngn(i); pn(i))$, and Euler characteristic estimates were calculated. Figure 10 shows the persistence diagram. In Fig. 31, the x-axis represents the birth time of a component (Birth) and the y-axis represents the death time of a component (Death). In the diagram, blue points correspond to zero-dimensional simplices, and orange points correspond to one-dimensional simplices.

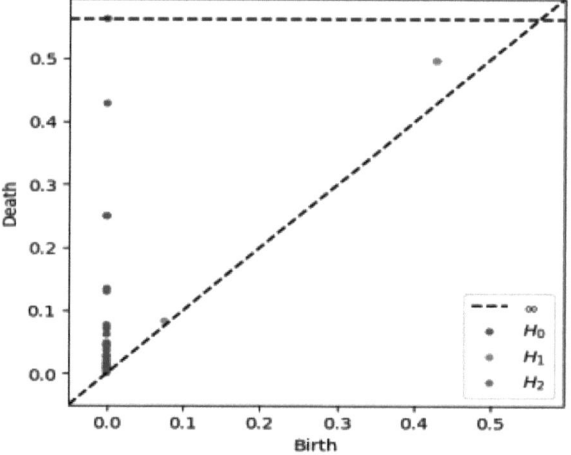

Fig. 10. Persistence diagram (DTA).

Points lying close to the diagonal line (dashed line) are interpreted as noise points and are not included in further calculations. Let $L(0)$ denote the lifespan of zero-dimensional simplices ($H0$)components, $L(1)$ the lifespan of one-dimensional simplices ($H1$) components, and B the Euler characteristic estimate. Then, according to [30,31]:

Figure 11 shows the graph of Euler characteristic estimates (B) and the normalized distances between the positive and negative series,

$$DW = dvn(pn, ngn)/m \tag{1}$$

where m is the maximum value of $dvn(pn, ngn)$. In Fig. 11, the x-axis represents the month number, and the y-axis represents the normalized values of the dv and B series. The graphs indicate that the behavior of the series is very similar, despite being constructed using different methods (DWT and DTA). Therefore, by introducing threshold values hw (for the DW series) and ht (for the B series), several clusters can be identified. In this study, we identified two clusters: the first cluster consists of points for which

$$DW < h(w), B < h(t), \qquad (2)$$

and the second cluster consists of points that do not meet this condition.

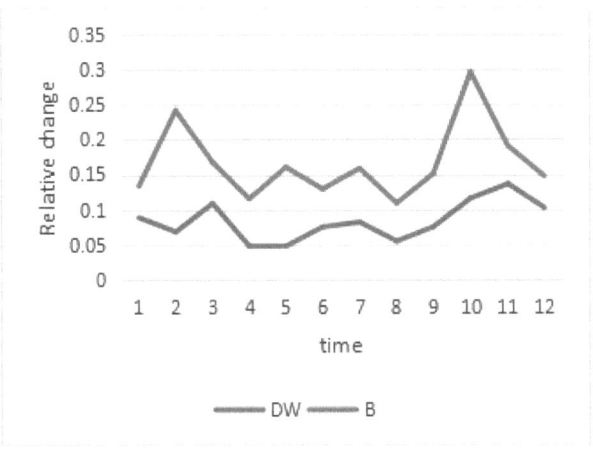

Fig. 11. Dynamics of estimates of the distance between rows and Euler's characteristics.

C. Time Series Forecasting Clustering time series allowed the construction of adequate predictive models for the time series data. A general mathematical model for the time series $x(t)$ is given by:

$$x(t) = u(t) + v(t) \qquad (3)$$

where $u(t)$ is the deterministic sequence or systematic component, and $v(t)$ is the random component. The goal of mathematical modeling of time series is to study the dynamics and predict the values of $x(t)$ several steps ahead. In this work, regression models were used to estimate $u(t)$, and autoregressive models were used to estimate $v(t)$.

$$ngn(t) = -0.04ngn(t-1) + 0.56ngn(t-2) - 0.34ngn(t-3) \qquad (4)$$

2.3 Conclusion

In the course of this study, a model integrating mathematical and neural network semantic algorithms was developed and tested. The resulting model was used to analyze the current situation and predict its development, assess well-being, and identify the presence or absence of social tension surrounding the implementation of a major urban development project in a metropolis using social media data.

The study analyzed the sentiment of different source types, digital platforms, and the aggregated dataset, and compiled rankings of actors generating positive and negative content. The most significant messages that garnered the most user interest were ranked. A list of key themes and semantic accents that were most important for actors in approving and criticizing the project implementation was formed after summarization, content analysis, thematic structure extraction, semantic network construction, and semantic core analysis.

To solve the problem of quantitative forecasting, an algorithm for clustering time series using a composition of DTW and TDA methods was developed. This approach allowed for a method that is independent of metric choice and robust to noise in the data. Clustering revealed two system states and enabled the construction of adequate predictive models for each cluster. The results of the calculations, obtained using the constructed mathematical models, confirmed a stable level of well-being and the absence of actual and potential social tension around the project.

All algorithms produced identical results, indicating a positive attitude of residents towards the project implementation and the absence of social tension. Additionally, data analysis allows for the prediction of a favorable situation if construction continues, with no serious potential conflicts with residents that could trigger conflict escalation. The development of the situation around the project confirmed the obtained results.

In future research, it is planned to expand the application of new algorithms by utilizing LLMs.

References

1. Zhao P., Lyu D.: Lifestyle Change and Transport in China. Springer, Singapore. https://doi.org/10.1007/978-981-19-4399-7
2. Ho, E., Schneider, M., Somanath, S., Yu, Y., Thuvander, L.: Sentiment and semantic analysis: urban quality inference using machine learning algorithms. iScience **27**(7), 110192 (2024)
3. Kakwani, N., Son, H.H.: Measuring social tension. In: Social Welfare Functions and Development. https://doi.org/10.1057/978-1-137-58325-3-3
4. Hayes, R.A., Smock, A., Carr, C.T.: [4]Face[book] management: self-presentation of political views on social media. Commun. Stud. **66**, 549–568 (2015)
5. Yang, W., et al.: Marginal gains to maximize content spread in social networks. IEEE Trans. Comput. Soc. Syst. **6**(3), 479–490 (2019)
6. Williams, H.T.P., McMurra, J.R., Kur, T., Lamber, F.H.: Network analysis reveals open forums and echo chambers in social media discussions of climate change. Glob. Environ. Chang. **32**, 126–138 (2015)

7. Kramer A.D.I., Guillory J., Hancock J.T: Experimental evidence of massive-scale emotional contagion through social networks. In: Proceedings of the National Academy of Sciences, vol. 32, pp. 126–138 (2015)
8. Shi, Q.H., Wang, C., Chen, J.: Post and repost: a holistic view of budgeted influence maximization. Neurocomputing **338**, 92–100 (2019)
9. Yang, W., Chen, S., Gao, S., Yan, R.: Boosting node activity by recommendations in social networks. J. Comb. Optim. **40**(3), 825–847 (2020). https://doi.org/10.1007/s10878-020-00629-6
10. Bharathi, S., Kempe, D., Salek, M.: Competitive influence maximization in social networks. In: Deng, X., Graham, F.C. (eds.) WINE 2007. LNCS, vol. 4858, pp. 306–311. Springer, Heidelberg (2007). https://doi.org/10.1007/978-3-540-77105-0_31
11. Walter, R., Lawrinenko, A.: A characterization of optimal multiprocessor schedules and new dominance rules. J. Comb. Optim. **40**(4), 876–900 (2020). https://doi.org/10.1007/s10878-020-00634-9
12. Reina, J.C., Cuaresma, J.C., Fenz, K., Zellmann, J., Yankov, T., Taha, A.: Gravity models for global migration flows: a predictive evaluation. Popul. Res. Policy Rev. **43**(29), 1–17 (2024)
13. D'Iorio, S., Forzani, L., Arancibia, R.G., Girela, I.: Predictive power of composite socioeconomic indices for targeted programs: principal components and partial least squares. Qual. Quant. (2023)
14. Zhu, X., Shek, D.: The predictive effect of depression on self-injury: positive youth development as a moderator. Appl. Res. Qual. Life **18**, 2877–2894 (2023)
15. Podbielska, H., Kapalla, M. (eds.): Predictive, Preventive, and Personalised Medicine: From Bench to Bedside. Springer, Cham (2023)
16. Seifert, M. (ed.): Judgment in Predictive Analytics. International Series in Operations Research, Springer, Cham (2023)
17. Alvarenga, V.O. (ed.): Basic Protocols in Predictive Food Microbiology. Humana New York, New York (2023)
18. Mahalle, P.N., Hujare, P.P., Shinde, G.R.: Predictive Analytics for Mechanical Engineering: A Beginners Guide. Springer Briefs in Computational Intelligence, Springer, Singapore (2023)
19. Tan, Y., Shi, Y.: Data Mining and Big Data. In: 8th International Conference, DMBD 2023, Sanya, China, December 9–12, 2023, Proceedings, Part I, II. Springer, Singapore (2024)
20. Zhan, J., Matwin, S.: Secure Data Mining. Springer, New York (2024)
21. Yang, X., et al.: Advanced data mining and applications. In: 19th International Conference, ADMA 2023, Shenyang, China, August 21–23, 2023, Proceedings, Part I, II. LNCS, vol. 14177.Springer, Cham (2023)
22. Olson, D.L., Lauhoff, G.: Deskriptives Data-Mining. Springer, Cham (2023)
23. Yang, D.-N., et al. (eds.): Advances in Knowledge Discovery and Data Mining, 28th Pacific-Asia Conference on Knowledge Discovery and Data Mining, PAKDD 2024, Taipei, Taiwan, May 7–10, 2024, Proceedings, Part I, II, III. LLNCS, vol. 14645. Springer, Singapore (2024)
24. Kharlamov, A., Pilgun, M. (eds.): Neuroinformatics and Semantic Representations. Theory and Applications, Cambridge Scholars Publishing, Newcastle upon Tyne (2020)
25. Sikos, L.F., Choo, K.-K.R. (eds.): Data Science in Cybersecurity and Cyberthreat Intelligence. Springer, Cham (2021)

26. Borovikov, V.P., Ivchenko, G.I.: Forecasting in the statistica system in the windows environment. In: Finance and Statistics, Fundamentals of Theory and Intensive Practice on a Computer, Moscow (2006)
27. Rani, S., Sikka, G.: Recent techniques of clustering of time series data: a survey. Int. J. Comput. Appl. **52**(15), 1 (2012)
28. Aghabozorgi, Y.S., Shirkhorshidi, A.S., Wah, T.Y.: Time-series clustering-a decade review. Inf. Syst. **53**, 16–38 (2015)
29. Al-Naymat, G., Chawla, S.J.: Taheri Sparse DTW: A Novel Approach to Speed Up Dynamic Time Warping, Wayback Machine. Accessed 13 Oct 2019
30. Edelsbunner, H., Letscher, D., Zomorodian, A.: Topological persistence and simplification. Discrete Com-put. Geom. **28**, 511–533 (2002)
31. Carlsson G., Zomorodian A.: Computing persistent homology. In: Proceedings of the 20th Annual Symposium Computation Geometry, pp. 347–356 (2004)

The Principle of Trajectories Separation for the Optimal Control Problem in a Feedback System

A. N. Daryina[1,2]

[1] Lomonosov Moscow State University Faculty of Computational Mathematics and Cybernetics, GSP-1, Leninskie Gory, 119991 Moscow, Russia
[2] Federal Research Center Computer Science and Control of RAS, Vavilova str., 44-2, 119333 Moscow, Russia
anna.daryina@gmail.com

Abstract. Ensuring autonomous navigation of mobile robots remains an important and relevant challenge in robotics. A variety of approaches have been proposed in the scientific community to address this problem. This paper provides an overview of existing robot navigation methods and focuses on the problem of optimal control of a mobile robot's movement. It is assumed that during its motion, the robot periodically receives data about its current position in a given coordinate system from an onboard positioning system. However, these data may contain measurement errors – meaning that the actual coordinates are determined with a certain degree of approximation. The paper presents an algorithm for correcting the robot's trajectory, which ensures reaching the final destination with a specified level of accuracy, even in the presence of measurement inaccuracies. A corresponding theorem is formulated and proven, justifying the correctness and effectiveness of the proposed approach. The foundation of the developed method is based on two well-known theoretical frameworks are the classical theory of optimal control synthesis for second-order systems and the graph-based method for finding the shortest path in a connected graph. Thus, the article offers an efficient solution to the problem of mobile robot navigation under conditions of uncertain localization – a crucial aspect for real-world applications.

Keywords: mobile robot control problem · trajectory correction algorithm · approximate positioning

1 Introduction

Nowadays, there are a number of different navigation methods used for autonomous unmanned robots. If we consider a navigation system as a combination of hardware, software, and algorithms that allow a device to determine its position in space, then satellite navigation is one of the most widely used approaches. However, it has certain limitations in practical application.

For example, indoors, satellite signals often fail to reach devices due to obstacles such as concrete and metal structures. Moreover, even when the signal is available, its accuracy may be insufficient for precise localization. The frequency of updating location and speed data 1 Hz, it can be increased to 5 10 Hz. The accuracy of determining location, for example by using GPS, is on average about 15 m. This is due to various factors such as signal inaccuracies, the influence of the atmosphere on radio signal propagation, and the quality of quartz generators in receivers. However, with the help of correction methods – such as base stations – it is possible to significantly improve the accuracy of position determination.

Global navigation systems, in addition to satellite navigation, also include radio beacons (placement in the robot's coverage area of radio signal sources, which are processed by an onboard microprocessor. The limitation is that radio beacons are placed at fixed points of some route; the device loses the ability to bypass obstacles or choose an alternative route). An example of such a system that can operate indoors is the Marvelmind Starter Set. The system determines its own coordinates based on the delay in the return of ultrasonic waves upon reaching stationary beacons (the trilateration principle), the measurement accuracy is ± 2 cm, which are updated at a frequency 16 Hz.

One of the primary methods for determining a robot's position, which does not require external signals or reference points, is odometry. It can be obtained through various means – such as wheel encoders, computer vision systems, inertial measurement units (IMUs), or a combination of these. However, each of these sensor systems has its own limitations and potential sources of error.

Visual inertial odometry algorithm [1–3] shows good results in robot navigation, but its implementation requires good illumination and a large number of landmarks in space, and it does not work in conditions of poor visibility (in the dark). In the dark, active sensors (for example, a strobe light source) or lidars can be used, although they also fail in fog or smoke. A new method of thermal inertial odometry has recently been proposed that uses data on the temperature of objects [4–6]. It is also not optimal for indoor use, since temperature gradients are often very small, due to which the studied space contains few landmarks, and it is necessary to use new methods to ensure a robust state assessment. Suitable sensors for such challenging environments are radars, including frequency modulated continuous wave (FMCW) radars, which are often used in automotive applications because they are less susceptible to external influences [7].

In environments that are unknown or not well explored, the Simultaneous Localization and Mapping (SLAM) technique stands out as one of the most promising navigation approaches. This method enables a robot to construct a map of an unfamiliar space while simultaneously determining its own position within that space. SLAM often integrates data from sources such as visual-inertial odometry and place recognition algorithms to align the developing map and improve location accuracy through a process called loop closure. Today, SLAM is extensively applied in robotics to support autonomous navigation and movement. The best-known algorithm is ORB-SLAM 2 [8], which operates at a frequency 20 Hz and gives approximately two percent deviation from the distance traveled in unfamiliar terrain.

2 Problem Statement

Let us assume that the speed problem in controlling a mobile robot moving from some initial point to a terminal point is solved using the principle of separating admissible trajectories proposed by V.A. Bereznev in [9].

In recent years, robotic devices have occupied new areas of activity from industrial and agricultural production to defense. Along with the development of robots themselves, methods for controlling them are also developing. This paper proposes an unconventional approach to solving the problem of controlling a mobile robot in R^2, the formal statement of which is reduced to the well-known problem of optimal speed with phase constraints.

The traditional approach to solving this problem is to use optimal control methods based on L. S. Pontryagin's maximum principle (see [10,11]). The behavior of a robot that must move from a certain point $x(0) = x^0$ to a terminal point $x(T) = \hat{x}$ in a minimum time T is described by second-order differential equations

$$\ddot{x}(t) = f(t, x(t), \dot{x}(t), u(t)), \qquad (1)$$

where $x(t) \in R^2$ is the desired optimal trajectory, and $u(t) \in R^2$ is an admissible control parameter that obeys the condition

$$u_{\min} \leq u(t) \leq u_{\max}, \quad u_{\min}, u_{\max} \in R^2. \qquad (2)$$

It is assumed that the component $u_1(t)$ specifies the acceleration (positive when picking up speed and negative when braking), and $u_2(t)$ specifies the position of the robot's steering wheel. It is also natural to assume that the robot's speed is limited, i.e. for any t the inequalities are satisfied

$$0 \leq \dot{x}(t) \leq \hat{v}. \qquad (3)$$

The condition of non-intersection of the robot's trajectory with circular obstacles means that for any t the curve $x(t)$ must satisfy the inequality

$$\|x(t) - C_k\| \geq R_k^+, \quad k \in \overline{1, K}, \qquad (4)$$

where K is the total number of obstacles, C_k denotes the coordinates of the center of the k-th obstacle.[1]

These conditions imply that the set of feasible robot trajectories is non-convex. For this reason, the use of optimal control methods, as well as mathematical programming methods (see, for example, [12], [13]) becomes very labor-intensive.

---

[1] we allow a non-strict inequality in (4), since in practical calculations we assume $R_k^+ = R_k + \rho$, where R_k is the actual radius of the k-th circular obstacle, and $\rho > 0$ is the minimum value that ensures the robot's movement without touching the obstacle, taking into account its dimensions.

3 A Principle of Trajectories Separation

The proposed results are based on the well-known theory of optimal control synthesis in second-order systems (see, for example, [11]), as well as on the method of constructing the shortest path on a connected graph. In particular, the proposed approach is as follows.

The perimeters of the circular obstacles are marked with specific points that serve as nodes $v_i, i \in \overline{1,n}$. These nodes constitute a connected graph $\Gamma(S,V)$, in which $V = \{v_i\}$ represents the set of vertices and $S = \{s_{ij}\}$ is the set of edges connecting pairs of vertices $i, j \in \overline{1,n}$. The weight assigned to each edge $s_{ij} \in S$ corresponds to the time it takes the robot to move from vertex v_i to vertex v_j.

We will limit our analysis to the linear case, where the acceleration is determined by the control input $\ddot{x}(t) = u_1(t)$. It is assumed that the path between points v_i and v_j is free of obstacles; otherwise, the corresponding graph edge s_{ij} is assigned an infinite length $+\infty$. Furthermore, we assume that the object begins its motion at v_i with an initial velocity $\dot{x} = q_i > 0$, and must arrive at the final point v_j with a specified velocity $\dot{x} = q_j > 0$. In these cases, as follows from [11], the controlled object belongs to a class for which the maximum principle is not only a necessary but also a sufficient condition for optimality, and, in addition, the statement is true that (see [11], p. 282) each control element that performs a transition from any starting point to any ending point takes only the values $u_1(t) = u_{\min,1}$ or $u_1(t) = u_{\max,1}$ and has no more than one switching.

At this point, it is worth emphasizing an important consideration. The earlier discussion regarding the number of control switches assumes that there are no restrictions on the maximum value of $\dot{x}(t)$. However, if a constraint 3 (such as a limit on velocity or control input) is introduced, the number of possible switches will depend on factors like the magnitude of $u_{\max,1}$ and the length of the edge s_{ij}. In such cases, either one or two switches between the control values $u_{\max,1}, 0, u_{\min,1}$ may occur.

Let us now address the question of how vertices are assigned. In their construction, a natural smoothness condition is imposed on the robot's trajectory. Specifically, the robot's path around a circular obstacle should start and end at points where it touches the boundary circles of that obstacle. The total set of vertices (excluding intermediate points along the obstacle boundaries) naturally includes the initial point $v_0 = x^0$ and the final target point $v_n = \hat{x}$.

Let $A(x_{01}, x_{02})$ be an arbitrary point on the plane, and let $C(x_{j1}, x_{j2})$ denote the center of the j-th circular obstacle. The coordinates of the contact points – that is, the points where the robot's trajectory touches the boundary of the obstacle – are then determined as solutions to the following nonlinear system of equations:

$$\begin{cases} (x_{01} - x_1)(x_{j2} - x_2) + (x_{02} - x_2)(x_{j1} - x_1) = 0, \\ (x_1 - x_{j1})^2 + (x_2 - x_{j2})^2 = R_j^2, \end{cases} \quad (5)$$

where R_j is the radius of the j-th circular obstacle.

Using straightforward algebraic manipulations, it can be shown that the solutions (x_1, x_2) of this system correspond to the roots of a quadratic equation of the form $ax^2 + bx + c = 0$, where the coefficients a, b, and c are derived from system (5).

In the special case when $x_{02} = x_{j2}$, the component x_1 can be computed directly using the formula $x_1 = w/(x_{01} - x_{j1})$, and the corresponding value of x_2 is then obtained from the second equation of the system (5).

When constructing tangents to two circles, it is possible to determine a specific point A, which represents the unique intersection point of the tangents and the line connecting the centers of the two circles. Once this point is determined, the problem reduces to the previously discussed case.

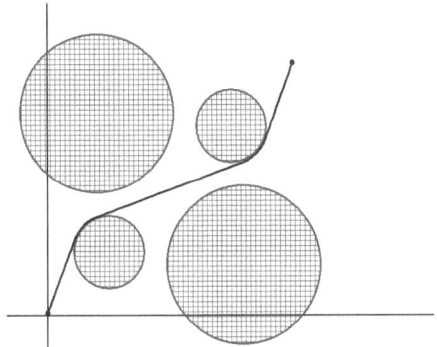

Fig. 1. Optimal trajectory

Consequently, any robot trajectory consists of a sequence of segments. Some of these segments are straight-line paths between two graph vertices with no obstacles in between, while others correspond to motion at constant speed along circular arcs. For a given robot, such trajectories define a finite set of possible states, from which one can choose the path that minimizes travel time. To achieve this, one can apply, for example, Dijkstra's algorithm [14], commonly used to find the shortest path in a graph. The optimal trajectory obtained through this method is illustrated in Fig. 1.

Let us consider an arbitrary rectilinear fragment of the obtained trajectory $x(t)$. Without loss of generality, we can assume that the fragment starts at the point $x^0 = (0,0)$ and ends at the point $\hat{x} = (\hat{x}_1, 0)$. We will also assume that the initial position of the robot at the point x^0 is known exactly and that the robot's velocity is constant over the entire interval $[x^0, \hat{x}]$, i.e. $\dot{x}(t) = v = const.$ We will also assume that control is carried out using some positioning system that, at each time interval τ, transmits to the controlled object its coordinates $y(t_i) = (y_1(t_i), y_2(t_i))$ with some small error δ_i, where $|\delta_i| < \varepsilon$, where $\varepsilon > 0$ — known positioning accuracy.

By $z(t) = (z_1(t), z_2(t))$ we will denote the unknown current coordinates of the robot, i.e. $z(t_i) = y(t_i) + \delta_i$. The coordinates $z(t)$ are assumed to be unknown due to a variety of reasons. These can be all sorts of external influences on the controlled object (weather conditions, the state of the surface on which the robot moves, etc.), as well as technical (design) features of the robot itself. In other words, the robot can deviate from the trajectory $x(t)$ and move in a direction in which it will not be able to reach the terminal point $\hat{x}$. Consequently, at some point its movement must be corrected in such a way as to guarantee that it will hit at least some acceptable neighborhood of the point $\hat{x}$.

4 An Algorithm and its Justification

The essence of the algorithm proposed for this purpose is as follows. If the coordinates $y(t_i)$ transmitted by the positioning system do not go beyond the boundaries of the ε—band on both sides of the theoretical trajectory of the robot $x(t)$, there is no need to correct its motion, since in this case there is a probability that the object has not deviated from the specified trajectory, and the deviation $y(t_i)-x(t_i)$ is a consequence of the error of the positioning system. If, taking into account the speed v and the time interval τ, it is possible for the point $y(t_{i+1})$ to go beyond the specified band, then a trajectory-correcting control $u = \hat{x} - y(t_i)$ should be introduced. Thus, the step-by-step algorithm takes the following form.

Algorithm 1.

Step 1. Set $i = 0$, $t_i = 0$. Go to step 2.
Step 2. Set $i := i+1$, $t_i := t_{i-1} + \tau$, accept positioning system data $y(t_i)$. Go to step 3.
Step 3. If $|y_2(t_i)| + v\tau < \varepsilon$, then go to step 2. Otherwise, go to step 4.
Step 4. Reconfigure robot control to move along vector $u = \hat{x} - y(t_i)$. Go to step 5.
Step 5. If $|y_1(t_i) - \hat{x}_1| < \varepsilon$, then stop. Otherwise, go to step 2.
Thus, the following is true

Theorem 1. *Let the control be carried out in accordance with algorithm 1. Then for any t the inequality*
$$|z_2(t)| < 2\varepsilon$$
is valid.

Proof. Let $|z_2(t_k)| \geq 2\varepsilon$ for some k, and for all $i < k$ the inequality $|y_2(t_i)| + v\tau < \varepsilon$ is valid. Then

$$|z_2(t_k)| \leq |z_2(t_{k-1})| + v\tau = |y_2(t_{k-1}) + \delta_{k-1}| + v\tau \leq$$
$$\leq |y_2(t_{k-1})| + |\delta_{k-1}| + v\tau < |y_2(t_{k-1})| + \varepsilon + v\tau < 2\varepsilon.$$

The resulting contradiction proves the statement.

5 Conclusion

This paper addresses the problem of optimal control for a mobile robot, under the assumption that the robot periodically receives position data from a positioning system within a given coordinate frame. It is further assumed that these coordinates are subject to approximation errors with known bounds. Building upon the theory of optimal control synthesis for second-order systems and the method of finding shortest paths on a connected graph, an algorithm is proposed to correct the robot's trajectory in real time. This algorithm ensures that the robot reaches its target point within a specified accuracy.

The implementation and analysis of numerical experiments based on this approach are intended to be addressed in future research by the author.

References

1. Bloesch, M., Omari, S., Hutter, M., and Siegwart, R.: Robust visual inertial odometry using a direct EKF-based approach. In: 2015 IEEE/RSJ International Conference on Intelligent Robots and Systems (IROS), pp. 298–304. Hamburg, Germany, September 28 - October 02 (2015)
2. Sun, K., Mohta, K., Pfrommer, B., Watterson, M., Liu, S., Mulgaonkar, Y., Taylor, C.J., Kumar, V.: Robust stereo visual inertial odometry for fast autonomous flight. IEEE Robot. Automation Lett. **3**(2), 965–972 (2018)
3. Qin, T., Li, P., Shen, S.: VINS-Mono: A robust and versatile monocular visual-inertial state estimator. IEEE Trans. Rob. **34**(4), 1004–1020 (2018)
4. Khattak, S., Papachristos, C., Alexis, K.: Keyframe-based thermal-inertial odometry. J. Field Robot. **37**(4), 552–579 (2020)
5. Saputra, M.R.U., et al.: DeepTIO: a deep thermal-inertial odometry with visual hallucination. IEEE Robot. Automation Lett. **5**(2), 1672–1679 (2020)
6. Zhao, S., Wang, P., Zhang, H., Fang, Z., and Scherer, S.: TP-TIO: a robust thermal-inertial odometry with deep thermalpoint. In 2020 IEEE/RSJ International Conference on Intelligent Robots and Systems (IROS), pp. 4505–4512. Las Vegas, NV, USA, October 25-29 (2020)
7. Dickmann, J., Klappstein, J., Hahn, M., Appenrodt, N., Bloecher, H., Werber, K., Sailer, A.: Automotive radar – The key technology for autonomous driving: From detection and ranging to environmental understanding. In 2016 IEEE Radar Conference (RadarConf), pp. 1–6. Philadelphia, PA, USA, May 2-6 (2016)
8. Mur-Artal, R., Tardos, J.D.: ORB-SLAM2: an open-source SLAM system for monocular, stereo, and RGB-D cameras. IEEE Trans. Rob. **33**(5), 1255–1262 (2017)
9. Bereznev, V.A.: The principle of dividing feasible trajectories in a robot control problem. Procedia Comput. Sci. **186**, 456–459 (2021)
10. Pontryagin, L.S., Boltyanskii, V.G., Gamkrelidze, R.V., Mishchenko, E.F.: Math. Theory Pptimal Processes. Interscience Publishers John Wiley and Sons, Inc (1962)
11. Boltyanskii, V.G.: Mathematical methods of optimal control. Nauka, Moscow (1968). In Russian
12. Karmanov, V.G.: Mathematical programming. Fizmatlit, Moscow (2000). In Russian

13. Izmailov, A.F., Solodov, M.V.: Numerical methods of optimization. Fizmatlit, Moscow (2003). In Russian
14. Dijkstra, E.W.: A note on two problems in connection with graphs. Numer. Math. Springer Science + Business media **1**(1), 269–271 (1959)

Fault Detection and Isolation for USV with INS Using Genetic Algorithm

Dmitry Bazylev(✉), Alexey Margun, and Radda Iureva

ITMO University, Saint Petersburg, Russia
`bazylevd@itmo.ru`

Abstract. This study is addressed to a problem of fault detection and isolation (FDI) of an inertial navigation system that is used by an unmanned surface vessel (USV). The inertial navigation system (INS) includes two sensors that measure angular and linear velocity. It is assumed that all parameters of the second order USV model are unknown. The developed algorithm is based on the parameter identification with a guaranteed finite time convergence, second order directional generators of residual signals for sensor faults detection and genetic algorithm for tuning design parameters. Simulation results demonstrate the effectiveness of the proposed approach providing accurate and timely fault identification of each sensor.

Keywords: Fault detection and isolation · Unmanned surface vessel · Parameter identification · Genetic algorithm

1 Introduction

Fault detection and isolation (FDI) is an important aspect aimed at increasing the reliability of complex technical systems. It involves determining the occurrence of faults and identifying their specific location in the system. Using inaccurate values of system parameters can significantly degrade the quality and efficiency of these processes. Therefore, parameter identification plays a crucial role in increasing the accuracy and robustness of FDI methods.

Among the widely used methods for fault diagnosis, the following key approaches can be distinguished: i) model-based techniques, ii) data-driven methods and iii) hybrid approaches.

The first direction of research is useful for accurate fault identification by comparing sensor measurements against predictions from precise mathematical models. These models can adapt to changes in system dynamics, offering flexibility and adaptability. However, developing accurate models, particularly for systems with nonlinearities or uncertainties, can be a complex and time-consuming endeavor. Furthermore, model inaccuracies or unforeseen system changes can lead to unreliable identification of faults, resulting in false alarms or missed detection.

On the other hand, data-driven methods eliminate the need for complex models by using historical system data. This makes them applicable to a wider range of systems and operating conditions, and easily scalable to handle large data sets and complex systems. However, their dependence on data quality is a significant challenge. Possible noise and missing information can compromise their effectiveness and interfere with the validity of the results.

Reliable fault detection and isolation in complex systems often requires finding trade-offs between model-based and data-driven methods. Hybrid techniques can leverage the strengths of both approaches and enhance robustness by addressing their limitations. However, such methods often require integration of several algorithms that may result in complexity of practical implementation.

Known approaches of estimation of system parameters that are used in FDI include: adaptive and robust observer-based methods, extended Kalman filter (EKF), model-based identification algorithms, machine learning techniques, etc. Such methods can be applied for fault detection of various sensors of technical systems. In this study, an unmanned surface vessel (USV) with an inertial navigation system (INS) is considered. INS of the USV is usually equipped with microelectromechanical accelerometer and gyroscope which correct measurements are strictly required for directional stability.

Using erroneous readings of INS sensors in a control system leads to generation of improper control signals that can cause unstable behavior of USV. In this regard, many methods for fault diagnosis of INS sensors have been proposed [1–3].

In [4] optimal filtering methods are used to effectively handle uncertainties arising from probabilistic measurement errors. A wide range of machine learning strategies are proposed for fault detection in [5–8]. Authors of [5] consider the use of autoencoders, specifically exploring their application in representation learning and monitoring tactics for FDI.

The study [6] demonstrates the effectiveness of generative adversarial networks (GAN) in detecting faults in the Tennessee Eastman process. GANs are powerful tools for generating synthetic data and identifying deviations from expected patterns, making them suitable for complex systems with potential anomalies. Papers [9,10] explore the combination of fuzzy logic systems and metaheuristic optimization algorithms for fault-tolerant control. This approach leverages the ability of fuzzy logic to handle uncertainties and the optimization capabilities of metaheuristics to find efficient solutions for systems with actuator faults and external disturbances. In particular, the authors present a fuzzy-based harmonic search metaheuristic technique for robust control under these conditions. Observer-based methods from [4,11] apply estimators of model state variables that can be adapted to diagnose specific input and output faults. However, practical implementation of such methods may be inefficient if the USV has unknown or inaccurately known parameters.

Improper settings of design parameters can lead to performance degradation of observer and parameter estimator (slow response, overshoot, oscillations and long-term estimation errors) that may result in low quality of FDI operation. The problem of optimal search of design gains in multidimensional space can be solved by various machine learning techniques, for instance, evolutionary algorithms (EAs). Such algorithms employ basic principles of biological evolution that usually include [12–14]: i) selection, ii) mutation and iii) reproduction processes. In [12] the structure of non-linear control is identified using genetic programming (GP). Also the possibility to tune proportional and integral gains of the controller is presented. Another examples of practical application of GAs are addressed to optimization of currents in the direct voltage controller [13] and design gain tuning of position observer of synchronous motors [14].

In this paper, fault detection and isolation algorithm is combined with parameter estimator for the USV with INS that ensures guaranteed finite time convergence of estimation errors to zero. GA is applied to find the optimal values of design gains for finite time parameter estimator. Presented estimation algorithm is based on filtering techniques, Kreisselmeier's dynamic extension of the regressor and ad-hoc procedure that results in a direct scalar solutions with a threshold. Detection and isolation of INS faults is based on reconstructed parameter values and full-order Luenberger observers, which output is used to design directional residual generators.

The remaining of the paper is organized as follows. The problem formulation and the state-space model of USV with unknown parameters and fault signals is described in Sect. 2. Section 3 presents the main result of the paper which includes finite time parameter estimator, FDI approach for INS sensors and GA implementation. Simulation results that illustrate the efficiency of the proposed approach are shown in Sect. 4. Finally, the paper is wrapped-up with concluding remarks in Sect. 5.

2 USV Model and Problem Formulation

Assume that the longitudinal speed of the USV is measurable and constant $u = u_0$. Then, the model of the USV is described by the second order Nomoto model [15] that is given by

$$M_R \dot{x} + N_R(u_0) x = B_R \delta_R, \tag{1}$$

where $M_R \in \mathbb{R}^{2\times 2}$ is a rigid body inertia matrix; $N_R \in \mathbb{R}^{2\times 2}$ is a matrix of Coriolis and centrifugal forces; $x \in \mathbb{R}^2$ is a velocity vector that includes transverse velocity v and angular speed r of the vessel course, $x = [v \quad r]^\top$; $B_R \in \mathbb{R}^2$ is a vector of external forces and moments and $\delta_R \in \mathbb{R}$ is a rudder angle.

Separate measurements of two sensors in the INS are used to obtain the linear v and angular velocities r. For example, r can be directly measured by the gyroscope and v can be calculated using accelerometer measurements.

The input-state-output form of the USV model (1) is represented as

$$\dot{x} = Ax + B\delta_R, \tag{2}$$
$$y = Cx, \tag{3}$$

where the matrices $A, C \in \mathbb{R}^{2\times 2}$ and $B \in \mathbb{R}^2$ are given by $A = -M_R^{-1} N_R(u_0) := \begin{bmatrix} a_1 & a_2 \\ a_3 & a_4 \end{bmatrix}$, $B = M_R^{-1} B_R := \begin{bmatrix} b_1 \\ b_2 \end{bmatrix}$, $C = \begin{bmatrix} 1 & 0 \\ 0 & 1 \end{bmatrix}$.

Taking into account sensor faults, the USV model (2)-(3) can be rewritten as

$$\dot{x} = Ax + B\delta_R, \tag{4}$$

$$y = Cx + I_i \begin{bmatrix} f_1 \\ f_2 \end{bmatrix}, \quad I_i = \begin{bmatrix} 1 & 0 \\ 0 & 1 \end{bmatrix}, \tag{5}$$

where $f_1, f_2 \in \mathbb{R}$ are failure signals from the accelerometer and gyroscope, respectively.

Consider the following assumptions imposed on the USV model with INS:

A1. The matrix of dynamical properties A and the input matrix B are *unknown*.
A2. Fault signals $f_1(t)$ and $f_2(t)$ are *unknown* time-varying functions.
A3. The states x_1, x_2 and the system output y are measurable with possible distortions due to faults.

The following three objectives are of interest in this study:

1. Estimation of all parameters of the matrices A and B of the model (2)-(3) in a finite time.
2. Detection and isolation of faults of the two sensors in the INS using parameter estimates.
3. Automatic tuning of design parameters in the developed approach using genetic algorithm.

3 Main Result

3.1 Finite Time Parameter Estimation Algorithm

The first step in the synthesis of the parameter estimator is to perform such transformation of the USV model (2)–(3) that results in a linear regression model. Introduce a linear time-invariant (LTI) filter

$$H_1(p) := \frac{h_1}{p + h_1}, \quad h_1 > 0 \tag{6}$$

where $h_1 > 0$ is a design parameter and $p = \frac{d}{dt}$ is a differentiation operator.

Hereinafter, $\frac{h_1}{p+h_1}\{\mu(t)\}$ and $H_1(p)\{\mu(t)\}$ denote the application of a filtering operation $H_1(p)\{\cdot\} = \frac{h_1}{p+h_1}\{\cdot\}$ to some time-varying signal $\mu(t)$.

Applying the filter $H_1(p)$ to (2) one can obtain

$$\frac{h_1}{p+h_1}\{\dot{x}_1\} = a_1 \frac{h_1}{p+h_1}\{x_1\} + a_2 \frac{h_1}{p+h_1}\{x_2\} + b_1 \frac{h_1}{p+h_1}\{\delta_R\}, \tag{7}$$

$$\frac{ph_1}{p+h_1}\{\dot{x}_2\} = a_3 \frac{h_1}{p+h_1}\{x_1\} + a_4 \frac{h_1}{p+h_1}\{x_2\} + b_2 \frac{h_1}{p+h_1}\{\delta_R\}, \tag{8}$$

Notice, that according to problem formulation the vector $\dot{x} = px$ is immeasurable while the filtered signals $\frac{h_1}{p+h_1}\{\dot{x}_{1,2}\} = \frac{ph_1}{p+h_1}\{x_{1,2}\}$ are known.

To simplify further notations, introduce the new signals $\bar{x}_1, \bar{x}_2 \in \mathbb{R}$ and $m \in \mathbb{R}^3$, $m := \begin{bmatrix} m_1 & m_2 & m_3 \end{bmatrix}^\top$ which are available for calculation

$$\bar{x}_1 = pH_1(p)\{x_1\}, \quad \bar{x}_2 = pH_1(p)\{x_2\}$$
$$m_1 = H_1(p)\{x_1\}, \quad m_2 = H_1(p)\{x_2\}, \quad m_3 = H_1(p)\{\delta_R\} \tag{9}$$

and the vectors of unknown constant parameters $\alpha, \beta \in \mathbb{R}^3$

$$\alpha = [\alpha_1\ \alpha_2\ \alpha_3]^\top = [a_1\ a_2\ b_1]^\top, \tag{10}$$
$$\beta = [\beta_1\ \beta_2\ \beta_3]^\top = [a_3\ a_4\ b_2]^\top. \tag{11}$$

Then, the linear regression models (7) and (8) can be represented as follows

$$\bar{x}_1 = m^\top \alpha, \tag{12}$$
$$\bar{x}_2 = m^\top \beta, \tag{13}$$

where the signals $\bar{x}_1, \bar{x}_2$ and m are known.

Next, one can use various methods to estimate unknown parameters in (12) and (13): standard gradient-descent algorithm, least-squares methods, dynamic regressor extension and mixing technique (DREM) [16, 17], etc. In this paper, the Kreisselmeier's dynamic regressor extension (KDRE) approach is applied for this task. The main reasons of such choice are i) better transient behavior compared to standard gradient algorithm and ii) lower number of design gains for tuning in contrast to DREM procedure.

According to KDRE, the models (12) and (13) are first extended multiplying equations from the left by the regressor function $m(t)$ keeping the excitation level

$$m\bar{x}_1 = mm^\top \alpha, \tag{14}$$
$$m\bar{x}_2 = mm^\top \beta. \tag{15}$$

The next step is to apply dynamic operator

$$H_2(p) := \frac{h_2}{p+h_2}, \quad h_2 > 0 \tag{16}$$

with the design parameter $h_2 > 0$ to the both sides of Eqs. (14) and (15). As a result, one can get

$$\bar{X}_1 = M\alpha, \tag{17}$$
$$\bar{X}_2 = M\beta, \tag{18}$$

where $\bar{X}_1, \bar{X}_2 \in \mathbb{R}^3$ and $M \in \mathbb{R}^{3\times 3}$ are known and given by

$$\bar{X}_1 = H_2(p)\{m\bar{x}_1\}, \quad \bar{X}_2 = H_2(p)\{m\bar{x}_2\}, \quad M = H_2(p)\{mm^\top\}.$$

Now, multiply the dynamically extended models (17) and (18) from the left by the adjoint matrix of M. Using such a mixing operation six scalar linear regressors are obtained

$$g_{1l} = q\alpha_l, \qquad (19)$$
$$g_{2l} = q\beta_l, \qquad (20)$$

where $l = 1, 2, 3$, $q \in \mathbb{R}$ and $g_1, g_2 \in \mathbb{R}^3$ are known signals

$$q = \det\{M\}, \quad g_1 = \mathrm{adj}\{M\}\bar{X}_1, \quad g_2 = \mathrm{adj}\{M\}\bar{X}_2, \qquad (21)$$

$\det\{M\} \in \mathbb{R}$ is the determinant of M and $\mathrm{adj}\{M\} \in \mathbb{R}^{3\times 3}$ is the adjoint matrix of M.

One of the possible solutions is to apply scalar standard gradient estimators to reconstruct the unknowns from (19) and (20). In paper [19], DREM procedure with such estimators is used ensuring two global convergence properties of estimation errors to zero: i) asymptotic, if $q(t)$ is not square integrable and ii) exponential, if $q(t)$ satisfies persistent excitation condition. However, verification of such conditions may be non-obvious and difficult in a number of technical systems. Moreover, the mentioned types of convergence operate with infinite time, thus creating uncertainty as to when the parameter estimates can be considered sufficiently accurate. To overcome these shortcomings, we propose to use a parameter estimator that is constructed using direct calculation from scalar Eqs. (19) and (20), adding a small threshold for the regressor function $q(t)$ to avoid its zero crossing. The finite time convergence properties of the estimator are ensured under the following assumption:

A4. Introduce a positive relatively small design parameter $\lambda > 0$. Let the finite time condition holds

$$\forall t \geq 0, \, \exists t_F > t : \, \det\left\{H_2(p)\left[mm^\top\right]\right\} \geq \lambda, \qquad (22)$$

where t_F is a finite time, $m(t)$ and $H_2(p)$ are given by Eqs. (16) and (9), respectively.

The following proposition establishes the finite time parameter estimator.

Proposition 1. *Let the assumptions A1, A3 and A4 hold. Consider the state-space model of USV given by (2)-(3). Introduce the functions (9) and (21) with filters $H_1(p)$ and $H_2(p)$ given by (6) and (16), respectively. The parameter estimator given by*

$$\hat{\alpha}_l = \frac{g_{1l}}{max\{q, \, \lambda\}}, \; l = 1, 3, \qquad (23)$$

$$\hat{\beta}_l = \frac{g_{2l}}{max\{q, \, \lambda\}}, \; l = 1, 3, \qquad (24)$$

with the design threshold $\lambda > 0$ ensures convergence of parameter estimation errors to zero in the finite time t_F

$$\tilde{\alpha}_l(t) = \hat{\alpha}_l(t) - \alpha_l = 0, \quad \tilde{\beta}_l(t) = \hat{\beta}_l(t) - \beta_l = 0 \quad \forall t \geq t_F$$

and accurate reconstruction of the matrices A and B of the USV model

$$\hat{A} = \begin{bmatrix} \hat{\alpha}_1 & \hat{\alpha}_2 \\ \hat{\beta}_1 & \hat{\beta}_2 \end{bmatrix}, \quad \hat{B} = \begin{bmatrix} \hat{\alpha}_3 \\ \hat{\beta}_3 \end{bmatrix}, \tag{25}$$

ensuring

$$\tilde{A} = \hat{A} - A = \begin{bmatrix} \hat{\alpha}_1 & \hat{\alpha}_2 \\ \hat{\beta}_1 & \hat{\beta}_2 \end{bmatrix} - \begin{bmatrix} a_1 & a_2 \\ a_3 & a_4 \end{bmatrix} = [0] \quad \forall t \geq t_F$$

and

$$\tilde{B} = \hat{B} - B = \begin{bmatrix} \hat{\alpha}_3 \\ \hat{\beta}_3 \end{bmatrix} - \begin{bmatrix} b_1 \\ b_2 \end{bmatrix} = [0] \quad \forall t \geq t_F.$$

Proof. The proof of the proposition is completed in two steps. First, consider the direct computation of parameter values from (19) and (20)

$$\alpha_l = \frac{g_{1l}}{q}, \quad \beta_l = \frac{g_{2l}}{q}, \quad l = 1, 3. \tag{26}$$

Next, recalling that $q = \det\{M\}$ and considering that the assumption A4 is fulfilled, one can get

$$\forall t \geq 0, \, \exists t_F > t : q \geq \lambda, \, \lambda > 0 \tag{27}$$

Finally, applying the latter condition to (23) and (24) we get

$$\max\{q, \lambda\} = q$$

and existence of parameter estimates calculated directly by (26).

3.2 Fault Detection Algorithm

The key idea of the presented fault detection and isolation approach is in the usage of parameter estimates generated by estimation device from Proposition 1 during synthesis of directional generators based on the residual signals. These residuals are calculated as differences between the measured faulty output and the estimates of output generated by the full-order Luenberger observers [18]. The full-order Luenberger observer for the USV model (2)-(3) with known A and B is given by

$$\dot{\hat{x}} = A\hat{x} + B\delta_R + K(y - \hat{y}), \tag{28}$$
$$\hat{y} = C\hat{x}, \tag{29}$$

where $\hat{x}(t)$ is the estimate of $x(t)$, $\hat{y}(t)$ is the estimate of $y(t)$ and $K \in \mathbb{R}^{2\times 2}$ is a design matrix. In this paper, two observers of the form (28)-(29) are applied to detect deviations of the system output signal caused by one or another sensor fault of INS. Sensitivity to each sensor failure is achieved by appropriately choosing the design matrix K based on two conditions that are related with isolability of faults and formulated below.

Let the assumption A4 hold. Then we get accurate estimates $\hat{\alpha}$ and $\hat{\beta}$ generated by the finite time estimator (23)-(24). Using the estimates from (25) introduce two observers for each sensor fault of the form (28)-(29)

$$\dot{\chi} = \hat{A}\chi + \hat{B}\delta_R + F(y - z), \tag{30}$$
$$z = C\chi, \tag{31}$$
$$\dot{\xi} = \hat{A}\xi + \hat{B}\delta_R + P(y - w), \tag{32}$$
$$w = C\xi, \tag{33}$$

where $z(t)$ and $w(t)$ are the estimates of $y(t)$ with accelerometer and gyroscope faults, respectively, $\chi(t)$ and $\xi(t)$ are the estimates of $x(t)$, $F, P \in \mathbb{R}^{2\times 2}$ are design matrices.

The model output (5) with separate sensor faults is given by

$$y_1 = Cx + \begin{bmatrix} 1 \\ 0 \end{bmatrix} f_1, \quad y_2 = Cx + \begin{bmatrix} 0 \\ 1 \end{bmatrix} f_2, \tag{34}$$

where $y_1, y_2 \in \mathbb{R}^2$.

Consider the state estimation errors $e_1, e_2 \in \mathbb{R}^2$

$$e_1 = x - \chi, \quad e_2 = x - \xi, \tag{35}$$

and the residual signals $r_1, r_2 \in \mathbb{R}^2$

$$r_1 := y - z = Cx - C\chi = Ce_1, \tag{36}$$
$$r_2 := y - w = Cx - C\xi = Ce_2. \tag{37}$$

Then, subtracting (30) from (2) and (31) from (3) one can get the dynamic model of the residual signal for the accelerometer fault

$$\dot{e}_1 = (\hat{A} - FC)e_1 + d_1 f_1, \quad r_1 = Ce_1, \tag{38}$$

where $d_1 \in \mathbb{R}^2$ is a vector that denotes the direction of accelerometer faults. Similarly, subtracting (32) from (2) and (33) from (3) the residual model for the gyroscope fault is obtained

$$\dot{e}_2 = (\hat{A} - PC)e_2 + d_2 f_2, \quad r_2 = Ce_2, \tag{39}$$

where $d_2 \in \mathbb{R}^2$ is a vector that denotes the direction of gyroscope faults.

The well-known condition of stability for the observers (30)-(31) and (32)-(33) is that the matrices $(\hat{A}-FC)$ and $(\hat{A}-PC)$ are of Hurwitz form.

In practice, sensor measurements are not ideal and often include noises (for example, zero drift of the gyroscope, high frequency noises of the accelerometer due to external vibrations). Therefore, two thresholds $\sigma_1 > 0$ and $\sigma_2 > 0$ with relatively small value are applied for each residual signal and switching functions are formed for correct detection of faults:

$$\bar{r}_1 = \begin{cases} r_1, & \text{if } |r_1| \geq \sigma_1, \\ 0, & \text{if } |r_1| < \sigma_1. \end{cases} \tag{40}$$

$$\bar{r}_2 = \begin{cases} r_2, & \text{if } |r_2| \geq \sigma_2, \\ 0, & \text{if } |r_2| < \sigma_2. \end{cases} \tag{41}$$

The localization of the failed sensor is ensured if the following conditions are satisfied [4]:

i) Cd_1 and Cd_2 are linearly independent vectors,
ii) $rank[d_1; (\hat{A}-FC)d_1] = 1$,
$rank[d_2; (\hat{A}-PC)d_2] = 1$.

The first requirement is related with separate localization of the sensor faults in a two-dimensional residual space. The second condition ensures changes of the residual signal r_1 in the direction of d_1 and r_2 in the direction of d_2 only. The latter requirement along with stability condition for observers define design matrices F and P that provide sensitivity for each sensor fault.

The final step is to compute the normalized values N_1 and N_2 of residuals' projections on the fault directions d_1 and d_2

$$N_{1,2} = \frac{|d_{1,2}^\top r_{1,2}|}{\left\|d_{1,2}^\top\right\|_2 \|r_{1,2}\|_2}. \tag{42}$$

3.3 Genetic Algorithm

In this section, a genetic algorithm (GA) is constructed to perform optimal tuning of design gains h_1, h_2 and λ ensuring global finite time convergence of parameter estimation errors to zero.

Genetic algorithm(GA) is a useful tool that can be applied to solve various optimization problems. It's application might be especially effective in multidimensional parameter spaces requiring considerably less computational resources in contrast to some other techniques, for example, brute force search.

In this paper, we consider GA that is based on the following evolution methods [12]: replication o_1, crossover o_2 and mutation o_3 operators. The task of the replication operator is to choose a single individual (agent) based on its fitness, allowing it to directly contribute to the next generation. The crossover operator combines the genetic material of two individuals, exchanging portions of their code. This process strengthens and refines successful strategies, facilitating the development of individuals in a next generation. The mutation operator introduces random changes to an individual's code. This encourages adaptation to the environment by allowing individuals to explore new possibilities and promotes diversity, thereby increasing exploration of the multidimensional parameter space.

Introduce the new denotations to construct the GA: $\rho(o_1)$ is the probability of replication, $\rho(o_2)$ is the probability of the crossover and $\rho(o_3)$ is the probability of the mutation. These quantities can be set to different values according to a specific optimization task. Here we construct the GA for the case when most agents with good fitness are contributed to the next generation using the crossover operation, then a smaller number of them are reproduced via the mutation operator and the agents with better fitness are replicated directly. Thus, the following conditions holds

$$\rho(o_1) + \rho(o_2) + \rho(o_3) = 1. \tag{43}$$
$$\rho(o_1) < \rho(o_3) < \rho(o_2). \tag{44}$$

Also denote two main tunable hyperparameters: N_P is a number of agents in a population and N_G is a number of generations. Each agent in the population is assigned a fitness value. And the more adapted agents have a higher probability of reproducing taking into account inequality (44).

The optimization problem itself is structured such that its solution can be represented as a set of genes (genotype). In general, these genes can take the form of bits, numbers, or other relevant data structures, effectively encoding the solution within the genetic framework. In this paper, we use the genotype with a fixed length that is represented by the set of the three tunable coefficients: h_1 and h_2 from the dynamic operators $H_1(p)$ and $H_2(p)$, respectively, and the threshold λ from (23) and (24). Therefore, the genotype is given by the vector

$$\nu = \begin{bmatrix} h_1 & h_2 & \lambda \end{bmatrix}^\top. \tag{45}$$

Invoking that all of these parameters must be strictly positive to ensure global finite time convergence of the estimators (23) and (24), the multidimensional space for optimal search is defined as follows

$$\nu \in \mathbb{R}_+^3 \Leftrightarrow h_1 > 0,\ h_2 > 0,\ \lambda > 0.$$

The parameter estimation errors are used as metrics for the cost (objective, fitness) function

$$\tilde{\alpha}_l(t) = \hat{\alpha}_l(t) - \alpha_l, \quad \tilde{\beta}_l(t) = \hat{\beta}_l(t) - \beta_l. \tag{46}$$

Finally, the cost function that should be minimized is formed in a quadratic form with respect to the mentioned estimation errors

$$J = \int_0^T \left(Q_\alpha ||\tilde{\alpha}||_2 + Q_\beta ||\tilde{\beta}||_2 \right) d\tau, \tag{47}$$

where T is the evolution time, $Q_\alpha, Q_\beta \in \mathbb{R}_+$ are weighting coefficients for norms of the vectors of the parameter estimation errors $\tilde{\alpha}$ and $\tilde{\beta}$, respectively.

4 Simulation Results

The first objective of simulations is to verify the performance and validate the optimal solution generated by the GA for the finite time parameter estimator. The second one is to investigate the efficiency of the proposed FDI algorithm that uses reconstructed values of USV parameters.

Initial states of the filters $H_1(p)$ and $H_2(p)$ and Luenberger observers are zero. The following parameter values are applied in the GA: $N_P = 20$, $N_G = 15$,

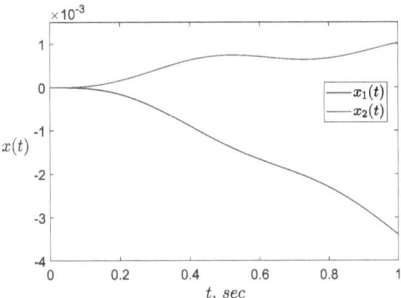

Fig. 1. Transients of system states $x(t)$ used in GA

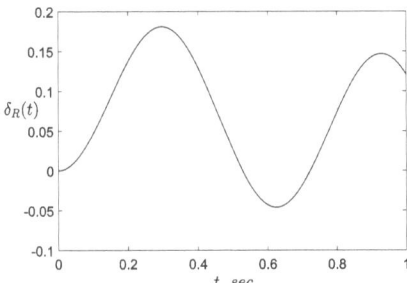

Fig. 2. Transients of rudder angle $\delta_R(t)$ used in GA

$Q_\alpha = Q_\beta = 0.5$. The condition for stopping the GA is reaching the limit on the number of generations. Figures 1 and 2 demonstrate the states of the USV model $x_1(t)$, $x_2(t)$ measured by INS and control signal $\delta_R(t)$ that are collected with a sampling interval of 0.0001 sec. This data is used as an input for parameter estimator. The reconstructed parameter values $\hat{\alpha}$ and $\hat{\beta}$ are then used in the cost function given by (47).

Figure 3 shows changes of the cost function J during evolution process. As can be seen from the figure, the values of $\log(J)$ decrease as the number of generations increases. Thus, the parameter estimation errors are minimized during the evolution process. Figure 4 illustrate the search of the design values in the parameter space from the early generations (red dots) to the best set (black dot) found by GA: $\nu_{opt} = \begin{bmatrix} 22.3821 & 3.5957 & 10^{-7} \end{bmatrix}^T$ with $J = 1.4209$.

Transients of parameter estimation errors are presented in Figs. 5 and 6. One can conclude, that the transient behavior is monotonic with small finite convergence time $t_F = 0.245$ sec. Finally, Figs. 7 and 8 show fault signals in sensors $f_{1,2}$ and normalized fault detection signals $N_{1,2}$ obtained by the residual generators with parameter estimates. As seen from Fig. 8, the proposed approach demonstrates correct and timely detection and isolation of INS sensor failures.

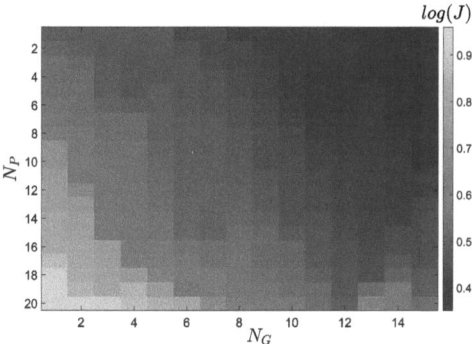

Fig. 3. The logarithm of the cost function $\log(J)$ depending on the population and the generation

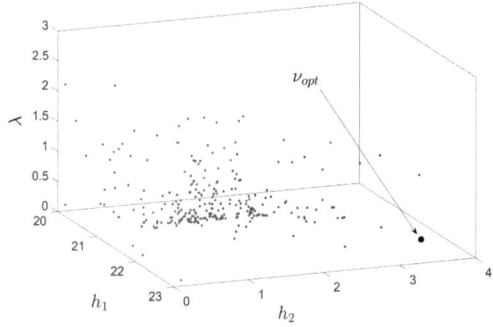

Fig. 4. Evolution process of individuals in the parameter space

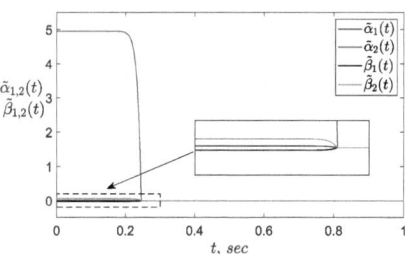

Fig. 5. Transients of estimation errors of parameters for matrix A

5 Conclusions

The problem of fault detection and isolation of the inertial navigation system that includes two sensors is considered in this paper. Both sensors are used for navigation of the unmanned surface vessel that is described by the second order dynamic model. The detection task is complicated by the assumption that all

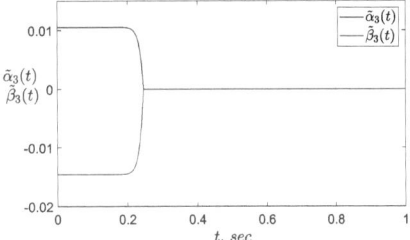

Fig. 6. Transients of estimation errors of parameters for matrix B

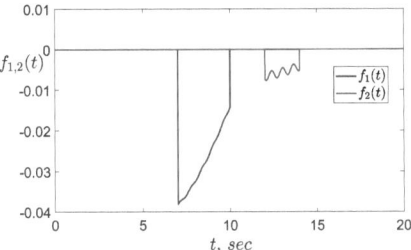

Fig. 7. INS sensor fault signals

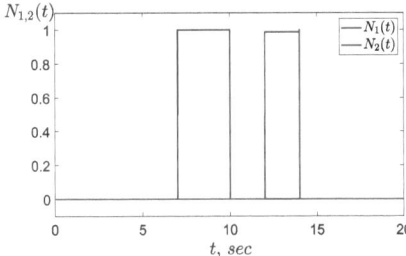

Fig. 8. Normalized fault detection signals

model parameters are unknown. This case is typical for many practical applications due to uncertain parameters, biases and deviations. Proposed detection strategy is based on three stages: i) finite time parameter identification, ii) FDI algorithm that uses directional generators of residual signals and parameter estimates and iii) genetic algorithm for tuning design gains of estimators.

It is shown that machine learning, in particular genetic algorithm, can be effectively used to find optimal design parameters of the analytically developed estimation algorithm. The advantages of the general approach, such as fast estimation of model parameters and accurate fault detection, are demonstrated using numerical simulations. Future research will focus on extending the presented approach to more complex and nonlinear USV models.

Acknowledgments. This work is supported by the Russian Science Foundation grant № 23-79-10071, https://rscf.ru/project/23-79-10071/.

References

1. Babaei, M., Shi, J., Abdelwahed, S.: A survey on fault detection, isolation, and reconfiguration methods in electric ship power systems. IEEE Access **6**, 9430–9441 (2018)
2. Perera, L.P.: Marine engine centered localized models for sensor fault detection under ship performance monitoring. IFAC-PapersOnLine **49**(28), 91–96 (2016)
3. Samy, I., Postlethwaite, I., Gu, D.W.: Survey and application of sensor fault detection and isolation schemes. Control. Eng. Pract. **19**, 658–674 (2011)
4. Chen, J., Patton, R.J.: Robust Model-based Fault Diagnosis For Dynamic Systems, p. 354. Kluwer Academic Publishers, Boston, MA, USA (1999)
5. Qian, J., Song, Z., Yao, Y., Zhu, Z., Zhang, X.: A review on autoencoder based representation learning for fault detection and diagnosis in industrial processes. Chemom. Intell. Lab. Syst. **231**(2), 104711 (2022)
6. Spyridon, P., Boutalis, Y. S.: Generative adversarial networks for unsupervised fault detection. In: 2018 European Control Conference (ECC), pp. 691–696. Limassol, Cyprus (2018)
7. Lo, N.G., Flaus, J.-M., Adrot, O.: Review of machine learning approaches in fault diagnosis applied to IoT systems. In: 2019 International Conference on Control, Automation and Diagnosis (ICCAD), pp. 1–6. Grenoble, France (2019)
8. Nie, G., Zhang, Z., Shao, M., Jiao, Z., Li, Y., Li, L.: A novel study on a generalized model based on self-supervised learning and sparse filtering for intelligent bearing fault diagnosis. Sensors **23**, 1858 (2023)
9. Patel, H.R.: Fuzzy-based metaheuristic algorithm for optimization of fuzzy controller: fault-tolerant control application. Int. J. Intell. Comput. Cybern. **15**(4), 599–624 (2022)
10. Patel, H., Shah, V.: Fuzzy logic based metaheuristic algorithm for optimization of type-1 fuzzy controller: fault-tolerant control for nonlinear system with actuator fault. IFAC-PapersOnLine **55**, 715–721 (2022)
11. Kolesnik, N.S., Margun, A.A., Kremlev, A.S., Zhivitskii, A.I.: Adaptive fault detection and isolation for DC motor input and sensors. In: Proceedings of the 19th International Conference on Informatics in Control, Automation and Robotics (ICINCO), pp. 703–710 (2022)
12. Brunton, S.L., Kutz, J.N.: Data-Driven Science and Engineering: Machine Learning, Dynamical Systems, and Control, p. 492. Cambridge University Press, Cambridge (2019)
13. Chaoui, H., Khayamy, M., Okoye, O., Gualous, H.: Simplified speed control of permanent magnet synchronous motors using genetic algorithms. IEEE Trans. Power Electron. **34**(4), 3563–3574 (2019)
14. Bazylev, D., Pyrkin, A., Dobriborsci, D.: Nonlinear state observer for PMSM with evolutionary algorithm. In: 31st Mediterranean Conference on Control and Automation (MED), pp. 581–586. Limassol, Cyprus (2023)
15. Do, K. D., Pan, J.: Control of Ships and Underwater Vehicles, p. 418. Springer London (2009)

16. Aranovskiy, S., Bobtsov, A., Ortega, R., Pyrkin, A.: Performance enhancement of parameter estimators via dynamic regressor extension and mixing. IEEE Trans. Autom. Control **62**(7), 3546–3550 (2017)
17. Bazylev, D.: Finite time parameter estimation algorithm for salient PMSM. In: 2022 European Control Conference (ECC), pp. 303–308. London, United Kingdom (2022)
18. Ellis, G.: Observers in Control Systems: A Practical Guide, p. 259. Elsevier Science (2002)
19. Bazylev, D., Margun, A., Dobriborsci, D.: Fault detection algorithm of INS of parametrically uncertain unmanned vessel. In: 32nd Mediterranean Conference on Control and Automation (MED), pp. 191–196. Chania - Crete, Greece (2024). https://doi.org/10.1109/MED61351.2024.10566125

The Necessity of Using Technical Analogues of Thinking and Consciousness for the Creation of AGI

Alexey Podoprosvetov[ID], Vladimir Smolin(✉)[ID], and Georgy Malinetsky[ID]

Keldysh Institute of Applied Mathematics, Miusskaya Square 4,
125047 Moscow, Russia
`{smolin,gmalin}@keldysh.ru`

Abstract. Data describe any objects and phenomena. Knowledge consists of data the system can use to form rational actions aligned with the current external environment's state. Information is extracted from data using knowledge to manage the process of forming actions and tracking the usefulness of continuing with a previously chosen plan or replacing it with another. Knowledge allows us to form rational actions, but its direct use is effective only when interacting with simple objects and phenomena. In complex situations, knowledge about the properties of simple objects enables preliminary modeling of event developments under various sequences of actions and helps select better actions than those obtained through direct approximation (without comparative modeling). The ability to use the same accumulated knowledge in two modes (for direct action generation and preliminary modeling) is the main idea for creating a developed system with mode control based on utilizing information extracted from the data. The processes of action formation and modeling must be organized hierarchically. The lower levels are always engaged in direct management of actions. The upper levels manage the lower levels, setting the goals of actions. While the lower levels successfully perform management of actions leading to the achievement of the goal, the upper levels can switch to the modeling mode. The operation modes without modeling and with modeling can be called technical intuition and thinking, while control over switching modes can be called technical consciousness (conscious control).

Keywords: Neural networks · analytical functions · neural nets training

1 Introduction

The understanding of intellectual activity has evolved throughout the history of civilization. Leibniz considered it a form of computation, while Turing saw it as the ability to outperform a human in a game. Today, machines far exceed humans in both computational power and winning "intellectual" games. However, even the most advanced neural network algorithms still rely heavily on knowledge acquired by humans. Is it possible to develop AGI (Artificial General Intelligence) that can independently acquire new knowledge without human intervention and use both this new knowledge and the knowledge accumulated through evolution and civilization to shape behavior in complex environments? And what exactly do we mean by knowledge, rational behavior, and a complex environment?

The term "Artificial Intelligence" was first proposed by John McCarthy during the 1956 Dartmouth Summer Research Project on AI [1]. Initially, the goal was to create AI systems capable of solving certain intellectual tasks. By the late 1980 s, many of these tasks had been addressed, though with limited success, leading to the idea that machines might one day approach human cognitive abilities. As noted in [2], "Strong AI claims that machines can be made to think on a level equal to humans". In the early 2000 s, Ben Goertzel [3] became one of the key figures popularizing this direction, which became known as AGI. Today, the development of AGI is a central goal of AI research in several major companies, such as OpenAI and Meta. By 2020, over 70 active AGI research and development projects were reported across 37 countries, and this number has since grown significantly.

The definitions of knowledge, rational behavior, and complex environments will be discussed below.

1.1 Scientific Methods for Acquiring New Knowledge

Since ancient times, acquiring new knowledge has involved using axioms and formal-logical transformations to determine "correct" actions. This approach was valued for justifying monarchical power and produced numerous theoretical and practical results. Despite its reliance on aligning axioms with real-world conditions, it remains influential in some sciences.

Modern scientific approaches focus on statistical significance, verifiability, and falsifiability, though they require a manageable number of parameters for accurate observation. Both axiomatic and experimental methods depend on human input to identify axioms, rules, and parameters from complex environments before applying these methods.

1.2 Success of Neural Networks in Approximating Complex Signal Transformations

The "miracle" of neural network algorithms lies in their ability to process complex signals (video, speech, text, etc.) that contain thousands or millions of parameters. In this case, human involvement to identify a small number of "essential" parameters is not required.

A simple calculation shows that even the vast datasets used for "large" neural network models are insufficient for building statistically reliable dependencies between such a large number of parameters.

This "miracle" can only be explained by the decomposition of complex signals in modern neural networks. This decomposition is carried out both by the properties of the back propagation method in deep neural networks and by numerous structural and algorithmic techniques embedded in the architecture and functions of modern neural network structures.

For example, the impressive success of large language models (LLMs), in the authors' opinion, is based on the use of human decomposition of the properties of a complex environment, achieved through its textual description (each word being a description of simple properties). This decomposition proves useful not only in text processing but also in other modalities of complex signals.

2 Data, Knowledge, and Information in Computers and Neural Networks

In everyday communication, these three concepts are often used as synonyms: we record data on information carriers and assume they contain knowledge. However, here we will differentiate between them:

- **Data** – the description of objects and phenomena in any form;
- **Knowledge** – the description of objects and phenomena in a form suitable for forming actions;
- **Information** – data that allows the selection of knowledge for forming actions.

Knowledge allows the transformation of input data into output and the extraction of information, which is used not to create output but to manage the transformation process.

2.1 Differences Between Data, Knowledge, and Information in Computers

In computers, data are stored in digital form in memory and processed using programs (knowledge about methods of processing and information about the addresses of the data and commands currently in use). While processors were once called ALUs (Arithmetic-Logic Units, i.e., a combination of a logic device with an arithmetic coprocessor), there is no fundamental distinction between data, knowledge, and information in this context, as they are all stored in the same memory and processed by the same processor (or multiple identical ones).

It can be argued that modern processors have different levels of memory (L1, L2, L3, etc.), and addressing is handled by a part of the processor distinct from the one that processes data. However, the proposed understanding of the differences between data, knowledge, and information is not as apparent (though easier to grasp for many) as it is in neural networks.

In the proposed definitions, all data operations executed linearly by a program rely on knowledge, while logical transitions extract information from data, based on the program's knowledge, to determine which knowledge will guide further processing. Knowledge defines data processing methods, including information extraction, and is stored long-term, whereas information resolves uncertainty at a specific moment. Information directs program execution after branching points but is quickly forgotten unless measures are taken to retain it.

2.2 Differences Between Data, Knowledge, and Information in Neural Networks of Higher Animals

Higher animals adapt their behavior in complex environments, similar to how computers process sensor input to extract knowledge. Unlike computers, where knowledge and data processing are separate, neural networks integrate both across all neurons.

In neural network models, data is the activation propagating through the network, and knowledge is represented by the weights of connections. While basic models show a clear separation between data and knowledge, advanced networks like GPT or AlphaZero not only transform data but also extract and apply different knowledge based on context.

In modern neural networks with gates and comparative modeling, the same input can be processed using different knowledge based on the network's history, similar to how humans and animals adjust their knowledge application according to past experiences and context.

3 Scalar and Vector Variables in Neural Networks

3.1 Scalar Variables O_j and w_{ji}

For millennia, philosophers have devised abstract, immaterial worlds to solve the problem of the emergence of abstract concepts, and even more so, variables – the parameters that characterize these concepts. In neural networks, variables that reflect the state of the external environment are a natural and inseparable property. The current state of each neuron in all neural networks is primarily characterized by its output activity. It is the output activity of neurons in animals that determines the structure of actions formed by neural networks (which, in animals, constitute the central nervous system).

A biological neuron is a complex living system with many properties, characterized by a large number of parameters that influence its output activity. In turn, the output activity of a biological neuron has a pulse (spike) nature, which prompts some researchers to search for hidden encoding in the sequence of spikes.

In modern neural network algorithms, an extremely simplified model of the biological neuron is used, originating from [4,5] and known as a formal neuron:

$$o_i = \phi \left(\sum_{j=0}^{n} w_{ji} o_j \right) \quad (1)$$

where o_i is the output activity of the i-th neuron, o_j is the output activity of neurons connected to its input synapses w_{ji}, and $\phi()$ is the non-linear transformation of the weighted sum (inside the parentheses).

The output activity of the formal neuron o_i models the spike generation rate, while the weight w_{0i} represents the threshold value of biological neurons (for this, a special neuron with constant unit activity is introduced in formal networks, $o_j \equiv 1$). Receptors (sensors) do not receive activation from other neurons, but instead generate their activity in response to external stimuli.

The variables o_i and w_{ji} and equation (1) do not exhaust the description of the functioning of formal neuron networks – there are also parameters that describe the network structure, variables that influence the stability of activity propagation, and a number of parameters and functions that define the rules for modifying not only the synaptic weights w_{ji} but also other auxiliary parameters. However, o_j and w_{ji} are the main variables characterizing the current state of a network of formal neurons.

The modern successes in AI, based on neural network algorithms relying on equation (1), demonstrate that solving complex "intellectual" tasks does not require the "complexity" of individual neurons but rather the rationality and size of the structures built from them.

3.2 Vector Variables and Their Distributed and Localized Description

When describing neural network structures (layers), scalar variables form vectors $\boldsymbol{O}^m = \{o_i\}$ and matrices $W^m = \{\boldsymbol{W}_i\}$ ($\boldsymbol{W}_i = \{w_{ji}\}$) that describe layer m. However, here we will consider only the vector description of the activity $\boldsymbol{O}^m$ (which, according to equation (1), is influenced by the parameters forming the matrix W^m).

The vector description of the activity of the input or previous hidden layer $\boldsymbol{O}^{(m-1)}$ can be transformed into the activity $\boldsymbol{O}^m$ in various ways, but we will highlight only two types of transformations: distributed and localized (Fig. 1).

If no special measures are taken and the activity $\boldsymbol{O}^m$ is calculated simply by equation (1), the activation of the layer will be distributed. To localize activation, certain conditions need to be created. In biological neural networks, localization of neuronal activity is carried out by mechanisms of feedback or lateral inhibition. In formal neural networks, algorithms such as WTA (winner-takes-all) or kWTA (k-winners-take-all) are commonly used for local representation of activity, ensuring that non-zero activity is maintained in only one or k neurons of the layer.

Both types of descriptions allow for approximations of broad classes of vector transformations. Since the dimensionality of the vector space is determined by the number of neurons in the layer (not the type of description), they offer comparable approximation accuracy (given the same number of neurons in the layer). Fault tolerance is also approximately the same: distributed descriptions achieve it by using a large number of activated elements in the layer, while localized descriptions rely on approximating with neighboring elements.

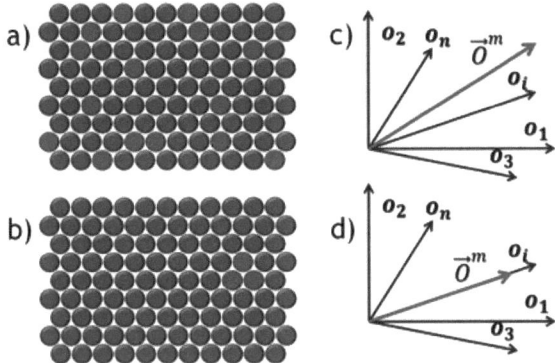

Fig. 1. Distributed (a) and localized (b) activation of layers and their vector description O^m (c and d).

The obvious advantage of distributed description is its simplicity of use, as it does not require additional costs for its formation. However, the costs of localization are not in vain; they lead to more efficient use of both the neural network's resources and the properties of the multidimensional state space of vectors O^m. Moreover, it is possible to combine the positive aspects of both distributed and localized descriptions of layer activity vectors, as discussed, for example, in [6].

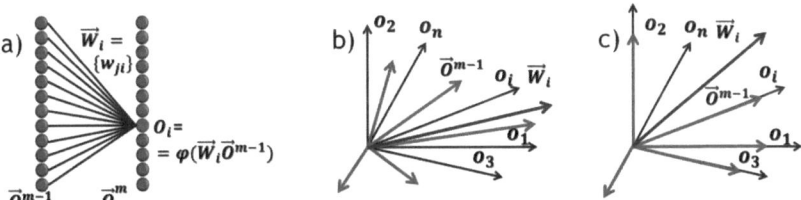

Fig. 2. Formation of activity of elements in layer m. Red arrows indicate various values of activity $\mathbf{O}^{(m-1)}$ in layer $m-1$ for different states of observed objects, blue arrows indicate $\mathbf{W}_i$. (a) – Activation scheme, (b) – "Nearly" orthogonal "distributed" $\mathbf{O}^{(m-1)}$, (c) – Strictly orthogonal "localized" $\mathbf{O}^{(m-1)}$. (Color figure online)

In high-dimensional spaces, two random vectors are highly likely to be orthogonal to each other. The weight vector of the input connections $\mathbf{W}_i$ of any element in the subsequent layer m can be represented as a sum of projections $\mathbf{W}_{ik}$ onto k activity vectors $\mathbf{O}_k^{(m-1)}$ of the previous layer:

$$\mathbf{W}_i = \sum_{k=0}^{n} \mathbf{W}_{ik} \qquad (2)$$

Thus, if layer $m-1$ is activated by various input signals and vectors of activity $\mathbf{O}^{(m-1)}$ are obtained, onto which the vector $\mathbf{W}_i$ is decomposed, the output activity of any element will be determined "almost" solely by the component $\mathbf{W}_{ik}$ for the distributed representation $0 < \alpha \ll 1$:

$$o_{ik} = \phi\left(\mathbf{O}_k^{(m-1)} \sum_{k=0}^{n} \mathbf{W}_{ik}\right) = \phi\left(\mathbf{O}_k^{(m-1)} \mathbf{W}_{ik} + \alpha \mathbf{O}_k^{(m-1)} \sum_{k=0}^{n} \mathbf{W}_{ik}\right), \quad (3)$$

Since random vectors $\mathbf{O}_k^{(m-1)}$ in high-dimensional space are "almost" orthogonal, the mutual influence of learning results from various input signals, represented by activation $\mathbf{O}_k^{(m-1)}$, is preserved, albeit with a small coefficient α. In the case of localized representation of activity, the vectors are orthogonal by construction, and $\alpha = 0$.

3.3 Short-Term Episodic Memory as Traces of Trajectories of Activity Centers

As noted in Sect. 2.2, in biological neural networks, unlike in computers, data (neuron activation) and knowledge (synaptic weights) have different physical natures. While data $\boldsymbol{O}^m$ can easily (though each time in a transformed form) propagate through the neural network, knowledge is formed in place, and there are no direct mechanisms for transmitting it (synaptic weights w_{ji}) elsewhere. Therefore, the knowledge accumulated in the synaptic weights must be utilized at the place of its formation in all processes, such as action formation and comparative modeling of various actions to select the best one for execution.

Comparative modeling of different action options requires short-term memory to avoid repeating previously considered actions. Short-term memory is also useful for repeatedly modeling the most successful (or, conversely, unsuccessful) actions to convert them into long-term knowledge. This short-term memory must also be implemented on the same neural layers where knowledge about data transformations is stored, as there are no mechanisms for transferring this knowledge to other structures.

Localized descriptions of activity vectors, based on topologically correct mapping [7], provide a good foundation for forming a short-term memory mechanism on the same structures that hold knowledge about data transformations. When the activity vectors $\boldsymbol{O}^{(m-1)}$ of input or modeled data change, localized descriptions $\boldsymbol{O}^m$ not only lead to the activation of a small number of neurons (the activity center, AC), but, according to the property of topological correctness, the AC shifts along the array of neurons in a continuous trajectory. If the dimensionality of the topologically correct map is greater than 2, the trajectory lies in the mapping space without self-intersections (with rare exceptions).

If the trajectory sweeps over an insignificant part of the mapping space, in each neuron activated during the AC's movement, a certain recent activation feature can simply be increased, which will return to its original state over time. This does not alter the knowledge stored in the synaptic weights, but, depending on the control commands issued, either complicates the reactivation of neurons included in the AC's trajectory or, conversely, stimulates the modeled movement of the AC along a previously realized trajectory.

Since the AC movement trajectories reflect the sequences of executed or modeled actions, short-term memory, as the memorization of traces of such trajectories in the layers performing transformations, can only be implemented in neural layers that map not just the states of activity vectors $O^{(m-1)}$ but the phase portraits of changes in their activity.

4 Tabular Transformations in Neural Networks

The weight vectors $\mathbf{W}_i = \{w_{ji}\}$ of the input connections of neurons (both natural and artificial) layer by layer form weight matrices $W^m = \{\mathbf{W}_i\}$. These matrices carry knowledge and define the transformations performed by neural network layers. Since a matrix is essentially a table, we can say that tabular transformations are carried out using the parameters of these tables. The properties of individual neurons and their interaction structure within a neural layer determine the rules for using the data from the table to perform transformations.

Discrete tables can approximate continuous transformations, as established in the Sampling Theorem [8]. With few independent parameters, any continuous multidimensional function with a limited frequency range can be represented by finite-domain tables.

However, as the number of independent parameters increases, a significant limitation emerges: while expanding frequency or domain along one parameter leads to a linear table size increase, adding independent parameters results in an exponential growth in table elements.

Fortunately, most real and virtual worlds consist of simple objects characterized by few parameters, making it feasible to use tabular descriptions for actions. Humans naturally simplify complex environments, yet this decomposition is often neglected in behavior analysis. This work posits that fundamental decomposition is essential for advancing AGI, as AI lacking this capability will not achieve human-like cognitive abilities.

4.1 Limitations on the Growth of Table Size

While the size of the human brain (and the tables of connection weights it contains) is genetically determined, artificial neural networks built on silicon have no such explicit limitations. What, aside from cost and power consumption, could serve as a barrier to the growth of table sizes describing transformations performed by neural networks?

This limitation arises from the challenge of populating the table with weight values w_{ji}, which represent knowledge of the external environment necessary for rational actions via tabular transformations. The Sampling Theorem enables the optimization of continuous transformations by tuning a finite set of parameters. However, in complex environments with many independent variables, the required number of parameters (weights w_{ji}) becomes impractically large, quickly exceeding "practical infinity", estimated between 10^{23} and 10^{26} for current computational technology.

"Practical infinity" can be observed in seemingly simple tasks like chess and Go, where the number of possible game scenarios is finite but immense — approximately 10^{47} for chess and 10^{171} for Go. If the computing power and memory continue to grow according to Moore's Law (a tenfold increase every five years), it would take over 100 years to move beyond "practical infinity" for chess and over 750 years for Go. In a complex real (or virtual) world, the number of independent variables vastly exceeds that of board games, potentially requiring billions of years to resolve them using brute-force methods under Moore's Law.

4.2 The Necessity of Decomposing a Complex Environment into Simple Objects and Phenomena

Many complex tasks, such as chess and Go, can be addressed by breaking them down into simpler tasks. These simpler components are characterized by a small number of independent variables, whose properties and transformations can be identified through statistical methods. Although the complexity of an environment and the uniqueness of its states hinder the accumulation of knowledge about all possible states, it is feasible to construct a model-based composition of the complex scene by leveraging knowledge of its simpler components. This enables a comparison between the constructed composition and the actual observed scene, facilitating assessments of how well actions align with a pre-established plan.

Identifying discrepancies between the plan and real outcomes can help refine models of objects and their interactions at various hierarchical levels. At lower levels, it is possible to gather knowledge that encompasses most properties of simple objects and phenomena. However, as one ascends the hierarchy, knowledge becomes limited to an increasingly smaller percentage of interaction scenarios involving more complex combinations, since the total number of combinations approaches infinity.

5 Goal-Driven Control

A hierarchical knowledge structure necessitates a matching hierarchical action scheme. In traditional control, higher levels set goals, while lower levels execute actions. Intermediate levels refine these goals, breaking them into sub-goals for lower levels.

In neural network hierarchical control, the focus extends beyond goal execution to also forming knowledge that improves goal attainment. Since higher levels guide lower levels by setting objectives, knowledge accumulation centers on defining increasingly efficient goals.

5.1 Feedback Loops

In cybernetics, the classical scheme features a feedback control loop to monitor the difference between the controlled system's state and the set goal. Lower hierarchy levels receive goals from higher levels and use sensors to monitor the system. The transfer function properties, learned and stored in connection weights, describe the controlled object's phase portrait.

At the same time, the activity of the lower level (denote it as $\mathbf{O}_0^m$) reflects the observed state of the controlled object, while the knowledge in the connection weights w_{ji} defines the actions for controlling the object. The control goal is set by the higher-level layer in the format of vectors $\mathbf{O}_0^m$.

For a higher level l, the object of control is the activity of the lower levels, which they represent by their activity $\mathbf{O}_l^m$. Knowledge about the control impact on the $l-1$ level is stored in the connections w_{ji} of the elements at level l, and this knowledge is in the format (dimension) of vectors $\mathbf{O}_{l-1}^m$.

Naturally, the formation of transfer functions based on observation (through trial and error) is effective only for the lower levels that control (relatively) simple objects. This is because both the duration of the impact on the object and the number of possible actions and their combinations are limited.

Building control laws for more complex objects and phenomena (goal formation) requires more advanced methods of designing the actions performed.

5.2 Adequacy of Accumulated Knowledge to the Set Goal

Can we predict if the knowledge in connection weights w_{ji} is sufficient for solving a control task and avoiding catastrophic failures? Higher animals monitor their offspring until they can achieve goals safely, while less-developed animals rely on high fertility to compensate. Historically, before medical advances, human families often had 5–10 children to offset high child mortality.

However, in a complex environment, one frequently encounters new, previously unseen situations in which parental control may not help.

This translation retains the technical nature of the original text, ensuring accuracy and clarity in conveying concepts related to hierarchical goal-driven control and feedback mechanisms in neural networks.

6 Modes of Operation of Hierarchical Levels: Technical Intuition and Thinking

Hierarchical layers represent the state of external environment components through their activity $\mathbf{O}_l^m$. Lower levels form simple sensory-based models, while higher levels extract properties from these models, enabling actions suited for complex scenarios.

The activity $\mathbf{O}_l^m$ across hierarchy levels reflects the environment's state, allowing $\mathbf{O}_l^m$ to serve as arguments in transformation functions that link environmental representations to actions.

6.1 Knowledge-Based Action Modeling

Even at the hierarchy's lowest levels, where activity $\mathbf{O}_0^m$ reflects the controlled object's observed state, accumulated knowledge about its dynamics—a model of its phase portrait—plays a key role. This model enables more accurate representation under noisy or interrupted sensory input, relying on $\mathbf{O}_0^m$ to guide actions when sensory data is unreliable.

Higher levels derive activities O_l^m from filtered representations of controlled object states. While lower levels manage directly observable objects, higher levels set goals for these lower levels, addressing objects beyond direct observation. Lower levels then model the dynamics of these objects in real time, enabling actions that align with the actual state, including obscured or out-of-view objects.

Actions based on real-time environmental knowledge can be termed **technical intuition**.

Knowledge of simple objects' dynamic properties enables complex situation modeling under various actions. Short-term memory and situational assessment allow comparison of action outcomes. Since no real actions are performed, changes in vectors $\mathbf{O}_l^m$ occur faster than real time.

In biological networks, action modeling occurs within the same structures that control objects through $\mathbf{O}_l^m$ activity, and can only happen when they are not in direct control, allowing higher levels to model at higher abstraction levels. Detailed modeling requires setting a long-term goal to free up lower levels.

In silicon-based systems, separate devices for control and modeling, with a unified knowledge format, allow seamless knowledge transfer. This facilitates modeling even during complex tasks with active goal generation. However, challenges remain in aligning modeling with control, effectively using modeling outcomes, and tracking environmental changes to guide modeling priorities.

Modeling outside of real-time, based on knowledge of object dynamics, can be termed as **technical thinking**.

6.2 The Need for Mode Control

The existence of two modes of knowledge use – technical consciousness and thinking – necessitates managing the switching of these modes and allocating information for the management process. Even when technical consciousness and thinking are implemented in separate blocks within silicon-based devices, issues of coordinating their activities remain.

Criteria for the possibility of switching certain levels to modeling mode (thinking) include: Approximating the activity vector of layer $\mathbf{O}_l^m$ to the specified target state or maintaining it; Correspondence to the expected list of scene components in a complex environment; Absence of a developed action plan with the required level of expected effectiveness.

Conditions for switching back are the opposite: Development of an action plan with the required level of expected effectiveness; Change in the expected list of scene components in a complex environment; Movement of the activity vector of layer $\mathbf{O}_l^m$ away from the specified target state.

More nuanced criteria for managing behavior generation involve assessing the possibility of creating conditions for thinking and selecting the direction of modeling. It should be understood that the management of the interaction between technical intuition and thinking in behavior construction is ongoing and similar to conscious control (based on understanding the state and development prospects of the situation). Hence, such management can be termed technical consciousness.

7 Three Components of Hierarchical Knowledge

The formation of actions using comparative modeling of different options presupposes the ability to perform:

- Modeling,
- Comparative evaluation, and
- Action formation.

Each of these operations has its own features, and the structures implementing them must possess specific, distinct properties.

7.1 The Necessity of Hierarchy for Holistic Perception of Complex Situations

When creating models of simple objects and phenomena during observation, it is essential to ensure their usability for describing complex scenes and to utilize such descriptions for comparative selection of action sequences.

A multi-level hierarchy provides the means to separate the variables that describe a complex scene not only by space and time but also by levels of abstraction in the representation of both objects and phenomena, as well as the properties of their interactions.

The complexity and uniqueness of environmental states prevent full knowledge accumulation for all possible conditions. However, constructing a model of a complex scene using knowledge of simpler components allows comparison with observed scenes and assessment of action alignment with the planned outcomes.

Discrepancies between the plan and reality provide a basis for refining object and interaction models across hierarchy levels. At lower levels, it is feasible to capture most properties of simple objects, but as hierarchy levels rise, knowledge becomes limited to a shrinking subset of possible interactions among increasingly complex combinations, whose total count approaches infinity.

7.2 Differences in Problems Addressed by the Three Components of the Hierarchy

The components listed at the beginning of Sect. 5 of the hierarchical structure aimed at action in a complex environment address different issues. The overall goal of the entire hierarchy is to ensure the preservation of knowledge about self-organization methods and to pass this knowledge to successors for continued evolution. This goal is a natural property of evolution – species and individuals that do not achieve this "goal", or achieve it with less success compared to competitors, are eliminated from the evolutionary process.

The hierarchical system for modeling environmental properties aims to align actions with real conditions and defined needs in the system of comparative assessments. This component was the focus of the previous text. However, there are two more components of the hierarchical system.

The system of comparative assessments, although it carries important knowledge about the requirements for maintaining homeostasis and more complex representations, such as aesthetic evaluation of a mate, must also adapt to the complex environment, as not all possible situations can be anticipated or evaluated in advance. Specialized algorithms for nonlinear aggregation of simpler (and initially available) assessments are necessary to build evaluations of complex situations. Assessments based on knowledge obtained from observing a complex environment allow the hierarchy to choose more rational behavior in complex situations.

The formation of actions to achieve goals also has its peculiarities. While goals are mostly independent, the execution of actions (movements) is often interdependent. The issues of compensating for these dependencies have their specifics, but this work, which is dedicated solely to the component of goal formation in a complex environment, does not address these issues.

8 Conclusions

Self-organization in living organisms and their products depends on knowledge. Initially, knowledge was preserved through RNA and DNA. Evolution led to neural network structures that augmented DNA-based knowledge with new, environment-related insights. Human-articulated speech accelerated knowledge exchange, and technological progress created effective data-processing systems. This paper examines AI as a system that acquires and applies knowledge to navigate complex environments.

Although neural networks now excel in many fields, the automation of knowledge acquisition is still incomplete and requires human involvement. Nonetheless, significant progress has been made in leveraging knowledge with the ability to combine it in various ways to solve practical problems. For example, Google DeepMind's AlphaFold 3 is not only capable of generating formulas for new organic compounds [9], but it has also enabled the creation of a database of over 2 million previously unknown stable organic compounds, demonstrating the effective use of knowledge about the properties of components. However, AlphaFold 3 has not yet reached the level of AGI, as it cannot independently acquire new knowledge about the chemical properties of the components it uses to create new organic substances without human input.

A key condition for the development of AGI is not only the search for effective solutions based on existing knowledge to form rational actions in complex environments but also the rapid and accurate new knowledge acquisition through the autonomous identification of simple, statistically observable components within the complex environment. As shown in the previous paper [6], a major path to solving this problem is the development of neural network methods for decomposing complex signals for their correspondence analysis with nonlinear compositions of identified components. Various aspects of these decomposition methods were also discussed in [10].

This report is primarily dedicated to describing methods of using decomposition techniques in complex hierarchical structures for the rapid acquisition and effective use of knowledge about complex environments. However, the limited scope of the report allows for the examination of only some of the simplest aspects of the operation of complex hierarchical systems for knowledge and information (as they defined in the report) extraction and their effective use.

The main conclusion is that in a complex, non-repeating environment, rational behavior depends on understanding its simple components. While hierarchical knowledge representation allows for considering component interactions, it alone isn't enough for choosing rational actions. A neural network must have systems for comparative modeling of actions and for managing resources and time between modeling and direct control. These systems are termed as "technical thinking" and "technical consciousness" in the report.

References

1. Solomonoff Grac. The Meeting of the Minds That Launched AI. IEEE Spectrum (2023). Retrieved 20 Oct 2024
2. Haugeland, J.: Artificial Intelligence: The Very Idea. MIT Press (1989). ISBN 978-0-2625-8095-3
3. Goertzel, B.: Artificial General Intelligence: Now is the Time. GoogleTalks Archive. Retrieved 20 Oct 2024–via YouTube (2007)
4. McCulloch, W.S., Pitts, W.: A logical Calculus of Ideas Immanent in Nervous Activity — Bull. Mathematical Biophysics (1943)
5. Widrow, B., Hoff, M.E.: Adaptive switching circuits. 1960 IRE WESTCON Convention Record. New York (1960)
6. Podoprosvetov, A., Smolin, V., Sokolov, S.: Vector analysis of deep neural network training process. In: Fred, A., Hadjali, A., Gusikhin, O., Sansone, C. (eds.) Deep Learning Theory and Applications. DeLTA. Communications in Computer and Information Science, vol. 2171. Springer, Cham (2024). https://doi.org/10.1007/978-3-031-66694-0_14
7. Kohonen, T.: Self-Organizing Maps. Third, Extended NY, USA, New York (2001)
8. Nyquist, H.: Certain topics in telegraph transmission theory. Trans. AIEE **47**, 617–644 (1928)
9. Abramson, J., Adler, J., Dunger, J., et al.: Accurate structure prediction of biomolecular interactions with AlphaFold 3. Nature **630**, 493–500 (2024)
10. Smolin, V., Sokolov, S.: AGI's hierarchical component approach to unsolvable by direct statistical methods complex problems. Eng. Proc. **33**, 67 (2023). https://doi.org/10.3390/engproc2023033067

Solving Boundary Value Problems Based on a Fractional Differential Equation of Diffusion Type with Arbitrary Values of the Orders of the Derivatives

Dmitry O. Zhukov and Konstantin K. Otradnov(✉)

Institute of Radio Electronics and Informatics, MIREA–Russian Technological University, 78 Vernadsky Avenue, 119454 Moscow, Russia
const.otradnov@yandex.ru

Abstract. This study introduces a generalized analytical framework for deriving the probability density function associated with observing a system's state x over time t, based on a fractional diffusion-type differential equation defined on an unbounded domain.

Employing methods from operational calculus and complex analysis-specifically, the residue theorem, we obtain an explicit closed-form expression for the amplitude distribution, capturing the temporal dynamics of time series over variable-length intervals.

In contrast to conventional approaches, which typically restrict the fractional time derivative β to the range $0 < \beta < 1$ and the spatial derivative α to $1 < \alpha < 2$, the proposed model accommodates arbitrary positive orders of differentiation. This generalization substantially enhances the versatility of fractional models in time series analysis.

The derived solution remains valid under the condition $0 < \beta/\alpha \leq 0.865$, thereby extending its applicability to a broader class of nonlocal, memory-dependent dynamical systems. Empirical validation using electoral time series data from the 2008, 2012, and 2016 U.S. presidential campaigns confirms the model's effectiveness and descriptive power.

Keywords: fractional differential equations · time series analysis · operational methods · electoral dynamics processes

1 Introduction

The human factor plays a pivotal role in the self-organization processes of complex systems, affecting both the persistence of memory associated with prior states and the development of emergent organized complexity. This phenomenon should not be interpreted simply as a cumulative effect of individual elements' properties; rather, it arises from the complex network of interactions within the system, accompanied by the adaptive redistribution of roles among its constituents.

Typically, researchers approach the analysis of original time series by decomposing them into three fundamental components: seasonal patterns, underlying trends, and stochastic fluctuations.

When examining time series, for example, you can first remove seasonal variations, then build forecasts based on filtered data [1], and then add the previously extracted seasonal component to the resulting model.

However, the use of seasonal adjustment during data preprocessing remains a contentious issue. Empirical studies have shown that seasonal behavior can change over time, and in many instances, it is difficult to clearly separate seasonal influences from non-seasonal ones, or to treat them as independent.

Accurately separating seasonal influences from other fluctuations remains a significant challenge, which has spurred the creation of specialized approaches tailored to periodic structures. In dynamical systems theory, such recurring seasonal behavior is typically illustrated using phase portraits akin to limit cycles [2].

The issue of stationarity in time series analysis is frequently discussed in the literature [3]. In practical applications, most observed time series exhibit non-stationary behavior, prompting the development of numerous methodologies aimed at transforming historical data into stationary sequences.

A frequently employed strategy consists in splitting a time series into two distinct elements: a dominant factor structure and a stochastic noise component [4]. It is worth noting that most models capturing the behavior of complex socio-economic systems are inherently non-stationary in nature.

Time series prediction models can, based on the methods used in them, be conditionally divided into: using statistical and stochastic methods and machine learning methods, as well as based on the decomposition of approximating functions, for example, into Fourier series [5], Bessel series [6], Volterra functional decomposition [7] and others. Statistical methods are based on the use of historical time series data, they predict future values based on regression methods (linear regression, exponential regression [8]), autoregressive dependencies, for example, such as the moving average model (ARMA) [9–11] and the autoregressive integrated moving average model (ARIMA) [12,13]. Stochastic prediction methods include Bayesian models, Gaussian models, use of beta distribution [14].

Other modern models of time series analysis and forecasting should also mention approaches based on the use of robust interval estimates of their parameters [15,16].

Machine learning has experienced a surge in adoption across numerous scientific and applied fields in recent years [17–20]. A notable example is the use of artificial neural networks (ANNs), which are particularly well-suited for capturing intricate nonlinear dependencies from historical observations. While shallow neural models are commonly utilized to detect such patterns [21], more advanced methodologies increasingly employ deep learning (DL) architectures to achieve greater modeling capabilities.

Recently, many new approaches have been developed to construct neural network topologies for nonlinear prediction of time series: graph neural networks

(GNN) [22], generative-adversarial networks (GAN) [23], long short-term memory networks (LSTM) [24]. However, the problem of their application is the problem of choosing a suitable neural network topology [25], as well as the exponential growth of modeling hyperparameters [26] as their topology becomes more complex.

Neural networks are now widely utilized for time series forecasting in a broad spectrum of application areas [27,28]. Their adaptability allows them to approximate diverse functional dependencies with considerable accuracy. A key advantage of these models lies in their data-driven nature: rather than depending on predefined assumptions about the generative mechanisms of the data, both their structure and parameters are learned directly from observed patterns.

Owing to their universality in function approximation, neural networks have found extensive practical application [24,29,30]. Nonetheless, their performance is highly sensitive to the size of the training dataset. When applied to small or short-term data samples, they often produce significant prediction errors, with actual values deviating noticeably from the forecast. Although modern neural network architectures commonly employed in tasks such as regression recovery can yield promising results in certain scenarios [31], they generally lack mechanisms to account for self-organization phenomena and memory effects typical of complex systems, instead adhering to Markovian assumptions.

The dynamic and nonlinear behavior typical of socio-economic systems poses significant challenges for conventional time series models. Methods like the integrated autoregressive moving average (ARIMA), known as the Box–Jenkins approach, frequently fail to capture the full complexity of these systems, leading to unreliable or misleading predictions.

Differential equations remain fundamental instruments for capturing the evolution and forecasting behavior of economic and social systems. Among these, the Fokker–Planck equation and its variants are extensively employed to analyze time series related to financial markets, commodity price dynamics, currency exchange rates, and other key economic indicators. A prominent example is the Black–Scholes model, which forms the theoretical basis for modern option pricing and remains one of the most widely used models in financial mathematics.

In their study, Jahanshahi et al. [32] propose a nonlinear model predictive control scheme to regulate a hyperchaotic economic system characterized by variable-order fractional dynamics. The model is defined by a set of four differential equations, each incorporating time-dependent fractional derivatives. The results confirm that this configuration not only stabilizes the system and suppresses chaotic responses but also provides resilience against bounded external perturbations.

In light of recent progress, we suggest employing fractional partial differential equations as a basis for studying time series data and the dynamics of intricate systems. The inclusion of fractional differentiation with respect to both temporal and state dimensions facilitates the modeling of long-term memory and naturally accommodates the emergence of self-organizing patterns.

Incorporating fractional operators in both temporal and spatial (state) dimensions enables the modeling of nonlocal behavior, wherein the evolution of a system depends not solely on its current or nearby states, but also on its complete past trajectory—thereby capturing essential memory effects. This type of temporal nonlocality frequently plays a pivotal role in driving self-organizing processes within complex systems.

This work focuses on analytically determining the probability density function that captures the extent of level fluctuations in time series, depending on the chosen observation interval. The approach relies on a fractional differential equation of the diffusion type, defined over an infinite domain and allowing for arbitrary, unrestricted orders of the fractional derivatives.

In the current literature, solutions to fractional differential equations are often derived under the assumption that the temporal derivative order β satisfies $0 < \beta < 1$, and the spatial derivative order α lies within $1 < \alpha < 2$. These limitations constrain the general applicability of fractional formulations, thereby reducing their potential to comprehensively describe the behavior of dynamic time series.

2 Fractional Partial Differential Equations

A diffusion-type differential equation that includes fractional partial derivatives can be expressed in the following general form:

$$\frac{\partial^\beta \phi(x,t)}{\partial t^\beta} = D \frac{\partial^\alpha \phi(x,t)}{\partial x^\alpha} \tag{1}$$

In this context, the parameters α and β are positive real numbers specifying the orders of the fractional derivatives in the Caputo sense, $\phi(x,t)$ denotes the probability density of the system being in state x at time t, and D represents a constant diffusivity parameter.

To illustrate, in diffusion scenarios, the function $\phi(x,t)$ quantifies the local concentration of the diffusing entity, and the parameter D governs the rate at which the diffusion process unfolds.

Let $G(z) \equiv \Gamma(z)$ denote the Euler gamma function.

In general, the fractional derivative of order ξ for the $\psi(x)$ function is defined as follows [33]:

$$\frac{d^\xi \psi(x)}{dx^\xi} = \begin{cases} \frac{1}{G(-\xi)} \cdot \int_a^x \frac{f(\xi)d\xi}{\{x-\xi\}^{\xi+1}}, & \xi < 0 \\ \frac{1}{G(1-\xi)} \cdot \frac{d}{dx} \int_a^x \frac{f(\xi)d\xi}{\{x-\xi\}^{\xi}}, & 0 \leq \xi < 1 \\ \frac{1}{G(2-\xi)} \cdot \frac{d^2}{dx^2} \int_a^x \frac{f(\xi)d\xi}{\{x-\xi\}^{\xi-1}}, & 1 \leq \xi < 2 \\ \frac{1}{G(3-\xi)} \cdot \frac{d^3}{dx^3} \int_a^x \frac{f(\xi)d\xi}{\{x-\xi\}^{\xi-2}}, & 2 \leq \xi < 3 \\ \ldots \end{cases}$$

Or:
$$\frac{d^\xi \psi(x)}{dx^\xi} = \frac{1}{G(1-[\xi])} \cdot \frac{d^{(max\{\xi\})}}{dx^{(max\{\xi\})}} \int_a^x \frac{f(\xi)d\xi}{\{x-\xi\}^{[\xi]}}, 2 \le \xi < 3$$

where $[\xi]$ – fractional part of the exponent ξ, $max\{\xi\}$ – rounding fractional key figure $[\xi]$ to the nearest largest integer.

Unlike the Riemann–Liouville formulation, the Caputo definition of a fractional derivative starts with an integer-order derivative of degree n, where $n = \lceil \xi \rceil$ is the smallest integer exceeding the fractional order ξ, and subsequently applies a fractional integral of order $n - \xi$.

Thus far, analytical solutions to the fractional diffusion Eq. (1) have been derived exclusively for cases where the parameters lie within the intervals $0 < \beta \le 1$ and $1 \le \alpha \le 2$ [34–41]. To the best of our knowledge, no exact solutions have been documented in the literature for values of α and β falling outside these specified ranges.

Due to the intrinsic mathematical complexity of these problems and the substantial difficulty of obtaining closed-form solutions, numerical techniques are commonly utilized in applied contexts.

For example, in [42], using numerical methods, the solution of the equation of the form:
$$\frac{\partial \phi(x,t)}{\partial t} = D \frac{\partial^\alpha \phi(x,t)}{\partial x^\alpha} \quad \text{at} \quad 1 < \alpha < 2$$

Reference [43] offers an analytical solution corresponding to a particular segment of an equation of the form:
$$\frac{\partial^\beta \phi(x,t)}{\partial t^\beta} = D \frac{\partial^2 \phi(x,t)}{\partial x^2} \quad \text{at} \quad 0 < \beta \le 1$$

The comparison of fractional differential Eq. (1) with the equation of ordinary diffusion is of interest: $\frac{\partial \phi(x,t)}{\partial t} = D \frac{\partial^2 \phi(x,t)}{\partial x^2}$. The resulting solution of this equation corresponds to the Poisson distribution law. $\phi(x,t) = \frac{1}{2\sqrt{\pi D t}} \cdot e^{-\frac{x^2}{4 \cdot D \cdot t}}$.

The function $\phi(x,t)$ can, for instance, be interpreted as the probability density of observing a specific value of the time series at time t, assuming that these levels evolve randomly over time.

In the case of a Poisson distribution, the dispersion $\sigma^2(t)$ follows a linear dependence on time, that is, $\sigma^2(t) \sim t$.

If slower growth occurs $\sigma^2(t)$ from t ($\sigma^2(t) \sim \sqrt[\beta]{t}$, where $0 < \beta < 1$), such a process qualifies as sub-diffusion [44,45] if faster growth is observed $\sigma^2(t)$ from t ($\sigma^2(t) \sim t^\beta$, where $1 < \beta < 2$), then such a process qualifies as superdiffusion [45].

The probability distribution characterizing subdiffusive phenomena can be derived by solving a time-fractional diffusion equation of the form: $\frac{\partial^\beta \phi(x,t)}{\partial t^\beta} = D \frac{\partial^2 \phi(x,t)}{\partial x^2}$, while superdiffusive processes are described by the equation: $\frac{\partial \phi(x,t)}{\partial t} = D \frac{\partial^\alpha \phi(x,t)}{\partial x^\alpha}$, where α and β are the exponents of the fractional derivatives, and In this context, the exponents α and β specify the fractional derivative orders, and D stands for a fixed coefficient that characterizes the rate of diffusion.

Abnormal diffusion processes can also be observed [46,47] described by a mixed-type fractional differential equation (see Eq. (1)).

Papers [48–52] consider the solution of the fractional Zener wave equation:

$$\nabla^2 \mathcal{U}(x,y,z,t) - \frac{1}{c_0^2}\frac{\partial^2 \mathcal{U}(x,y,z,t)}{\partial t^2} + \tau_\sigma^\alpha \frac{\partial^\alpha}{\partial t^\alpha}\nabla^2 \mathcal{U}(x,y,z,t) - \frac{\tau_\epsilon^\beta}{c_0^2}\frac{\partial^{\beta+2}\mathcal{U}(x,y,z,t)}{\partial t^{\beta+2}} = 0,$$

where $\mathcal{U}(x,y,z,t)$ in the case of elastic waves, the amount of displacement of the medium particles from the equilibrium position at the point with coordinates x, y, z at time t; c_0—wave velocity; τ_ϵ and τ_σ—some positive time constants. Here, the parameters α and β correspond to fractional orders of temporal derivatives, in contrast to the Laplacian ∇^2, which preserves its conventional integer-order structure.

Derived from the fractional Zener model, which governs the stress-strain relationship in viscoelastic media, this equation incorporates linearized conservation principles for mass and momentum. The resulting wave equation captures three attenuation regimes governed by power-law dynamics [48], and has found important applications in fields like acoustic wave modeling and medical elastography.

[50] provides a detailed description of the solution of the Zener equation for the case $\alpha = \beta$ (α and β can be derived fractional positive numbers specifying time derivatives t).

This result is particularly significant, as obtaining analytical solutions to fractional differential equations is known to be a challenging problem.

The solution of wave equations with a fractional derivative in time was further developed in [51] to describe the behavior of waves in non-Newtonian liquids.

Currently, the most developed and studied are fractional-differential models created to describe various kinds of physical problems, mainly devoted to the study of the processes of physical kinetics, diffusion and relaxation in various media [52–59], in [60] the analysis of time series of sociodynamic processes based on fractional-differential equations was considered.

The inclusion of fractional derivatives with respect to time t and space x facilitates the representation of non-local processes, where the likelihood of reaching a particular state x is shaped not solely by its immediate vicinity, but by the dynamics of the entire spatial domain.

This approach inherently accounts for memory effects, as it ties the system's present evolution to both its spatial configuration and complete temporal history. Such models, grounded in fractional calculus, provide a comprehensive analytical framework for capturing complex phenomena such as memory-driven evolution and self-organization.

It should be noted that in a number of works [61–66] models were developed that allow describing complex processes taking into account memory and self-organization, but without considering fractional differential equations.

3 Problem Statement and Mathematical Formulation

Consider a scenario in which a fractional differential equation of the diffusion type is provided. The objective is to analyze its behavior and derive meaningful insights under the assumption of non-integer differentiation orders.

$$\frac{\partial^\beta \phi(x,t)}{\partial t^\beta} = D \frac{\partial^\alpha \phi(x,t)}{\partial x^\alpha},$$

Here, $\phi(x,t)$ represents the probability density associated with observing a particular state x at time t, such as the intensity of level deviations in a time series across a defined interval. The constant D governs the diffusion rate, and the fractional derivative orders α and β are assumed to be strictly positive, i.e., $\alpha > 0$ and $\beta > 0$.

It is important to note that, unlike the common assumptions in the literature, the values of α and β are not restricted to the intervals $1 < \alpha < 2$ and $0 < \beta < 1$.

The following boundary conditions are imposed in the course of the analysis:

$$\phi(x,t)|_{x \to -\infty} = \phi(x,t)|_{x \to \infty} = 0,$$

supplemented by the initial condition given below:

$$\phi(x,0) = \delta(x-0),$$

Accordingly, the task is to solve a boundary value problem to determine $\phi(x,t)$, given the prescribed conditions, without imposing restrictions on the values of α and β.

4 Derivation of the Solution Under Boundary Constraints

The next step involves applying the Laplace transform in time t, leading to the frequency variable s, along with a Fourier transform in space x, introducing the variable k. As a result, we obtain the doubly-transformed form of the function, denoted by $\widetilde{\overline{H(k,s)}}$. $\phi(x,t)$:

$$\widetilde{\overline{H(k,s)}} = \frac{1}{s \cdot \left\{ 1 - \frac{(ik)^\alpha D}{s^\beta} \right\}} \quad (2)$$

Introducing the substitution $\xi = \frac{(ik)^\alpha D}{s^\beta}$, and under the condition $0 \leq \xi < 1$, we apply the binomial series expansion to facilitate further simplification.

$$\widetilde{\overline{H(k,s)}} = \frac{1}{s \left\{ 1 - \frac{(ik)^\alpha D}{s^\beta} \right\}} = \frac{1}{s\{1-\xi\}} = \frac{1}{s} \sum_{q=0}^{\infty} \xi^q =$$

$$= \frac{1}{s} \sum_{q=0}^{\infty} \left\{ \frac{(ik)^\alpha D}{s^\beta} \right\}^q = \sum_{q=0}^{\infty} \frac{\{(ik)^\alpha D\}^q}{s^{\beta q+1}}$$

Next, we invert the Laplace transform to express the solution in terms of the original temporal variable t:

$$\widehat{H(k,s)} = \sum_{q=0}^{\infty} \frac{\{(ik)^\alpha \cdot Dt^\beta\}^q}{G(\beta q + 1)} \qquad (3)$$

Note that $\widehat{H(k,s)}$ is a Mittag-Loeffler function.

To carry out the inverse Fourier transform of this function and find the distribution density, let's move from calculating the infinite sum to the integral representation of the Mittag-Loeffler function through the contour integral [67].

This is achieved through the use of the Hankel contour representation for the Euler gamma function:

$$\frac{1}{G(\tau)} = \frac{1}{2\pi i} \int_{C'} e^v v^{-\tau} du = -\frac{1}{2\pi i} \int_{C''} e^{-v}(-v)^{-\tau} du,$$

In this setting, C' is defined as a contour originating from $-\infty$, passing near the origin in a counterclockwise arc, and returning. The contour C'', in turn, begins at $+\infty$, loops around the origin counterclockwise, and returns. Applying this representation of the gamma function with contour C'', we derive:

$$\widehat{H(k,s)} = \sum_{q=0}^{\infty} \frac{\{(ik)^\alpha \cdot Dt^\beta\}^q}{G(\beta q + 1)} = \sum_{q=0}^{\infty} \frac{\xi^q}{G(\beta q + 1)} =$$

$$= -\frac{1}{2\pi i} \sum_{q=0}^{\infty} \{(ik)^\alpha \cdot Dt^\beta\}^q \int_{C''} e^{-v} \cdot (-v)^{-\beta q - 1} du = \qquad (4)$$

$$= -\frac{1}{2\pi i} \int_{C''} \frac{e^{-v} \cdot (-v)^{\beta - 1}}{(-v)^\beta - \{(ik)^\alpha \cdot Dt^\beta\}} du$$

Next, the inverse Fourier transform is employed to recover the function in the original spatial variable: $\frac{e^{-v} \cdot (-v)^{\beta-1}}{(-v)^\beta - \{(ik)^\alpha \cdot Dt^\beta\}}$:

$$\frac{e^{-v} \cdot (-v)^{\beta-1}}{2\pi} \int_{-\infty}^{\infty} \frac{e^{-ik|x|} dk}{(-v)^\beta - \{(ik)^\alpha \cdot Dt^\beta\}} \qquad (5)$$

The integral obtained may be computed using the residue theorem, provided that an appropriate analytic continuation of the function is carried out.

Consider the function

$$\frac{e^{-ik|x|}}{(-v)^\beta - [(ik)^\alpha Dt^\beta]},$$

and its representation along the imaginary axis:

$$\frac{e^{-ik|x|}}{(-y)^\beta - [(ik)^\alpha Dt^\beta]}.$$

To evaluate the corresponding integral, we construct a closed contour G in the complex plane that encloses the singularities of the integrand—specifically, the points at which the denominator

$$\frac{1}{(-y)^\beta - [(ik)^\alpha Dt^\beta]}$$

becomes undefined.

According to the residue theorem, the integral of a meromorphic function over the closed contour G is equal to the sum of the residues at the isolated singularities enclosed within G. Importantly, a portion of the contour G coincides with the real axis, which allows us to use this contour integration to compute the original integral by means of residue analysis.

The integrand exhibits a simple pole at a specific point in the complex plane, $k = -i\left\{\frac{(-v)^\beta}{Dt^\beta}\right\}^{1/\alpha}$.

Notably, when $k = 0$:

$$\widetilde{H(0,t)} = -\frac{1}{2\pi i}\int_{C''}\frac{e^{-v}\cdot(-v)^{\beta-1}}{(-v)^\beta}du = -\frac{1}{2\pi i}\int_{C''}e^{-v}\cdot(-v)^{-1}du = \frac{1}{G(1)} = 1$$

Consequently, the function $\widetilde{G(k,t)}$ satisfies the normalization condition, which permits its interpretation as a probability density function associated with a particular distribution. This probabilistic perspective suggests that, for the purpose of evaluating the integral $\int_{-\infty}^{\infty}\frac{e^{-ik|x|}dk}{(-v)^\beta-\{(ik)^\alpha\cdot Dt^\beta\}}$ it may not be necessary to perform an explicit computation over the full contour G, nor to separately integrate along each segment. Instead, the integral can be reformulated and represented as:

$$\int_{-\infty}^{\infty}\frac{e^{-ik|x|}dk}{(-v)^\beta - \{(ik)^\alpha\cdot Dt^\beta\}} := 2\pi i C \sum_l \text{res}\left[\frac{e^{-ik|x|}dk}{(-v)^\beta - \{(ik)^\alpha\cdot Dt^\beta\}}\right] =$$

$$= -2\pi C \frac{e^{-|x|\left\{\frac{(-v)^\beta}{Dt^\beta}\right\}^{1/\alpha}}}{\alpha\left\{\frac{(-v)^\beta}{Dt^\beta}\right\}^{1-1/\alpha}\cdot Dt^\beta} = -2\pi C \frac{e^{-|x|\left\{\frac{(-v)^\beta}{Dt^\beta}\right\}^{1/\alpha}}}{\alpha\cdot(-v)^{\beta-\beta/\alpha}\cdot Dt^{\beta/\alpha}},$$

The constant C is obtained by applying the normalization requirement to the function, yielding the following expression:

$$= -\frac{2\pi C}{\alpha\cdot(-v)^{\beta-\beta/\alpha}\cdot Dt^{\beta/\alpha}}\int_0^\infty e^{-|x|\left\{\frac{(-v)^\beta}{Dt^\beta}\right\}^{1/\alpha}}dx = -\frac{2\pi C}{\alpha(-v)^\beta} = 1$$

Furthermore, the normalization constraint is required to be satisfied for arbitrary t, leading to the following formulation:

$$\widehat{H(t)} = \widehat{H(0,t)} = -\frac{1}{2\pi i} \int_{C''} \left\{ -\frac{2\pi C}{\alpha(-v)^\beta} \cdot \frac{e^{-v}(-v)^{\beta-1}}{(-v)^\beta} \right\} du =$$

$$= \frac{C}{i\alpha} \int_{C''} e^{-v} \cdot (-v)^{-1} du = -2\pi \frac{C}{\alpha \cdot G(1)} = 1$$

This leads to the conclusion that $C = -\frac{\alpha}{2\pi}$. Incorporating all previous transformations, we obtain the following representation for $\phi_{\alpha,\beta}(x,t)$:

$$\phi(x,t) = \frac{1}{2\pi} \sum_{n=0}^{\infty} \frac{(-1)^n \left\{ \frac{|x|^\alpha}{Dt^\beta} \right\}^{n/\alpha}}{\{Dt^\beta\}^{1/\alpha} \cdot n! \cdot G\left[1 - \frac{\beta}{\alpha}(n+1)\right]}$$

When the argument of the gamma function is negative, one can utilize the reflection (or complement) formula to facilitate evaluation:

$$G\left[1 - \frac{\beta}{\alpha}(n+1)\right] = \frac{\pi}{G\left[\frac{\beta}{\alpha}(n+1)\right] \cdot \sin\left\{\pi\frac{\beta}{\alpha}(n+1)\right\}}$$

Thus, the final expression corresponds to the target probability density function $\phi_{\alpha,\beta}(x,t)$, and is given by:

$$\phi_{\alpha,\beta}(x,t) = \frac{1}{2\pi} \sum_{n=0}^{\infty} \frac{(-1)^n \left\{ \frac{|x|^\alpha}{Dt^\beta} \right\}^{n/\alpha} G\left[\frac{\beta}{\alpha}(n+1)\right] \cdot \sin\left\{\pi\frac{\beta}{\alpha}(n+1)\right\}}{\{Dt^\beta\}^{1/\alpha} \cdot n!} \quad (6)$$

The results of numerical analysis suggest that the series (6) converges provided the inequality

$$0 < \frac{\beta}{\alpha} \leq 0.865$$

holds for the fractional orders α and β.

5   Solution Analysis

It is important to note that the function $\phi_{\alpha,\beta}(x,t)$ fulfills the normalization requirement. Particularly, when $\beta = 1$ and $\alpha = 2$, it coincides with the standard form of the Poisson distribution:

$$\phi_{2,1}(x,t) = \frac{1}{2\pi} \sum_{n=0}^{\infty} \frac{(-1)^n \left\{\frac{|x|^2}{Dt}\right\}^{n/2} G\left[\frac{(n+1)}{2}\right] \cdot \sin\left\{\pi \frac{(n+1)}{2}\right\}}{\{Dt\}^{1/2} \cdot n!} =$$

$$= \frac{1}{2\sqrt{Dt}} \left\{ \frac{1}{\sqrt{\pi}} - \frac{\frac{x^2}{4Dt}}{\sqrt{\pi}} + \frac{\left\{\frac{x^2}{4Dt}\right\}^2}{2! \cdot \sqrt{\pi}} - \frac{\left\{\frac{x^2}{Dt}\right\}^3}{3! \cdot \sqrt{\pi}} + \cdots \pm \frac{\left\{\frac{x^2}{Dt}\right\}^n}{n! \cdot \sqrt{\pi}} \right\} =$$

$$= \frac{1}{2\sqrt{\pi Dt}} \sum_{n=0}^{\infty} (-1)^n \frac{\left\{\frac{x^2}{Dt}\right\}^n}{n!} = \frac{e^{-\frac{x^2}{4Dt}}}{2\sqrt{\pi Dt}}$$

We proceed to investigate the key statistical moments associated with the probability distribution $\phi_{\alpha,\beta}(x,t)$. In particular, we focus on the expectation value $\mu_{\alpha,\beta}(t)$ and the variance $\sigma^2_{\alpha,\beta}(t)$, which are defined by the integrals:

$$\mu_{\alpha,\beta}(t) = \int_0^\infty x \cdot \phi_{\alpha,\beta}(x,t)\, dx = \frac{D^{1/\alpha} \cdot t^{\beta/\alpha}}{G\left(\frac{\beta}{\alpha}+1\right)},$$

$$\sigma^2_{\alpha,\beta}(t) = \int_0^\infty x^2 \cdot \phi_{\alpha,\beta}(x,t)\, dx = \frac{2D^{2/\alpha} \cdot t^{2\beta/\alpha}}{G\left(2\frac{\beta}{\alpha}+1\right)}.$$

It should be noted that in the particular case when $\alpha = 2$ and $\beta = 1$, the quantities $\mu_{\alpha,\beta}(t)$ and $\sigma^2_{\alpha,\beta}(t)$ exactly match the mean and variance of the Poisson distribution, which represents the solution of the classical diffusion equation involving integer-order derivatives.

To illustrate the behavior of the derived model, let us perform numerical simulations using the parameter value $D = 0.05$ (in arbitrary units). We examine a range of representative cases by selecting various pairs of α and β such that the ratio satisfies the inequality $0 < \beta/\alpha \leq 0.865$.

Figure 1a illustrates the evolution of the probability density function $\phi_{\alpha,\beta}(x,t)$ for parameter values $\alpha = 1.70$, $\beta = 1.36$. The curves correspond to time points $t = 110$, 120, and 130, and are labeled as curves 1, 2, and 3, respectively.

Figure 1b presents the function profile computed at $\alpha = 1.70$, $\beta = 0.85$, with the distribution shown for $t = 85$, 100, and 150.

In Fig. 1c, the behavior of $\phi_{\alpha,\beta}(x,t)$ is shown for $\alpha = 1.70$, $\beta = 0.51$, at time instances $t = 50$, 100, and 200.

Figure 1d displays the function evaluated with $\alpha = 3.50$, $\beta = 2.80$, at $t = 60$, 65, and 70.

Figure 1e shows the evolution of the distribution for parameters $\alpha = 3.50$, $\beta = 1.75$, with snapshots taken at $t = 50$, 100, and 150.

Finally, Fig. 1f visualizes $\phi_{\alpha,\beta}(x,t)$ for $\alpha = 3.50$, $\beta = 1.05$, with time values $t = 100$, 300, and 500.

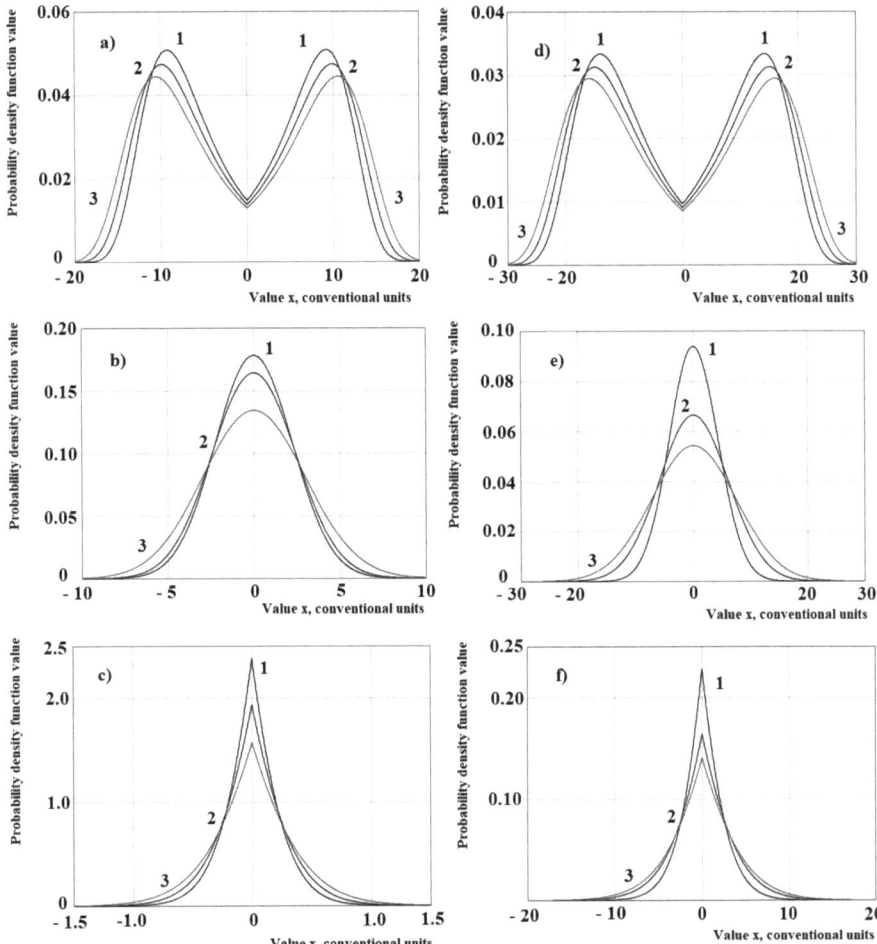

Fig. 1. Graphical representation of the probability density function $\phi(x,t)$ for various combinations of the fractional derivative orders α and β.

Visual inspection of the simulation results suggests a natural partitioning of the (α, β) parameter space into two regimes. The first, corresponding to $\beta/\alpha \in (0.50, 0.865]$, exhibits one type of behavior (Figs. 1a and 1d), while the second, defined by $\beta/\alpha \in (0, 0.50]$, demonstrates qualitatively different dynamics (Figs. 1c and 1f).

When $\beta/\alpha = 0.50$, the distribution $\phi_{\alpha,\beta}(x,t)$ exhibits behavior that is qualitatively similar to that of a Poisson distribution, as illustrated in Figs. 1b and 1e. However, it is crucial to point out that exact equivalence with the classical Poisson form occurs exclusively under the condition $\alpha = 2$ and $\beta = 1$.

A detailed comparison of Figs. 1a–1f shows that adjusting the parameters α and β predominantly affects the scale of the distribution curves, rather than their shape. The qualitative features of the probability density function remain largely unchanged within each parameter domain, a pattern particularly evident in the visual similarity between figure pairs 1a–1d, 1b–1e, and 1c–1f.

With increasing values of t, the height of the probability density curves decreases, accompanied by a widening of their support, indicating growing dispersion over time. The distributions preserve symmetry about the vertical axis through the origin throughout this process. However, in the parameter range $0.50 < \beta/\alpha \leq 0.865$, one observes a nontrivial effect: while the profiles remain symmetric in form, their maxima shift in opposite directions as time progresses—some moving rightward, others leftward (cf. Figs. 1a and 1d).

Another important observation is that $\phi_{\alpha,\beta}(x,t)$ tends rapidly toward zero as $|x|$ becomes large.

Although the parameters α and β may be integers, the function $\phi_{\alpha,\beta}(x,t)$ continues to exhibit the same qualitative characteristics as those shown in Figs. 1a–1f, as long as the ratio β/α falls within the identified regimes.

We now turn to an analysis of the time-dependent behavior of the mean value $\mu_{\alpha,\beta}(t)$ and the variance $\sigma^2_{\alpha,\beta}(t)$.

For this purpose, we carry out a numerical analysis using the fixed diffusion coefficient $D = 0.05$ and set $\alpha = 1.70$. The investigation focuses on two representative parameter intervals: $0 < \beta/\alpha \leq 0.50$ and $0.50 < \beta/\alpha \leq 0.865$.

Figure 2a presents the theoretical computation of the mathematical expectation based on the expression

$$\mu_{\alpha,\beta}(t) = \frac{D^{1/\alpha} \cdot t^{\beta/\alpha}}{G\left[\frac{\beta}{\alpha}+1\right]},$$

while Fig. 2b displays the corresponding variance, given by

$$\sigma^2_{\alpha,\beta}(t) = \frac{2D^{2/\alpha} \cdot t^{2\beta/\alpha}}{G\left[2\frac{\beta}{\alpha}+1\right]}.$$

In both figures, the individual curves are associated with specific values of the fractional ratio β/α: curve 1 corresponds to 0.3, curve 2 to 0.5, and curve 3 to 0.8, respectively.

Figure 2 illustrates in the interval $0 < \beta/\alpha \leq 0.50$, both the mean and the variance exhibit comparable growth behavior. In this regime, their increase tends to follow a smoother and more gradual trend, deviating from a strictly linear pattern—as can be observed, for instance, in curve 2 of Fig. 2b.

6 Comparison of Simulation Results with Observed Data

The simulation results illustrated in Fig. 2 suggest that comparable trends in the evolution of the mathematical expectation and variance may emerge in the analysis of real-world time series. One notable example is the modeling of complex sociodynamic phenomena, such as the dynamics of U.S. presidential election campaigns.

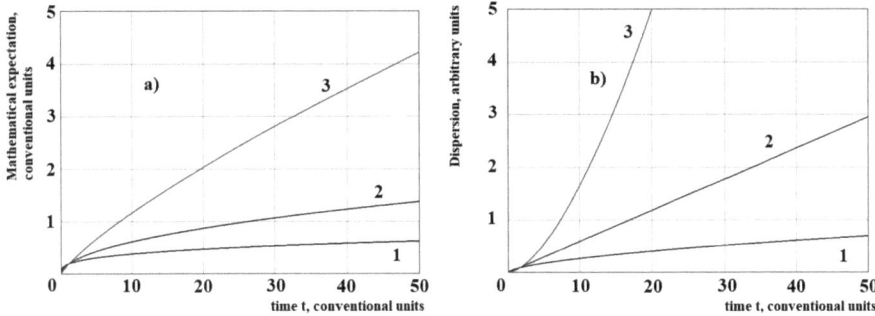

Fig. 2. Graphical representation of the simulation results for the mathematical expectation $\mu_{\alpha,\beta}(t)$ at various values of the ratio β/α.

To assess the adequacy of the proposed model, empirical time series data were obtained from the RealClearPolitics platform (realclearpolitics.com). This source offers comprehensive and systematically curated information on electoral campaign dynamics, including fluctuations in candidate popularity and trends in voter sentiment. Due to its reliability and structure, the dataset is frequently employed in political science and sociological studies for analyzing patterns of public opinion during election cycles.

The time series on RealClearPolitics are presented individually for each candidate and include daily support metrics based on various polls. Each data point in the time series corresponds to a specific date, and the associated value reflects the level of public support for a particular candidate or political party on that day. This structure enables researchers not only to track trends in popularity dynamics but also to identify potential patterns and compare changes in support for different candidates in relation to campaign events and stages.

To perform time series analysis, data retrieved from the RealClearPolitics website (realclearpolitics.com) were employed. The evaluation followed a sliding window approach, which proceeded as follows:

1. Begin by defining the width of the *moving time interval*, which specifies the temporal spacing between successive observations (e.g., one day, two days, three days, etc.). Using this interval, extract the corresponding overlapping segments from the time series for further analysis.
2. Second, compute the amplitude of variations in the time series values over each specified interval, as determined by the selected window size.
3. Third, for every interval length, sort the resulting set of amplitudes in ascending order (from negative to positive), and construct histograms that approximate the probability density functions of the deviation amplitudes.
4. Fourth, calculate the statistical moments such as the mean (mathematical expectation), variance, skewness, and kurtosis of each histogram corresponding to its respective window size.

5. Finally, derive functional relationships that describe how the expectation and variance of the amplitude deviations evolve as functions of the window length (i.e., the time interval used in the sliding window).

Figure 3 depicts how the mathematical expectation (see Fig. 3a) and the variance (see Fig. 3b) of time series amplitude fluctuations vary as a function of the observation interval length. The data used in the analysis pertain to the dynamics of voter support for the winning candidates during the U.S. presidential elections of 2008, 2012, and 2016. Specifically, curve 1 in both panels corresponds to Barack Obama in 2008, curve 2 to his re-election campaign in 2012, and curve 3 to Donald Trump's candidacy in 2016.

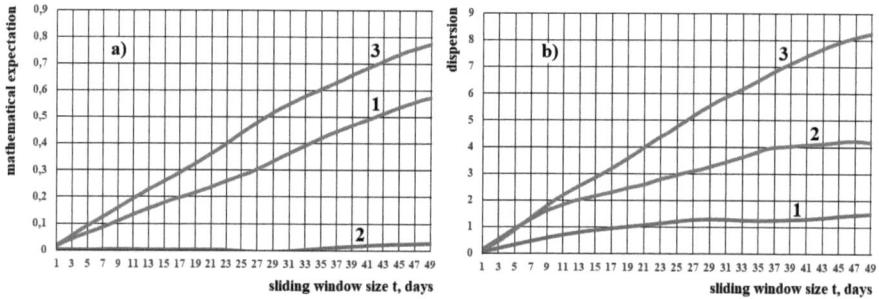

Fig. 3. Temporal evolution of the mean and variance of amplitude fluctuations in time series levels, evaluated over varying observation intervals. The results are based on voter preference data from the winning candidates in the U.S. presidential election campaigns of 2008, 2012, and 2016.

A comparison between Figs. 2 and 3 reveals a strong correspondence between the empirical trends and the theoretical predictions of the proposed model, particularly when the parameter ratio β/α falls within the range $0 < \beta/\alpha \leq 0.50$. Both the mathematical expectation and the variance were computed over the full duration of each presidential election campaign considered.

7 Conclusion

Fractional derivatives in both the temporal variable t and the spatial variable x establish a foundation for modeling non-local dynamics, wherein the system's evolution toward a particular state is governed by influences extending beyond local neighborhoods. Such a formulation captures the dependence on the entire spatial domain and the full temporal history of the process, thereby incorporating memory effects. Accordingly, fractional differential equations with time and space components constitute an effective tool for representing systems characterized by memory retention and emergent self-organization.

The investigation led to the development of a new analytical solution describing the probability density of observing a particular state x of a process—such as the amplitude of deviations in time series levels—at time t, representing the observation interval. This solution is constructed within the framework of a fractional diffusion equation defined over an infinite spatial domain.

Unlike conventional approaches, which typically confine fractional orders to $\beta \in (0, 1)$ for time and $\alpha \in (1, 2)$ for space, the solution developed in this work admits arbitrary values of α and β, provided that their ratio satisfies $0 < \beta/\alpha \leq 0.865$. This generalization significantly broadens the applicability of the model beyond the scope of prior studies.

The analysis of the proposed model indicates that variations in the parameters α and β predominantly influence the scaling of the simulated outcomes, while the qualitative structure of the resulting functional dependencies remains invariant across the explored parameter space. This structural consistency is especially evident when comparing the results shown in paired Figs. 1a with 1d, 1b with 1e, and 1c with 1f.

As the time parameter t grows, the probability distribution curves exhibit a reduction in peak height accompanied by an increase in their spread. Nonetheless, the distributions maintain symmetry with respect to the vertical line at $x = 0$. Within the parameter range $0.50 < \beta/\alpha \leq 0.865$, although the overall symmetry is preserved, the locations of the distribution maxima progressively shift in opposite directions along the x-axis as t increases (refer to Figs. 1a and 1d).

A notable property of the probability density function $\phi_{\alpha,\beta}(x, t)$ is its rapid attenuation as $|x|$ grows, with the function effectively vanishing in the asymptotic regions.

Variation in the diffusion coefficient D induces a consistent rescaling of the computed distributions, while preserving their fundamental qualitative structure within all examined parameter regimes.

Notably, the orders of the fractional derivatives, α and β, may also assume integer values.

Provided that the ratio β/α falls within the established range, the qualitative behavior of $\phi_{\alpha,\beta}(x, t)$ remains in agreement with the distributions presented in Figs. 1a–1f.

As illustrated in Figure, when β/α, both the expectation and variance curves demonstrate closely aligned behavior, characterized by a consistent upward trend. This progression is more regular and monotonic than a simple linear growth pattern, as can be seen in curve 2 of Fig. 2b.

The results of the analysis validate the internal coherence of the proposed modeling framework. Notably, the theoretical estimates for the behavior of the distribution's first and second moments are in close agreement with empirical observations derived from voter preference time series during the U.S. presidential election campaigns of 2008, 2012, and 2016. This concordance is most evident in the comparative analysis of Figs. 2 and 3, particularly in the regime defined by $0 < \beta/\alpha \leq 0.50$.

Acknowledgments. This research was supported by the Russian Science Foundation (RSF), grant no. 23-21-00153 «Analysis and modeling dynamics of transient time series of fractal processes with memory realization (aftereffection) and self-organization based on the use of differential equations with fractional derivatives».

References

1. Bandara, K., Bergmeir, C., Hewamalage, H.: LSTM-MSNet: leveraging forecasts on sets of related time series with multiple seasonal patterns. IEEE Trans. Neural Netw. Learn. Syst. **32**, 1586–1599 (2020)
2. Taniguchi, M., Kato, S., Ogata, H., Pewsey, A.: Models for circular data from time series spectra. J. Time Ser. Anal. **41**, 808–829 (2020)
3. Martiny, E.S., Jensen, M.H., Heltberg, M.S.: Detecting limit cycles in stochastic time series. Phys. A: Stat. Mech. Its Appl. **605**, 127917 (2022)
4. Hsu, N.J., Sim, L.H., Tsay, R.S.: Testing for symmetric correlation matrices with applications to factor models. J. Time Ser. Anal. **44**, 622–643 (2023)
5. Zinenko, A.V.: Prognozirovanie finansovy'x vremenny'x ryadov s ispol'zovaniem singulyarnogo spektral'nogo analiza [Forecasting financial time series using singular spectral analysis]. Biznes-informatika. **17**(30), 87–100 (2023)
6. Chaudhary, P.K., Pachori, R.B.: Automatic diagnosis of glaucoma using two dimensional Fourier-Bessel series expansion based empirical wavelet transform. Biomed. Signal Process. Control. **64**, 102237 (2021)
7. Son, J.H., Kim, Y.: Probabilistic time series prediction of ship structural response using Volterra series. Mar. Struct. **76**, 102928 (2021)
8. Wu, S., Xiao, X., Ding, Q., Zhao, P., Wei, Y., Huang, J.: Adversarial sparse transformer for time series forecasting. Adv. Neural Inf. Process. Syst. **33**, 17105–17115 (2020)
9. Yao, Z., Wann, J., Chen, J.: Generalized maximum entropy based identification of graphical ARMA models. Automatica **141**, 110319 (2022)
10. Entezami, A., Sarmadi, H., Behkamal, B., Mariani, S.: Big data analytics and structural health monitoring: a statistical pattern recognition-based approach. Sensors **20**, 2328 (2020)
11. Montagnon, C.E.: Forecasting by splitting a time series using singular value decomposition then using both ARMA and a Fokker Planck equation. Phys. A Stat. Mech. Its Appl. **567**, 125708 (2021)
12. Shi, Q., et al.: Block Hankel tensor ARIMA for multiple short time series forecasting. In: Proceedings of the AAAI Conference on Artificial Intelligence, vol. 34, pp. 5758–5766 (2020)
13. Khan, S., Alghulaiakh, H.: ARIMA model for accurate time series stocks forecasting. Int. J. Adv. Comput. Sci. Appl. **11**, 524–528 (2020)
14. Guan, Y., Li, D., Xue, S., Xi, Y.: Feature-fusion-kernel-based Gaussian process model for probabilistic long-term load forecasting. Neurocomputing. **426**, 174–184 (2021)
15. Nikulchev, E., Chervyakov, A.: Development of trading strategies using times series based on robust interval forecast. Computation. **11**(5), 99 (2023)
16. Nikulchev, E., Chervyakov, A.: Symmetric seasonality of time series in interval prediction for financial management of the branch. Symmetry **15** (2023). https://doi.org/10.3390/sym15122100

17. Lara-Benitez, P., Carranza-Garcia, M., Riquelme, J.C.: An experimental review on deep learning architectures for time series forecasting. Int. J. Neural Syst. **31**, 2130001 (2021)
18. Sezer, O.B., Gudelek, M.U., Ozbayoglu, A.M.: Financial time series forecasting with deep learning: a systematic literature review: 2005–2019. Appl. Soft Comput. **90**, 106–181 (2020)
19. Lim, B., Zohren, S.: Time-series forecasting with deep learning: a survey. Philos. Trans. R. Soc. A Math. Phys. Eng. Sci. **379**, 20200209 (2021)
20. Meisenbacher S., et al.: Review of automated time series forecasting pipelines. Wiley Interdiscip. Rev. Data Min. Knowl. Discov. **12**, e1475 (2022)
21. Jahangir, H., Golkar, M.A., Alhameli, F., Mazouz, A., Ahmadian, A., Elkamel, A.: Short-term wind speed forecasting framework based on stacked denoising autoencoders with rough ANN. Sustain. Energy Technol. Assess. **38**, 100601 (2020)
22. Cheng, D., Yang, F., Xiang, S., Liu, J.: Financial time series forecasting with multimodality graph neural network. Pattern Recogn. **121**, 108218 (2022)
23. Diqi, M., Hiswati, M.E., Nur, A.S.: StockGAN: robust stock price prediction using GAN algorithm. Inter. J. Inform. Technol. **14**(5), 2309–2315 (2022)
24. Hu, J., Wang, X., Zhang, Y., Zhang, D., Zhang, M., Xue, J.: Time series prediction method based on variant LSTM recurrent neural network. Neural Processing Lett. **52**, 1485–1500 (2020)
25. Wu L., Perin G., Picek S.: I choose you: automated hyperparameter tuning for deep learning-based side-channel analysis. IEEE Trans. Emerging Topics Comput. (2022). Early Access. https://doi.org/10.1109/TETC.2022.3218372
26. Ishaq, A., Asghar, S., Gillani, S.A.: Aspect-based sentiment analysis using a hybridized approach based on CNN and GA. IEEE Access. **8**, 135499–135512 (2020)
27. Hewamalage, H., Bergmeir, C., Bandara, K.: Recurrent neural networks for time series forecasting: Current status and future directions. Int. J. Forecast. **37**, 388–427 (2021)
28. Smyl, S.: A hybrid method of exponential smoothing and recurrent neural networks for time series forecasting. Int. J. Forecast. **36**, 75–85 (2020)
29. Li, Z., Han, J., Song, Y.: On the forecasting of high-frequency financial time series based on ARIMA model improved by deep learning. J. Forecast. **39**, 1081–1097 (2020)
30. Chen, Z.Y., Xiao, F., Wang, X.K., Deng, M.H., Wang, J.Q., Li, J.B.: Stochastic configuration network based on improved whale optimization algorithm for nonstationary time series prediction. J. Forecast. **41**, 1458–1482 (2022)
31. Demidova, L.A., Gorchakov, A.V.: Application of bioinspired global optimization algorithms to the improvement of the prediction accuracy of compact extreme learning machines. Russian Technol. J. **10**(2), 59–74 (2022). https://doi.org/10.32362/2500-316X-2022-10-2-59-74
32. Jahanshahi H., Sajjadi S.S., Bekiros S., Aly A.A.: On the development of variable-order fractional hyperchaotic economic system with a nonlinear model predictive controller. Chaos, Solitons & Fractals **144**, 110698 (2021); Miller K.S., Ross B.: An Introduction to the Fractional Calculus and Fractional Differential Equations. John Wiley & Sons. Inc., New York (1993)
33. Mainardi, F., Luchko, Y., Pagnini, G.: The fundamental solution of the space-time fractional diffusion equation. Fractional Calculus Appli. Analy. **4**(2), 153–192 (2001)
34. Gorenflo, R., Iskenderov, A., Luchko, Y.: Mapping between solutions of fractional diffusion-wave equations. Fractional Calculus Appli. Analy. **3**(1), 1-13 (2000)

35. Luchko, Y., Gorenflo, R.: Scale-invariant solutions of a partial differential equation of fractional order. Fractional Calculus Appli. Analy., 1–17 (1998)
36. Gorenflo, R., Mainardi, F.: Fractional calculus: integral and differential equations of fractional order. In: Carpinteri, A., Mainardi, F. (eds.) Fractals and Fractional Calculus in Continuum Mechanics. Springer (1997). https://doi.org/10.1007/978-3-7091-2664-6_5
37. Buckwar, E., Luchko, Yu.: Invariance of a partial differential equation of fractional order under the Lie group of scaling transformations. J. Math. Anal. Appl. **227**, 81–97 (1998)
38. Engler, H.: Similarity solutions for a class of hyperbolic integro differential equations. Differential Integral Eqn-s **10**(5), 815–840 (1997)
39. Fujita, Y.: Integrodifferential equation which interpolates the heat and the wave equations. Osaka J. Math. **27**(309–321), 797–804 (1990)
40. Gorenflo, R., Luchko, Yu., Mainardi, F.: Analytical properties and applications of the Wright function. Fractional Calc. Appli. Analy. **2**, 383–414 (1999)
41. Goloviznin, V., Kiselev, V., Korotkin, I., Yurkov, Y.: Some features of computing algorithms for the equations fractional diffusion. Preprint (textnumero) IBRAE-2002-01, p. 57 . Nuclear Safety Institute RAS, Moscow (January 2002)
42. Bondarenko, A.N., Ivaschenko, D.S.: Numerical methods for solving boundary problems of anomalous diffusion theory. Siberian Electron Math. Rep. **5**, 581–594 (2008)
43. Mainardi, F.: Waves and Stability in Continuous Media. In: Rionero S., Ruggeri T. (eds.) World Scientific, Singapore (1994)
44. Wyss, W.J.: The fractional diffusion equation. J. Math. Phys. **27**, 2782 (1986). https://doi.org/10.1063/1.527251Math
45. Schneider, W.R., Wyss, W.J.: Fractional diffusion and wave equations. J. Math. Phys. **30**, 134 (1989). https://doi.org/10.1063/1.528578Math
46. Ilic, M., Liu, F., Turner, I., Anh, V.: Numerical approximation of a fractional-inspace diffusion equation. I. Fractional Calc. Appli. Analy. **8**(3), 323–341 (2005)
47. Oldham, K.B., Spanier, J.: The Fractional Calculus. Academic Press, New York (1974)
48. Näsholm, S.P., Holm, S.: On a fractional Zener elastic wave equation. Fractional Calc. Appli. Analy. **16**(1), 26–50 (2013). https://doi.org/10.2478/s13540-013-0003-1
49. Holm, S., Näsholm, S.P.: A causal and fractional all-frequency wave equation for lossy media. J. Acoust. Soc. Am. **130**, 2195–2202 (2011)
50. Näsholm, S.P., Holm, S.: Linking multiple relaxation, power-law attenuation, and fractional wave equations. J. Acoust. Soc. Am. **130**, 3038–3045 (2011)
51. Pandey, V., Holm, S.: Connecting the grain-shearing mechanism of wave propagation in marine sediments to fractional order wave equations. J. Acoust. Soc. Am. **140**(6), 4225–4236 (2016). https://doi.org/10.1121/1.4971289
52. Zelenyi, L.M., Milovanov, A.M.: Fraktal'naya topologiya i strannaya kinetika [Fractal topology and strange kinetics]. Uspexi fizicheskix nauk T. 174, (textnumero) 8. S. 809–852 (2004)
53. Uchaikin, V.V.: Avtomodel'naya anomal'naya diffuziya i ustojchivy'e zakony' [Self-similar anomalous diffusion and stable laws]. Uspexi fizicheskix nauk. T. 173, (textnumero) 8. S. 847–874 (2003)
54. Nakhushev, A.M.: Drobnoe ischislenie i ego primenenie [Fractional calculus and its application]. M.: Fizmatlit, 272 s (2003)

55. Samko, S.G., Kilbas, A.A., Marichev, O.I.: Integraly' i proizvodny'e drobnogo poryadka i nekotory'e ix prilozheniya [Integrals and derivatives of fractional order and some of their applications]. Minsk.: Nauka i texnika, 688 s (1987)
56. Chukbar, K.V.: Stoxasticheskij perenos i drobny'e proizvodny'e. Zhurnal e'ksperimental'noj i teoreticheskoj fiziki [Stochastic transfer and fractional derivatives]. T. 108, vy'p. 5(11). S. 1875–1884 (1995)
57. Kobelev, V.L., Romanov, E.N.: Kobelev Ia.L. i dr. Nelinejnaya relaksaciya i diffuziya v fraktal'nom prostranstve [Nonlinear relaxation and diffusion in fractal space]. DAN. T. 361, (textnumero) 6. S. 755–758 (1998)
58. Kochubei, A.N.: Diffuziya drobnogo poryadka. Differencial'ny'e uravneniya [Fractional order diffusion. Differential equations]. T. 26, (textnumero) 4. S. 660–672 (1990)
59. Metzler, R., Gockle, W.G., Nonnenmacher, T.F.: Fractional model equation for anomalous diffusion. XXPhys. A **211**, 13–24 (1994)
60. Demidova, L.A., Zhukov, D.O., Andrianova, E.G., Sigov, A.S.: Modeling sociodynamic processes based on the use of the differential diffusion equation with fractional derivatives. Information (Switzerland), **14**(2), 121 (Q2, SCOPUS) (2023). https://doi.org/10.3390/info14020121
61. Zhukov, D., Khvatova, T., Aleshkin, A., Schiavone, F.: Forecasting news events based on the model accounting for self-organisation and memory. In: Proceeding of the 2021 IEEE Technology and Engineering Management Conference - Europe, TEMSCON-EUR 2021, Article number 94886342021
62. Zhukov, D., Andrianova, E., Trifonova, O.: Stochastic diffusion model for analysis of dynamics and forecasting events in news feeds. Symmetry **13**(2), Article number 257, 1–23 (2021)
63. Zhukov, D., Khvatova, T., Zaltsman, A.: Stochastic dynamics of influence expansion in social networks and managing users' transitions from one state to another. In: Proceedings of the 11 th European Conference on Information Systems Management, ECISM 2017, The University of Genoa, Italy, 14 -15 September, pp. 322–329 (2017). ISBN: 978-191121852-4
64. Sigov, A.S., Zhukov, D.O., Khvatova, T.Y., Andrianova, E.G.: A model of forecasting of information events on the basis of the solution of a boundary value problem for systems with memory and self-organization. J. Commun. Technol. Electron. **63**(12), 1478–1485 (2018). https://doi.org/10.1134/S1064226918120227
65. Zhukov, D., Khvatova, T., Istratov, L.: A stochastic dynamics model for shaping stock indexes using self-organization processes, memory and oscillations. In: Proceedings of the European Conference on the Impact of Artificial Intelligence and Robotics, ECIAIR 2019, 31 October–1 November 2019, Oxford, UK, pp. 390–401, E-Book ISBN: 978-1-912764-44-0, Book version ISBN: 978-1-912764-45-7
66. Zhukov, D., Khvatova, T., Istratov, L.: Analysis of non-stationary time series based on modelling stochastic dynamics considering self-organization, memory and oscillations. In: ITISE 2019 International Conference on Time Series and Forecasting. Proceedings of Papers, 25-27 September 2019, Granada (Spain), vol. 1, pp. 244 -254. ISBN: 978-84-17970-78; Hurst, H.E.: Long - term storage capacity of reservoirs. Trans. Am. So. Civil Eng. **116**, 770 (1951)
67. Djrbashian, M.M.: Integral Transforms and Representations of Functions in the Complex Plane. Nauka, Moscow (1966)

Automation Systems for Scientific Calculations for Multiscale Modeling of Composite Materials

K.K. Abgaryan[(✉)] and E.S. Gavrilov

Federal Research Center "Computer Science and Control" of Russian Academy of sciences, Moscow, Russia
kabgaryan@frccsc.ru

Abstract. Currently, systems for automating scientific calculations are being actively developed, the use of which makes it possible to carry out calculations using high-performance resources, control the parallel execution of an application in a distributed computing environment, and carry out multivariate calculations. The paper deals with the issues of building automation systems for scientific services for multi-scale modeling of composite materials. The presented methods, typical architecture and software tools make it possible to integrate the author's software module or application software package into a platform for multi-scale modeling of composite materials with little effort. The constructed software solutions can be used to speed up and reduce the cost of searching for new composite materials with desired properties.

Keywords: multiscale modeling · composite materials · integration platform · software package · distributed system · automation of scientific calculations

1 Introduction

The solution of applied problems related to multiscale modeling [1,2] of the structure and properties of composite materials involves conducting large-scale computational experiments with complex combinations of various tools and computational modules, which in large distributed systems are traditionally referred to as resources. The task of effective resource management requires the use of dynamically changing datasets that interact with computational modules. The specifics of this task are associated with the complexity of its formulation and approaches to its solution, the variety of modeling methods and techniques, large volumes of data, and the heterogeneity of their types and properties. In addition, the semantics of preparing informational content and the diversity of result processing methodologies play an important role. Developers of information support tools and the implementation of the integration framework for the software system face the challenge of providing specialists in the subject area, particularly related to modeling the properties of composite materials [3], with a convenient, understandable, universal, and scalable set of tools for preparing, executing, and analyzing tasks of varying complexity and execution time. This work contains materials describing approaches to creating software tools for the integration platform of multiscale modeling [4–6], which allow for increased performance of computational experiments, enhanced flexibility of developed software components, and accelerated integration of new computational modules into the system.

2 Calculation Modules

Let's consider the concept of a calculation module and the existing types of calculation modules. By a calculation module, we mean a program that implements an algorithm for calculating the structure or properties of a material according to a physical-mathematical model at a specific scale level. The program can be executed locally on the user's computer or on a computing cluster using mechanisms for parallel execution of calculations, utilizing specific resources (CPU cores, GPUs, libraries), in a Docker container.

In practice, the following two types of modules are the most common, and their integration fundamentally differs from each other:

1. Highly specialized proprietary program:
   – in most occasions, only the author knows how to work with it, therefore, knowledge about the module or even the module itself can be lost;
   – there are plenty of modules, but they usually solve a narrow problem with a single algorithm;
   – inevitably, different authors use various technological stacks, which complicates integration.
2. Software package (VASP, Quantum Espresso, LAMMPS, and others):
   – implements a variety of algorithms for solving a class of problems;
   – requires user training; effective use depends on the availability of documentation, educational materials, and specialists;
   – some packages require purchasing a license for use and/or publication of calculation results;
   – the package may be presented as a set of specialized modules with a common ideology;
   – the package may have its own internal configuration mechanism or programming language.

To solve problems of multiscale modeling, a combination of modules of both types is used, with the number of modules for a complete solution to a single problem typically ranging from 5 to 7. If we generalize solutions for classes of problems, then tens to hundreds of modules will be required. Therefore:

1. A service-oriented architecture is necessary, with loose coupling, that would allow to scale the development, integration, and further support of computational modules.
2. Each service should provide a somewhat unified programming interface for integration with the scenario service, while the degree of unification should

not overly restrict the authors of the scenario in their ability to use the module features (there should be the possibility of flexible configuration of module-specific parameters).
3. A program presented as a service can be published in a unified service catalog for reuse.

Microsoft Azure Machine Learning and Amazon Machine Learning cloud services are good examples of implementation.

Module Requirements For Platform Integration

Let's consider the main requirements for the computational module so that it can be implemented as a service:

1. Operation in batch (non-interactive) mode.
2. Transmission of input and output data through files or other OS means available for automation.
3. Ability to control the calculation process:
   - start,
   - monitoring the calculation process,
   - forced termination of the operation.
4. Producing an output log (logging) (Fig. 1)

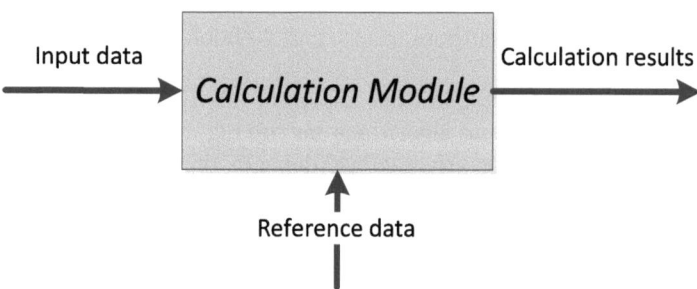

Fig. 1. Calculation module schema

3 Software Package

Calculation Module Integration Architecture

To implement the computational module as a service, the approach of "microservice architecture" is used, which provides such key non-functional properties as

- *loose coupling* of services (computational module services are usually not interconnected, and the scenario service uses the computational module services without special knowledge of their functionality);

- *independent development* – allows for parallel development/improvement of services in terms of organizing work, for example, in the form of students' course or diploma projects;
- the ability to use *different technologies* (programming languages, platforms). This allows for the use of more modern technologies for new services, maintaining the interest of developers and students in the project;
- *scalability*: when there is an increased load on a particular service, it can be scaled horizontally across several nodes;
- *fault tolerance*: in the event of a failure or restart of the service, ongoing calculations should not be lost, i.e., the service should automatically restore its state after a restart;
- support for *parallel execution* of multiple calculations;
- being *resistant to network failures*, since calculations are performed directly on various remote nodes connected via the Internet using the SSH protocol, if the connection is lost, the service should automatically restore it without losing calculation results (Fig. 2).

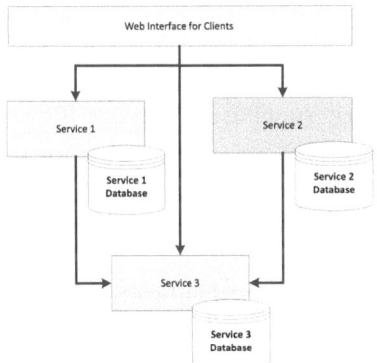

Fig. 2. Basic microservice architecture

Thus, the computational module service can be represented as several blocks (Fig. 3).

1. The module controller provides:
   - starting the calculation,
   - monitoring the execution of the calculation,
   - reading the output log,
   - stopping the calculation.
2. The local database is responsible for storing:
   - module-specific reference data (dictionaries),

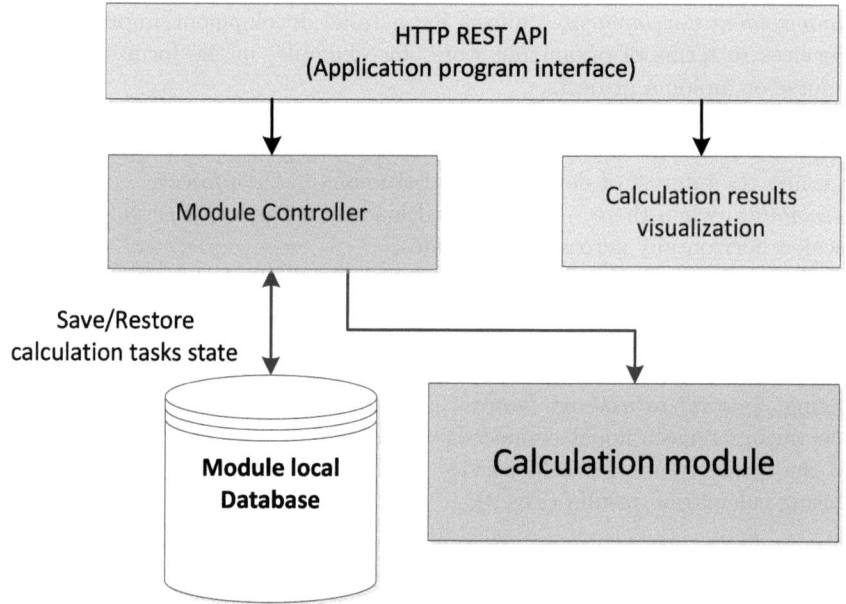

Fig. 3. Typical calculation module service architecture

- the state of currently running calculations (task ID, process PID, status, start time, and others),
- calculation results.

The programming interface provides access to the controller's functions and visualization through the HTTP protocol.

Visualization (optional) contains module-specific logic for visualizing calculation results.

For full operation in a concurrent multi-user environment, computational module services must met the following non-functional requirements:

- The ability to run one module multiple times in parallel with different parameters.
- The ability to execute modules in various physical locations (clusters).

At the same time, ensuring calculation resources balancing, tracking clusters free resources and organizing queues are not within the responsibility of the service. This functionality should be implemented on the computing cluster side, and necessary commands (queueing, task tracking) included in the service command line settings.

Architecture of the Computational Module Service Framework

Due to the quite high non-functional requirements described above in Sect. 3, the implementation of general functionality has been moved to a framework library for quick creation of new services in the future. Although using the framework is not mandatory, it significantly reduces costs for implementing new services. Let's consider the implementation of the framework in Fig. 4 below.

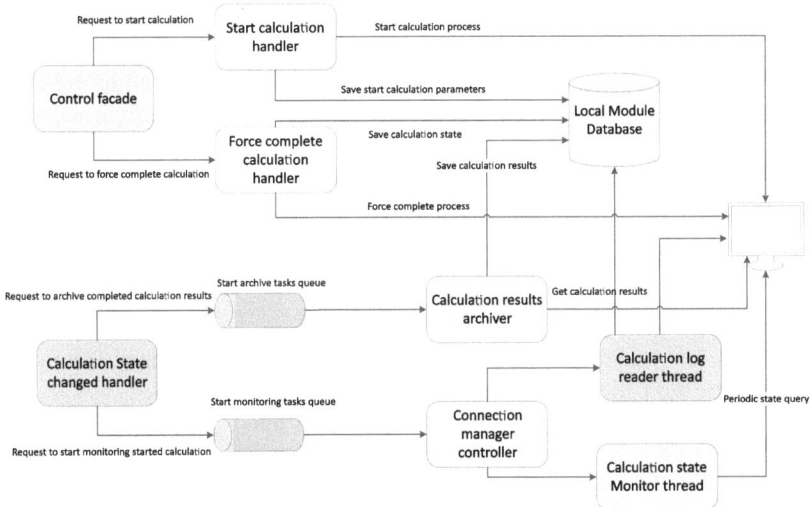

Fig. 4. Calculation module service framework architecture

The control facade is the integration point and central entry point for all operations of the controller. It is used by the module controller to send commands to start calculations, force termination, and other operations. The remaining blocks perform operations corresponding to their names.

The architecture of the framework is based on the following principles.

1. The principle of separation of responsibilities between the framework and the calculation module service: the service is responsible for data models, methods for converting input data, reading calculation results from files, and converting them into a common model. As a result, implementing the service on the framework takes 1-2 d for the developer.
2. Modularity and inversion of control: the configuration and initialization of the control facade is handled by the service implementation, which allows for the replacement of any component and strategy in the framework library with a custom implementation.
3. Asynchronous operations with files and external processes: since all potentially heavy operations are executed asynchronously through task queuing, service calls are completed in a fixed time, and heavy operations are limited by the resource size of the thread pool.
4. When the service starts, the contents of the job processing queues are restored by reading states from a local database. This ensures recovery after failures and, combined with request balancing between multiple instances of the service, provides fault tolerance.

Template Method for Calculation Service

The approach described above using the framework library is sufficient in most cases. However, in some advanced calculation packages (e.g., LAMMPS), there are no clearly defined input and output data. The package represents a specialized development environment with its own specialized programming language. A program in LAMMPS consists of one or more text files interpreted by LAMMPS. Thus, developing a single service for LAMMPS with a fixed configuration is not feasible, and creating a new service for each molecular dynamics calculation task is impractical in terms of development and maintenance effort.

A compromise is to add "templates" for the operation of a single LAMMPS service that possess sufficient flexibility to solve various tasks. A "template" includes:

- configuration of input and output data,
- procedure for generating input files from text templates,
- procedure for parsing LAMMPS output results into a standard format,
- set of static text files containing the program for LAMMPS,
- entry point into the LAMMPS program with a fixed name.

Thus, to add new LAMMPS calculation type one should add a new "template". Standardizing the LAMMPS program and preparing a new template may take 2-3 h, which is significantly less than developing a new module and has minimal impact on the cost of further support.

Calculation Service for Running External Scripts

A special category of calculation modules consists of small scripts that solve a very specialized task and usually frequently changed by the author. Implementing a separate service for each script is also impractical: changes in the input and output data types in the script would require modifications, recompilation, and deployment of a new version of the service. For such cases, even templates do not provide enough flexibility: the configuration can be moved to the settings of the script itself. In this case, adding or modifying a script will only require changes in the script, which allows for greater creativity among researchers with programming skills, for example, in Python. The calculation service in this case exists as a single instance and accepts configuration from the script in the form of:

- executable file path on the computing server,
- list of input and output data types.

> For example: "meta": {
> "inputs": ["layer_width", "elastic_constants;"],
> "outputs": ["composite_layer_angle_elastic"] ,
> "profile": "CCAS_CpuGold",
> "script": "/home/crystal/scripts/source/unidir.py"
> }

In this case, the script is responsible for reading the input data and writing the output data in a standardized format.

Of course, in this way, all computational modules could be implemented as scripts, but this is only advisable for rapidly changing implementations. For stable implementations on a compiled platform and ready-made packages, using wrapper scripts leads to difficulties in testing and support. The code in Kotlin, which is used to develop the services, is strongly typed and less prone to errors compared to dynamic languages.

4 Conclusions

The paper presents a typical architecture and software tools that can be used in the creation of systems for automating scientific calculations for multiscale modeling of composite materials.

The proposed approaches allow for the integration of an author's software module or a package of application programs for calculating material properties into the multiscale modeling platform of composite materials with minimal efforts. The developed software solutions can be used to accelerate and reduce the costs of the processes for searching for new composite materials with specified properties.

References

1. Abgaryan, K.K.: Mnogomasshtabnoe modelirovanie v zadachakh strukturnogo materialovedeniya [Multiscale modeling in material science problems], 284 p. MAKS Press, Moscow
2. Abgaryan, K.K.: Informatsionnaya tekhnologiya postroeniya mnogomasshtabnykh modeley v zadachakh vychislitel'nogo materialovedeniya [Information technology is the construction of multiscale models in problems of computational materials science]. Sistemy vysokoy dostupnosti [High Availability Systems] **15**(2), 9–15 (2018)
3. Naffakh M. et al. Morphology and thermal properties of novel poly (phenylene sulfide) hybrid nanocomposites based on single-walled carbon nanotubes and 8 inorganic fullerene-like WS 2 nanoparticles. J. Mater. Chem. **22**(4), 1418-1425 (2012)
4. Abgaryan K.K., Gavrilov E.S. Raspredelennaya informacionnaya sistema dlya rascheta strukturny'x svojstv kompozicionny'x materialov. Informatika i ee primenenie **15**(4). S. 50–58 (2021). https://doi.org/10.14357/19922264210407 [Informatics and its application]
5. Gavrilov, E.S.: Integrirovannyj interfejs k modulju sploshnosrednogo vzaimodejstvija [Integrated interface for continuum interaction module]. RF Computer program registration (textnumero) 2021681058 (2021)
6. Gavrilov, E.S.: Programmnye sredstva dlja hranenija i obmena dannymi v zadacah modelirovanija kompozitnyh materialov [Software for storage and data exchange in composite material modeling tasks]. RF Computer program registration No 2021681762 (2021)
7. Ilyushin, G.Ya., Sokolov, I.A.: Organizaciya upravlyaemogo dostupa pol'zovatelej k raznorodny'm vedomstvenny'm informacionny'm resursam [Organization of users' manageable access to heterogeneous departmental Informational resources]. Informatika i ee primenenie [Informatics and its application] **4**(1), 24-40 (2010)

Automated Generation of Educational Video Material

B. S. Ksemidov(✉) and K. K. Abgaryan

Federal Research Center Computer Science and Control of the Russian Academy of Sciences, Moscow, Russian Federation
sokboriswork@yandex.ru

Abstract. The purpose of this study is to develop an intelligent tutoring system (ITS) using the modern machine learning algorithms to automate the most labor-intensive part of teachers' work. Existing systems require preliminary preparation of educational material of the required format, which is a very labor-intensive process. This paper presents an ITS, the peculiarity of which is the availability of functionality to help the teacher in creating video and text material, as well as in editing and structuring it in an automated mode. This approach allows students to receive material not only in text form, but also in video format, thereby reducing the amount of work on preparing educational material.

Keywords: machine learning · intelligent tutoring system · domain model · knowledge base · speech synthesis · data analysis

1 Introduction

Intelligent tutoring systems are very popular due to the flexibility of learning, which is achieved by adapting the system to the individual characteristics of students, as well as the possibility of remote interaction, thanks to which you can learn from anywhere in the world, which speaks of the great importance of modern intelligent tutoring systems in our time [1].

The purpose of the research is to develop an adaptive ITS using the modern machine learning methods to automate the most labor-intensive part of the teachers' work by using the following functions:

- automated preparation of educational material;
- providing educational materials to students;
- providing answers to students' questions;
- drawing up individual learning plans in accordance with the students' academic performance;
- analysis of statistics on student performance and evaluation of educational materials (for example, the degree of difficulty of the course, as well as the relationship between practical tasks and theoretical material) to inform the teacher;

- notifying other teachers of the course about updating the subject-matter teaching material with developments from other courses;
- updating educational materials using version control systems.

It is assumed that this development will be implemented as part of the course on programming, data analysis and machine learning.

2 Intelligent Tutoring Systems

An intelligent tutoring system (ITS) is an e–learning system with elements of artificial intelligence that solves the tasks of drawing up individual plans for the educational process in accordance with the student's academic performance and includes intelligent decision analysis, assistance in completing tasks and intelligent monitoring of the learning process [14]. Thus, ITS can provide students with the following in order to increase the effectiveness of their training:

- personalized materials;
- practical tasks and tests.

The organization of the educational process with the help of ITS consists in the consistent implementation of the following steps:

- a preparation of educational material;
- setting up interactive support for the process of solving practical problems.

3 Domain Model

A domain model is a set of knowledge that a student needs to learn. The following approaches exist for domain modeling [8]:

- rule-based systems;
- frame-based models;
- formal logical models;
- ontology-based models.

3.1 Rule-Based Models

Rule-based models represent knowledge in the form of a set of rules of the type <if a condition, then an action>, where the condition is a sentence that is searched in the knowledge base, and the action is an action performed upon successful search result [8]. The search is performed using an output machine that iterates through the rules using forward output or reverse. Most often, rule-based models are used in expert systems.

The advantages of rule-based models [8, 12]:

- a high modularity;

- a simplicity of creating, modifying, and understanding rules;
- a simplicity of the logical inference mechanism.

The disadvantages:

- an inconsistency of the knowledge structure of the system with the structure of human knowledge (does not allow to describe meta-knowledge and fuzzy logic);
- an absence of the possibility of describing the mutual relations of the rules.

3.2 Frame-Based Models

Frame-based models (Fig. 1) represent knowledge in the form of a frame name and a set of slots with a value for each slot (listing 1). The slot value can be the name of another frame, so you can create networks of frames. The frame may also contain fields for determining the slot type and special processing procedures [8].

Listing 1. Frame structure.

```
<frame name>:
<the name of the first slot> : <the value of the first slot>,
<the name of the second slot> : <the value of the second slot>,
----
<the name of the n-th slot> : <the value of the n-th slot>.
```

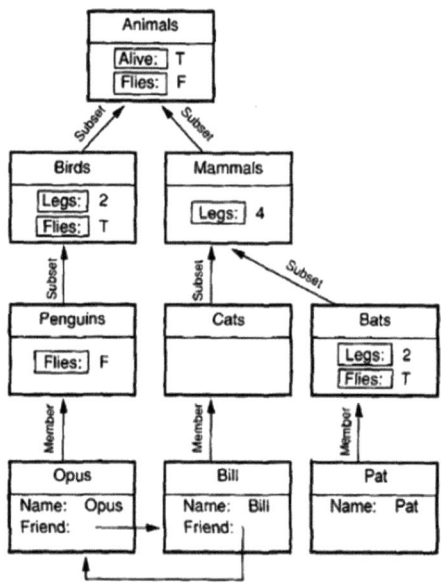

Fig. 1. Example of a frame model.

Frame-based models have the following advantages:

- an effective implementation of logical inference procedures;
- a possibility of non-monotonic output;
- a possibility of combining different models of knowledge representation to combine their advantages and compensate for disadvantages.

The disadvantages of frame-based models:

- deleting or including a new frame is a very laborious operation, as it involves removing all the component parts that may be components of other frames;
- the representation of time processes is a very complex and time-consuming procedure.

3.3 Formal Logical Models

Formal logical models represent knowledge in the form of a set of axioms, which imposes large restrictions on the subject area, so this approach is usually used in research expert systems. Formal logical models use first-order predicate logic [8].

3.4 Ontology-Based Models

Ontology-based models use the following components to represent knowledge:

- a concept (a concept or essence of a subject area);
- relationships between concepts;
- additional restrictions (axioms).

Semantic Networks. For ontology-based models, models are distinguished in the form of semantic networks (Fig. 2), representing ontologies in the form of a directed graph in which nodes correspond to concepts of ontology, and arcs correspond to relations between concepts [8].

The following groups of relationships can be distinguished in semantic networks [8]:

- logical;
- quantitative;
- hierarchical;
- functional;
- spatial;
- temporary;
- attributive.

The advantages of the semantic network [12]:

- a universality achieved through flexibility in choosing a set of possible relationships between concepts;
- the correspondence of the knowledge structure of the system to the structure of human knowledge (allows you to describe meta-knowledge and fuzzy logic),

The disadvantages:

- updating the knowledge base and modifying it is a very laborious operation.

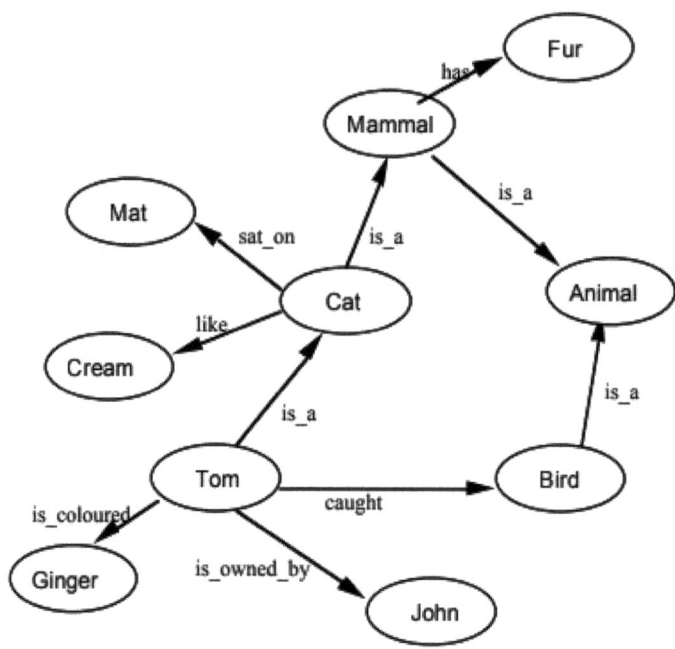

Fig. 2. An example of a semantic network.

4 Analysis of Existing Solutions

During the study of existing solutions, the following ITS were analyzed for compliance with the above functionality (systems for teaching technical sciences were considered):

- ZOSMAT (it is an ITS, primarily designed for teaching mathematics, but it can be adapted to technical sciences);
- OATutor;
- Thesis-ITS;
- GnuTutor;
- GuruTutor.

All the considered solutions do not have all the necessary functionality, in particular:

- there is no automated preparation of educational material or it is severely limited;
- there is no or limited module for submitting answers to questions;
- there is no version control system for educational materials;
- statistics on student academic performance are available, but there are no general statistics for a specific element of the educational materials of the domain model, and in addition, there is no assessment of the complexity of the teaching material;

Table 1. Comparative table of ITS.

Functional	ZOSMAT	OATutor	Thesis-ITS	GnuTutor	GuruTutor
Automated preparation of educational material	-	±	-	-	-
Providing educational material to students	+	+	+	+	+
Providing answers to students' questions	-	-	-	±	±
Preparation of individual educational plans	+	+	-	-	-
Analysis of student performance statistics and evaluation of educational materials	±	±	±	-	-
Notifying other teachers of the educational course about updating the thematic educational material with developments from other courses	-	-	-	-	-
Preparation of updated educational materials using version control systems	-	-	-	-	-

- there is no notification about the update of the associated educational material for teachers.

The results of the comparison can be seen in the Table 1 [4,9–11].

5 Architecture of an Intelligent Tutoring System

The architecture of the system under development (Fig. 3) consists of the following main components [5,6]:

- a domain model (storing elements of course materials);
- a student model (representation and storage of information about an individual student, including his knowledge, skills, learning preferences, misconceptions and other relevant characteristics);
- a mentor model (preparation of individual educational plans)
- a user interface model (a web interface for ITS interaction with a student).

In addition to the main components, a new component can be identified in the form of a module for working with educational material, responsible for:

- automated preparation of educational text and video materials;
- preparation of updated educational materials using version control systems;
- analysis of student performance statistics and evaluation of educational materials.

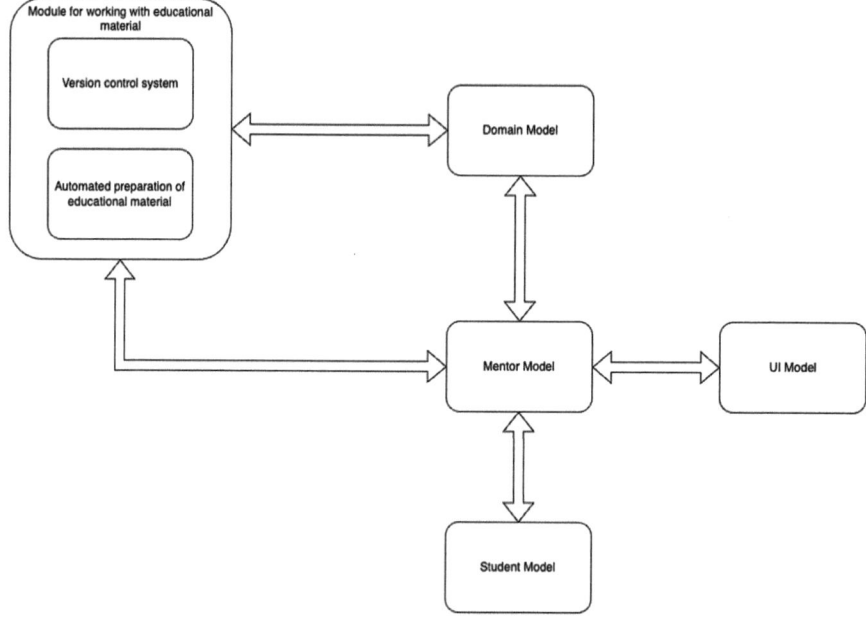

Fig. 3. ITS Architecture.

6 Automated Video Preparation

Automated video preparation allows you to generate video lectures based on a given text description. To prepare a text description, the Markdown [2] markup language is proposed. Using this language, presentation slides are described, which will then be automatically voiced using a neural network model.

The following tools were used for the software implementation of this component:

- Python programming language version 3.8 for the implementation of the CLI utility for generating video lectures;
- Silero TTS open neural network model for voice generation;
- Marp for generating presentation slide images;
- ffmpeg for overlaying voiceover on presentation slides.

7 Software Implementation of Automated Video Preparation

The process of preparing the video (Fig. 4) occurs as follows: the author of the course prepares the educational material in text form, sets the structure of the presentation, and then describes his presentation using a small dialect of the Markdown markup language, which was specially developed for this software.

This language variant is slightly different from the standard Markdown. Let's look at some differences:

– each title describes its own slide, the slides are separated from each other using triple hyphens;
– special control instructions are available for each slide, at the time of writing this article, only one instruction is available:
  • «/speech» - this control instruction allows you to set a speech for a specific slide to be voiced.

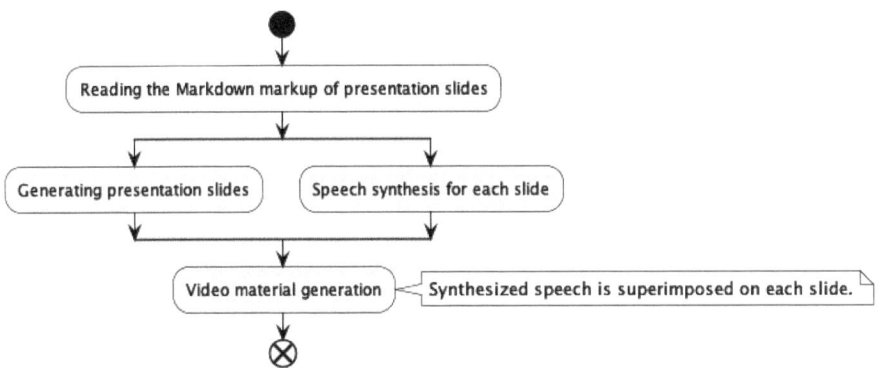

Fig. 4. UML diagram of video material generation activity

An example of presentation markup is a listing 3.

Listing 2. Markdown Example.

```
# Python

'''
print('Hello, world')
'''

/speech{This slide shows a simple program
written in the Python programming language. This program
simply outputs the specified words to the terminal.}

---

# Python

'''
a = 2
b = 4
print(a * b)
'''

/speech{And here is another program
that multiplies the number two by the number four.}
```

Listing 3. Markdown Example.

```
# Machine Learning

- Supervised Learning;
- UnSupervised Learning.

/speech{Let's look at two machine learning approaches}

---

# Training

X = (x1, x2, x3)
x1 = (1, 2, 3)
y = (0, 0, 1)

/speech{And here is example of dataset.}
```

Various approaches have been considered for voicing slides:

- Coque TTS—Python library for generating speech from a given text (in particular, the neural network model xtts_v2 participated in testing);
- Google Text-To-Speech—Google's Internet service for speech generation;
- Silero TTS—a pre-trained neural network model for speech generation [3], presented by Silero.

To compare the approaches, testing was conducted on the perception of speech by students. As a result, Silero TTS was chosen as a less robotic voice compared to others and one of a small number of models that was trained for speech synthesis in Russian. Silero TTS is based on the Tacotron 2 architecture, which is a modified sequence to sequence [13] architecture by the Accentor approach, which solves the problem of stress placement by manually controlling the stress placement by the user [7].

The implemented software provides two types of interface (Fig. 5) for automatic generation of video lectures:

- CLI;
- Web-interface (an example of working with the Web interface is shown in Fig. 6).

8 Conclusion

As a result of the work, the following was done:

- a comparative analysis of the existing ITS for compliance with the required functionality has been carried out;
- the system architecture has been developed;
- the component of automated video material generation is implemented programmatically.

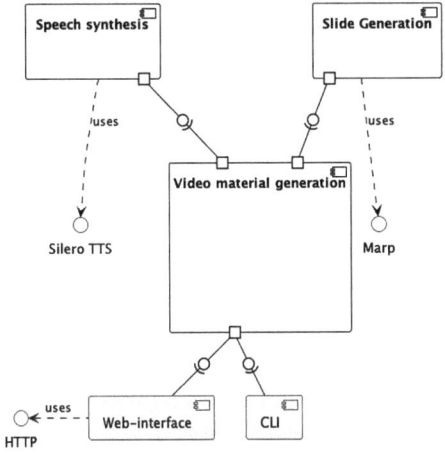

Fig. 5. UML diagram of system components.

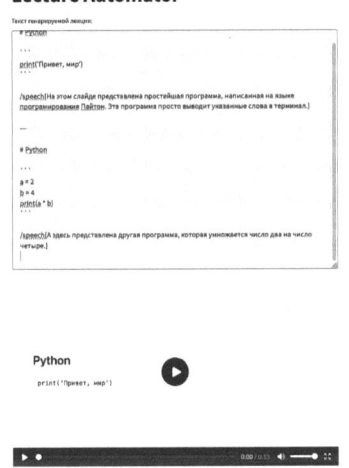

Fig. 6. Screenshot of video generation in the web interface.

For further development of the system, in addition to the implementation of the ITS components, it is planned:

- implementation of new control instructions of Markdown markup;
- implementation of automated preparation of practical tasks.

References

1. How intelligent tutoring systems are changing education. https://medium.com/@roybirobot/how-intelligent-tutoring-systems-are-changing-education-d60327e54dfb, Accessed 10 June 2023
2. Lecture automator. https://github.com/CapBlood/lecture-automator, Accessed 30 May 2023
3. Silero models. https://github.com/snakers4/silero-models, Accessed 10 June 2023
4. Thesis-its. https://github.com/robertoguazon/Thesis-ITS/wiki, Accessed 10 June 2023
5. Alesheva: Intellektual'nye obuchayushchie sistemy. Vestnik universiteta (1), 149–155 (2018)
6. Alkhatlan, A., Kalita, J.: Intelligent tutoring systems: a comprehensive historical survey with recent developments. arXiv preprint arXiv:1812.09628 (2018)
7. Geneva, D., Shopov, G., Garov, K., Todorova, M., Gerdjikov, S., Mihov, S.: Accentor: an explicit lexical stress model for TTS systems
8. Karpenko: Model'noe obespechenie avtomatizirovannyh obuchayushchih sistem. obzor. Mashinostroenie i komp'yuternye tekhnologii (7), 12 (2011)
9. Keleş, A., Ocak, R., Keleş, A., Gülcü, A.: Zosmat: web-based intelligent tutoring system for teaching-learning process. Expert Syst. Appl. **36**(2), 1229–1239 (2009)
10. Olney: Gnututor: an open source intelligent tutoring system based on autotutor. In: 2009 AAAI Fall Symposium Series (2009)
11. Pardos, Z.A., Tang, M., Anastasopoulos, I., Sheel, S.K., Zhang, E.: Oatutor: an open-source adaptive tutoring system and curated content library for learning sciences research. In: Proceedings of the 2023 Chi Conference on Human Factors in Computing Systems, pp. 1–17 (2023)
12. Popov: iskusstvennyj intellekt: modeli i metody. Radio i svyaz' (1990). https://books.google.ru/books?id=B7HTwAEACAAJ
13. Shen, J., et al.: Natural TTS synthesis by conditioning wavenet on mel spectrogram predictions. In: 2018 IEEE International Conference on Acoustics, Speech and Signal Processing (ICASSP), pp. 4779–4783. IEEE (2018)
14. Yurkov: Intellektual'nye komp'yuternye obuchayushchie sistemy. Penza: Izd-vo PGU (2010)

Application of Stochastic Petri Nets in a Network-Centric Approach to the Organization of the Emergency Response Process

Viktor A. Drogovoz[✉]

Federal Research Center "Computer Science and Control" of the Russian Academy of Sciences, Moscow, Russia
`vdrog@mail.ru`

Abstract. Initially, the use of network-centric approaches in process management appeared in the military sphere, and then spread to public administration and other sectors of the economy, including healthcare. The introduction of network-centric approaches in the interaction of emergency response processes places additional requirements on interoperability at the organizational, semantic and technical levels. Therefore, the optimization of interaction between participants in the elimination of the consequences of an emergency situation remains an urgent problem. The paper presents a stochastic model of emergency response, analyzes various states of the model, and provides recommendations for improving the effectiveness of both the process itself as a whole and information interaction between participants in the process.

Recommendations are presented on the key parameters of the stochastic model of emergency response to support decision-making in conditions of limited time and resources.

Keywords: Stochastic Petri net · emergency · network-centric system · interoperability · telemedicine system · disaster

1 Introduction

The network-centric approach to operations was tested in the military sphere by the countries of the NATO bloc and then scaled up to other sectors of the economy. The theory of a network-centric approach to warfare is based on three basic principles [1], the essence of which is to increase the information interaction of forces and, as a result, a sharp increase in the effectiveness of combat missions. Scaling this approach to other sectors of the economy, including emergency services and healthcare, actually boils down to informatization and improving the efficiency of information exchange. The works [2, 3] present solutions to improve the process of victim assistance using mobile telemedicine complexes. At the same time, the problem of decision-making support in the processes of eliminating the consequences of an emergency situation remains an urgent scientific and technical task.

Network-centric campaigns in crisis management.

In [4–7], some aspects of the application of network-centric approaches in an emergency situation are presented, namely:

- forces and means to counter disasters and risks should be combined into a single system that carries out operations;
- implementation of parallel strategies and formation of parallel operational actions to ensure their dispatch and synchronization.

Network-centric approaches in the activities of emergency services.

The paper [8] presents the main indicators of the quality of functioning of the network-centric management system of the Ministry of Emergency Situations of Russia. As the main indicator, it is proposed to use the indicator of minimizing the management cycle for the elimination of the consequences of an emergency situation while maximizing the information transmitted to participants in the liquidation of an emergency situation and minimizing the costs of completing the management cycle.

In [9], various concepts of applying network-centric approaches to the organization of emergency services are proposed, where information interaction plays a leading role both between different groups or within the same departmental subordination.

In addition to ensuring information interaction (interoperability) between the participants in the process, one of the ways to implement the network-centric approach is the integration of information resources, as shown in [10].

Network-centric approaches in healthcare.

In [11], an example of network-centric provision of medical units is presented by creating a single information space for medical support in order to heal the wounded and sick as quickly as possible. There are also known approaches to the implementation of network-centric elements in the management of medical support for troops [12].

The experience of using Petri nets in modeling processes related to disaster response.

Petri nets have proven themselves well enough as a tool for modeling discrete event processes and parallel structures. In relation to the field under consideration, the works [13–15] consider stochastic Petri nets for the analysis of processes related to the allocation of resources in the aftermath of disasters. The difference of this study is that it considers a network-centric approach to organizing the process of eliminating the consequences of an emergency situation, where such a criterion as information interaction of participants comes first. Also in this paper, the hypothesis is formulated that with a network-centric approach, the process management cycle is shortened, including due to interoperability between information systems.

In [16], a generalized network-centric control model based on a Petri net of a discrete event system is considered.

Thus, despite the work in the field of modeling on Petri nets, the task of modeling the process of eliminating the health consequences of an emergency using network-centric approaches was not considered independently.

2 Materials and Methods

2.1 A Network-Centric Model of Emergency Response Based on a Stochastic Petri Net

It was shown in [17] that the use of telemedicine consultations in the focus of an emergency situation significantly increases the accuracy of diagnosis and reduces the time to provide qualified medical care.

However, as shown in [18], in difficult cases, it is necessary to consult either with the regional center or with a federal organization, since "narrow" expert doctors are required. In this case, it is necessary to connect to other telemedicine centers in conditions of severe shortage of time resources.

Moreover, taking into account information interaction, in parallel with the provision of pre-medical and first aid in the center of an emergency, hospital facilities are being deployed for subsequent patient admission.

Figure 1 shows the process of emergency response using telemedicine consultations in the form of a Petri net.

There may be several parallel processes, to simplify calculations, the "branch" of transmitting information to the situation center (operational headquarters, government authority), interaction with the media, etc., was not specified, because this greatly increases the calculation part in possible markings, but is not essential for understanding the essence of the model. If the model is refined by information systems of other departments, it is also necessary to add positions responsible for the state of interoperability at levels (technical, semantic and organizational), as well as transitions with the intensity of ensuring interoperability between various heterogeneous information systems operating in parallel in order to eliminate the consequences of an emergency situation.

Model positions:

P1 – Victim detected.

P2 – Formation of an electronic medical record of the victim with the formulation of a request for telemedicine consultation.

P11 – Preparing a response to a telemedicine consultation request.

P12 – Data from a clarifying telemedicine consultation has been received.

P4 – Preparing the victim for evacuation to the hospital.

P3 – Waiting for first aid.

P5 – Waiting for evacuation measures.

P6 – Monitoring of vital signs during evacuation.

P7 – Arrival of the evacuation team at the hospital.

P8 – The hospital received information about the evacuation of the victim to them.

P9 – The hospital is ready to receive the victim.

P10 – The evacuation medical team issued a report on the victim.

Table 1 provides a description of the T transitions and the average transition intensities λ. A number of intensities regarding the time of telemedicine consultation were previously obtained in the work [17], the remaining parts are from standards for medical care and empirical data on working with medical information systems.

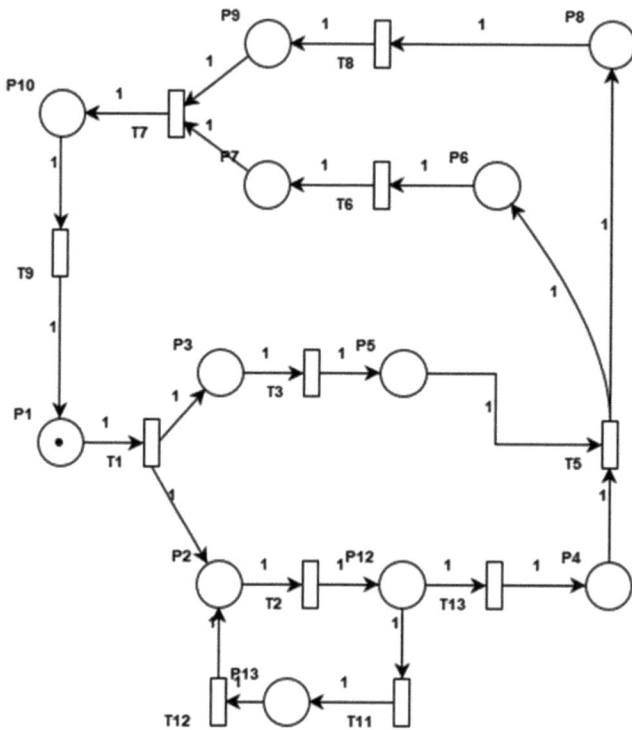

Fig. 1. The process of eliminating the consequences of an emergency situation using telemedicine consultations in the form of a Petri net.

Table 1. Transitions and transition intensities.

Transition, T	The functionality being performed	Transition intensity, λ (1/min)
T1	Transfer of the victim's service to the medical team	1/1
T2	Forming and sending a request for a telemedicine consultation	1/8
T11	Conducting a clarifying telemedicine consultation in a higher-level medical organization (if necessary in difficult cases)	1/10
T12	Sending the results of a clarifying telemedicine consultation to doctors working in the center of an emergency	1/2
T13	Application of the recommendations received from telemedicine consultations in the care of the victim in the center of an emergency situation	1/10

(*continued*)

Table 1. (*continued*)

Transition, T	The functionality being performed	Transition intensity, λ (1/min)
T3	Conducting first aid in the center of an emergency situation	1/10
T5	Evacuation of the victim to the hospital	1/20
T6	Transfer of data on the condition of the victim and the required treatment facilities to the hospital	1/5
T8	Preparation of the required capacities in the hospital for the maintenance of the victim	1/10
T7	Moving the victim from the evacuation team to the hospital service	1/2
T9	The return of the evacuation team to the place of emergency	1/20

3 Results and Discussion

Considering that M. Molloy's theorem [19] on the isomorphism of stochastic Petri nets and Markov chains with continuous time expands the possibilities of quantitative analysis of stochastic Petri nets in steady-state (stationary) mode, this will allow us to evaluate the parameters of efficiency, productivity of individual process elements, the architecture of a network-centric system, to assess the impact of time ensuring interoperability between various participants in the process on the duration of the liquidation of the consequences of an emergency situation.

The graph of the ergodic Markov chain corresponding to the stochastic Petri net from Fig. 1 is shown in Fig. 2.

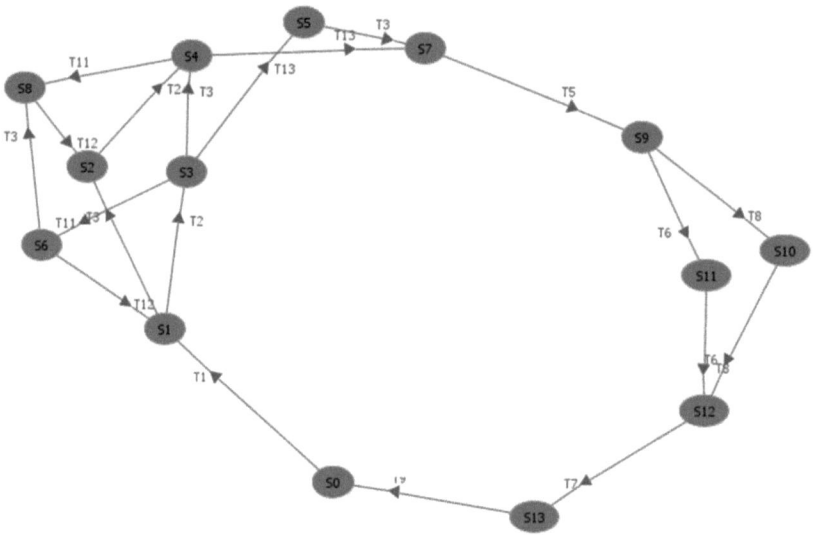

Fig. 2. Graph of an ergodic Markov chain corresponding to a constructed Petri net.

Using the T1–T9 transition intensities specified in Table 1 and the Markov chain analysis apparatus, a matrix of arc intensities corresponding to the graph in Fig. 2 is constructed.

$$M = \begin{pmatrix} -\lambda 1 & \lambda 1 & 0 & 0 & 0 & 0 & 0 & 0 & 0 & 0 & 0 & 0 & 0 & 0 \\ 0 & -(\lambda 2+\lambda 3) & \lambda 3 & \lambda 2 & 0 & 0 & 0 & 0 & 0 & 0 & 0 & 0 & 0 & 0 \\ 0 & 0 & -\lambda 2 & 0 & \lambda 2 & 0 & 0 & 0 & 0 & 0 & 0 & 0 & 0 & 0 \\ 0 & 0 & 0 & -(\lambda 13+\lambda 11+\lambda 3) & \lambda 3 & \lambda 13 & \lambda 11 & 0 & 0 & 0 & 0 & 0 & 0 & 0 \\ 0 & 0 & 0 & 0 & -(\lambda 11+\lambda 13) & 0 & 0 & \lambda 13 & \lambda 11 & 0 & 0 & 0 & 0 & 0 \\ 0 & 0 & 0 & 0 & 0 & -\lambda 3 & 0 & \lambda 3 & 0 & 0 & 0 & 0 & 0 & 0 \\ 0 & \lambda 12 & 0 & 0 & 0 & 0 & -(\lambda 12+\lambda 3) & 0 & \lambda 3 & 0 & 0 & 0 & 0 & 0 \\ 0 & 0 & 0 & 0 & 0 & 0 & 0 & -\lambda 5 & 0 & \lambda 5 & 0 & 0 & 0 & 0 \\ 0 & 0 & \lambda 12 & 0 & 0 & 0 & 0 & 0 & -\lambda 12 & 0 & 0 & 0 & 0 & 0 \\ 0 & 0 & 0 & 0 & 0 & 0 & 0 & 0 & 0 & -(\lambda 8+\lambda 6) & \lambda 8 & \lambda 6 & 0 & 0 \\ 0 & 0 & 0 & 0 & 0 & 0 & 0 & 0 & 0 & 0 & -\lambda 6 & 0 & \lambda 6 & 0 \\ 0 & 0 & 0 & 0 & 0 & 0 & 0 & 0 & 0 & 0 & 0 & -\lambda 8 & \lambda 8 & 0 \\ 0 & 0 & 0 & 0 & 0 & 0 & 0 & 0 & 0 & 0 & 0 & 0 & -\lambda 7 & \lambda 7 \\ \lambda 9 & 0 & 0 & 0 & 0 & 0 & 0 & 0 & 0 & 0 & 0 & 0 & 0 & -\lambda 9 \end{pmatrix} \quad (1)$$

Considering that mathematician A.N. Kolmogorov confirmed that to describe a homogeneous Markov process, a system of differential equations can be used to calculate the probability of states

$$\frac{dP(t)}{dt} = P(t)M \qquad (2)$$

Since the $P_\phi(t)$ vector does not change in the stationary state of the system, the derivative is zero, and the probability vector of finding the system can be found, we solve a system of differential equations.

$$P_\phi M = 0, \qquad (3)$$

Application of Stochastic Petri Nets in a Network-Centric Approach 241

M – the matrix of arcs of intensities.

One of the equations of the system must be replaced by the equation of the sum of all probabilities of finding the states of the system, which is equal to one.

For the example under consideration, the initial system of Kolmogorov equations will have the form:

$$\begin{cases} -p_0 + 0.05p_{13} = 0 \\ p_0 - 0.225p_1 + 0.5p_6 = 0 \\ 0.1p_1 - 0.125p_2 + 0.5p_8 = 0 \\ 0.125p_1 - 0.3p_3 = 0 \\ 0.125p_2 + 0.1p_3 - 0.2p_4 = 0 \\ 0.1p_3 - 0.1p_5 = 0 \\ 0.1p_3 - 0.6p_6 = 0 \\ 0.1p_4 + 0.1p_5 - 0.05p_7 = 0 \\ 0.1p_4 + 0.1p_6 - 0.5p_8 = 0 \\ 0.05p_7 - 0.3p_9 = 0 \\ 0.1p_9 - 0.2p_{10} = 0 \\ 0.2p_9 - 0.1p_{11} = 0 \\ 0.2p_{10} + 0.1p_{11} - 0.5p_{12} = 0 \\ 0.5p_{12} - 0.05p_{13} = 0 \end{cases} \qquad (4)$$

Replacing the last equation with the linear independence condition, we obtain the following system for solving

$$\begin{cases} -p_0 + 0.05p_{13} = 0 \\ p_0 - 0.225p_1 + 0.5p_6 = 0 \\ 0.1p_1 - 0.125p_2 + 0.5p_8 = 0 \\ 0.125p_1 - 0.3p_3 = 0 \\ 0.125p_2 + 0.1p_3 - 0.2p_4 = 0 \\ 0.1p_3 - 0.1p_5 = 0 \\ 0.1p_3 - 0.6p_6 = 0 \\ 0.1p_4 + 0.1p_5 - 0.05p_7 = 0 \\ 0.1p_4 + 0.1p_6 - 0.5p_8 = 0 \\ 0.05p_7 - 0.3p_9 = 0 \\ 0.1p_9 - 0.2p_{10} = 0 \\ 0.2p_9 - 0.1p_{11} = 0 \\ 0.2p_{10} + 0.1p_{11} - 0.5p_{12} = 0 \\ p_0 + p_1 + p_2 + p_3 + p_4 + p_5 + p_6 + p_7 + p_8 + p_9 + p_{10} + p_{11} + p_{12} + p_{13} = 1 \end{cases}$$
(5)

Solving the system of Eqs. (5), for example, by the method of inverse matrices [20] in the MATLAB environment, we obtain a vector of probabilities of the process of eliminating the consequences of an emergency in a stationary state.

$$P = \{0.012;\ 0.062;\ 0.127;\ 0.026;\ 0.092;\ 0.026;\ 0.004;\ 0.236;\ 0.019;\ 0.039;\ 0.02;\ 0.079;\ 0.024;\ 0.236\} \tag{6}$$

To estimate the average time spent by the victim in the system, it is possible to apply Little's law, adapting it to the considered model of a stochastic Petri net:

$$L = \lambda W \tag{7}$$

In our case, L is the average number of tokens in the system, W is the average time a token stays in the system, and λ is the average frequency of tokens entering the system.

Let T be the average time the token stays in the system. Then, using Little's law, we write down the formula (8) to calculate the average time spent on servicing the victim by the participants in the process (including the time of the evacuation team's return to the site) in the simulated emergency response process:

$$T = \frac{\overline{N}}{V(t, p)} = \frac{\sum_j j \times P[M\,(p_i) = j]}{W(t, p) U(t) \lambda} \tag{8}$$

W(t,p) is the weight of the Petri net arc from position t to position p (everywhere equal to 1).

λ - the frequency of the input stream from the transition U(t1), which is equal to 1 min.

U(t) is the bandwidth (workload) of the input junction, that is, U(t1).

The transfer capacity can be calculated using the formula (9):

$$U(t) = \sum_{M \in E} P[M] \tag{9}$$

Table 2 shows the values of the probability of the system positions being occupied according to Eq. (10):

$$P[M\,(p) = i] = \sum_j P[M_j],\ M_j \in [M_0 > M_j(p) = i \tag{10}$$

Table 2. The average number of tokens in the positions of the system.

Position number	The probability of a position being occupied by a token
P1	0,01178
P10	0,23569
P11	0,11785

(*continued*)

Table 2. (*continued*)

Position number	The probability of a position being occupied by a token
P12	0,02357
P2	0,18855
P3	0,11784
P4	0,2615
P5	0,47362
P6	0,05892
P7	0,10213
P8	0,11785
P9	0,04321

Using formula (8) and data from Table 2, we calculate the average time spent by the participants of the process on servicing the victim in the initial stages of emergency response:

$$T = \frac{1.74073}{0.01178 \times 1} = 147.7 \text{ minutes} \qquad (11)$$

The calculation using formula (10) does not take into account the probability of employment in the initial marking M0.

3.1 Practical Application of the Developed Model

By analyzing the probabilities of position occupancy and the throughput of transitions (loading) with chips, it is possible to find those transition intensities that most strongly affect the total time spent by the chip in the system.

This can be used to optimize and identify the most sensitive transitions whose throughput needs to be increased. This is achieved both by parallelizing chip maintenance and reducing chip processing time in the transition.

Another area of application may be the application of the theory of fuzzy calculations, since the initial data on the intensity of the transition are not always mathematically strictly calculated according to the laws of distribution and the average value. In this case, it is necessary to represent the transition intensities as fuzzy variables, which will make it possible to calculate the probabilities of the system staying in a stationary state as well as a fuzzy function, preferably of a triangular nature.

Since the article considers a network-centric approach to the organization of emergency response, the proposed model can be supplemented by the positions and transitions of other participants in the process (for example, authorities, situation centers, law enforcement agencies). In this case, it is necessary to introduce positions and transitions responsible for the process of ensuring interoperability, that is, positions responsible for organizational, semantic and technical levels of interoperability and transitions that ensure the achievement of a certain indicator of interoperability between different participants according to a certain criterion (for example – low, medium, high).

Then, the obtained fuzzy variables, including the location of the system in certain states of interoperability, can be correlated with a range of indicators of the levels of interoperability of information systems used in the aftermath of an emergency. Thus, knowing the indicators of system interoperability, it is possible to simulate the probability of finding a system in various positions responsible for the levels of interoperability.

For example, if the probability of finding a system in a position responsible for ensuring organizational interoperability (let's say - P15) lies in a certain range according to the theory of fuzzy sets, this will make it possible to estimate the required indicator of interoperability for the information resources used. In the future, to increase information interaction, it is necessary to refine the information system by adding software agents, semantic mediators, and gateways to ensure high-quality information interaction.

In the case of using a network-centric approach, the optimization of the emergency response process is complicated by the fact that it is necessary to reduce the time needed to eliminate the consequences of an emergency, but at the same time increase the awareness of participants and not significantly increase the costs of organizational and technical support for the network-centric approach. In the case of an overabundance of awareness, on the contrary, the management cycle will lengthen.

4 Conclusions

In the future, it is advisable to rank and optimize transition intensity indicators to minimize the load of resource-intensive model positions, for example, using fuzzy triangular functions. This will make it possible to optimize the total time in the model to eliminate the medical and sanitary consequences of an emergency.

To assess the qualitative and quantitative indicator of information interaction (interoperability), it is advisable to detail the states and transitions of the model, supplementing it with states of interoperability (at the technical, semantic, organizational levels) and mechanisms for its provision (instructions, semantic mediators, hardware gateways).

The proposed methodology can be scaled to assess information interaction with a heterogeneous interdepartmental group working in the center of an emergency situation.

In the case of a network-centric approach to emergency response, it is possible to use the considered model and an assessment of typical scenarios for emergency response (for example, based on retrospective statistical data) in order to assess the indicator of interoperability required for the effective functioning of a heterogeneous grouping.

In the future, the obtained model data on interoperability indicators at three levels (organizational, semantic, technical) can form the basis for improvements to the information systems used to operate the latter in a network-centric emergency response scenario.

References

1. Department of Defense. The Implementation of Network-Centric Warfare. Washington, D.C., 83 p. (2005). https://archive.org/details/DTIC_ADA446831. Accessed 27 Nov 2024

2. Drogovoz et al.: A model of a mass medical service system for emergency recovery based on mobile telemedical complexes. Biomed. Eng. **43**(1), 1–5 (2009)
3. Drogovoz, V.A., Berkovich, Yu.A., Orlov, O.I.: Choosing the optimal configuration of a mobile telemedicine complex for the tasks of eliminating the medical and sanitary consequences of emergencies according to the criterion of competitiveness. Aerosp. Environ. Med. **43**(N1), 57–62 (2009)
4. Trahtenherts, E.A.: Network-centric methods of computer counteraction to disasters and risks. UBS. 2013. No.41. https://cyberleninka.ru/article/n/setetsentricheskie-metody-kompyuternogo-protivodeystviya-katastrofam-i-riskam. Accessed 27 Nov 2024
5. Shershakov, V.M., Trakhtengerts, E.A., Kamaev, D.A.: Network-Centric Methods of Computer Support for Disaster Management. 1st edition. LENAND, Moscow, 160 p. (2015)
6. Trakhtenherts, E.A., Pashchenko, F.F.: Network-centric Management Methods in Large-Scale Networks. LENAND, Moscow, 200 p. (2016)
7. Trahtenherts, E.A.: The use of the network-centric principle of self-synchronization in management. Open Education, no. 2 (2015). https://cyberleninka.ru/article/n/ispolzovanie-setetsentricheskogo-printsipa-samosinhronizatsii-v-upravlenii. Accessed 27 Nov 2024
8. Maksimov, A.V., Ivanov, A.N.: Substantiation of the choice of quality indicators for the functioning of the network-centric management system of the Ministry of Emergency Situations of RUSSIA. In: Natural and Man-Made Risks (Physico-Mathematical and Applied Aspects), no. 1, pp. 5–10 (2014). https://journals.igps.ru/ru/nauka/article/68573/view. Accessed 27 Nov 2024
9. Ryazanov, V.A.: A network-centric approach in the management of fire protection forces. Fires and Emergencies, no. 4 (2010). https://cyberleninka.ru/article/n/setetsentricheskiy-podhod-v-upravlenii-silami-pozharnoy-ohrany. Accessed 27 Nov 2024
10. Barkovsky, S.A., Spiridonov, A.V., Ovsyanik, A.I., Semikov, V.L., Belkin, K.A., Godlevsky, P.P.: A network-centric approach to the prevention and elimination of forest fires based on the integration of information resources and systems. Technosphere Safety Technologies **3**(89), 98–109 (2020). https://doi.org/10.25257/TTS.2020.3.89.98-109
11. Krainyukov, P., Kuandykov, M., Singilevich, D., Abashin, V.: Informatization of medical supply of the armed forces in military conflicts based on the concept of netcentric control. Russ. Military Med. Acad. Reports **39**, 31–38 (2020). https://doi.org/10.17816/rmmar78240
12. Kalachev, O.V., Kraynyukov, P.E., Stolyar, V.P., Ovcharov, O.M., Voloshko, D.I., Katulin, A.N.: Possibilities of implementing a network-centric approach in managing medical support for troops in new type wars. Military Med. J. **343**(5), 4–16, 16 May 2022. https://doi.org/10.52424/00269050_2022_343_5_04
13. Sun, H., Liu, J., Han, Z., Jiang, J.: Stochastic petri net based modeling of emergency medical rescue processes during earthquakes. J. Syst. Sci. Complexity **34**, 1–24 (2021). https://doi.org/10.1007/s11424-020-9139-3
14. Zhang, Q., Wang, X., Jin, Y., Luo, X.: Modeling and analysis of postdisaster aviation medical rescue process using SPN-MC. J. Adv. Transp. (2022). https://doi.org/10.1155/2022/7123594
15. Xiao, L., Shi, Y., Wang, Z., Zhang, W.: Modified stochastic petri net-based modeling and optimization of emergency rescue processes during coal mine accidents. Geofluids 1–13 (2021). https://doi.org/10.1155/2021/4141236
16. Ambartsumyan, A.: Network-centric control based on Petri nets in the structured discrete-event system. Autom. Remote Control **73** (2012). https://doi.org/10.1134/S000511791207 0120.
17. Drogovoz, V.A.: Improving the process of servicing victims in emergency situations using mobile telemedicine complexes : dissertation ... Candidate of Technical Sciences : 05.26.02 / Drogovoz Viktor Anatolyevich; [Place of protection: Institute of Medical Biology. problems], Moscow, 130 p. (2009)

18. Jabbar, A., Muhammad, Hussain, M.: Formal Modeling and Analysis of Integrated Healthcare System using Colored Petri Nets, **10**, 211–226 (2022)
19. Molloy, M.K.: Performance analysis using stochastic petri nets. IEEE Trans. Comp. **C–3**(9), 913–917 (1982)
20. Electronic resource. http://solidstate.karelia.ru/p/tutorial/meth_calc/files/matlab2.shtml. Accessed 27 Nov 2024

A Self-organization Intelligent Model in Network-Centric Systems Based on Coalition-Structural Synthesis

V. A. Serov[(⊠)], D.A. Kozlov, O.V. Trubienko, S.A. Skaev, and S.A. Tepsikoev

MIREA – Russian Technological University, Moscow, Russia
ser_off@inbox.ru

Abstract. The article formulates the problem statement of an optimal heterogeneous coalition structure forming in a network-centric system at the form of a discrete multi-criteria optimization problem. A method of multicriteria lexicographic coalition-structural synthesis for a network-centric system has been developed. The main idea of the proposed method is to introduce priorities on a set of tasks to be solved and transform the initial problem into a sequence of p[roble ms for forming individual coalitions. A hypervector effectiveness criteria of the coalition structure has been formed, which includes vector effectiveness criteries of heterogeneous coalitions. Graph-theoretic metric of betweeness centrality and game-theoretic metric of centrality by Shapley vector are used to calculate the components of vector effectiveness criteria for a heterogeneous coalitions. Solving the problem of an optimal heterogeneous coalition structure forming makes it possible to implement an intelligent mechanism of self-organization in network-centric systems under situational uncertainty.

Keywords: The intellectual model of self-organization · network-centric system · lexicographic coalition-structural synthesis · multicriteria optimization · Shapley vector · betweenness centrality

1 Introduction

One of the key problems of the group control methodology is the models of structural adaptation development of group control systems (GCS), providing the ability to solve a wide class of tasks under situational uncertainty [1–8]. In [1,4–12], the perspective and universal nature of game approaches are noted, which allow taking into account various types of uncertain factors: a purpose uncertainty , the external environment uncertainty and conflict uncertainty in the models and methods development of GCS structural adaptation. At the same time, within the framework of game approaches, the mechanisms of coalition-structural adaptation can naturally be implemented in the form of models based on the formation of heterogeneous coalition structures of the GCS [1,8,13,14], focusing on a set of tasks solving under uncertainty of the external environment.

Currently, the methodology of network-centric systems (NCS) is being actively introduced into the theory and practice of GCS [15–20], based on the following principles:

- formation of a unified network information and communication space;
- collection, integration, and up-to-date maintenance of heterogeneous information by using the maximum number of available sources of primary and operational information;
- formation of an environment of situational analysis, forecasting, decision-making and control in real time.

The implementation of the above principles within the framework of combining game-theoretic and graph-theoretic approaches makes it possible to form an intellectual model of self-organization of the NCS [13–15], which provides the opportunity:

- formation and changes in the composition of heterogeneous coalitions and the coalition structure of the NCS based on situational analysis;
- taking into account the various types of conflict interaction between coalitions within the coalition structure of NCS;
- combining of conflict equilibrium principles in order to form stable and effective group control strategies;
- optimal communication NCS resources distribution between coalitions in real time.

The problem of a NCS heterogeneous coalition structure forming is one of the base ones, because its solution allows you to implement a self-organization mechanism in the NCS. At the same time, the effectiveness of each agent-member of the coalition and the coalition as a whole should be assessed by a vector criterion, the components of which should take into account their competence and communication characteristics.

The competence aspect characterizes the specialization of the coalition and its importance, determined by the set of competencies of its members. The communication aspect characterizes the information connectivity of the coalition members, as well as the degree of integration of the coalition into the information infrastructure of the NCS.

To estimate the competence capabilities of participants in a heterogeneous coalition and the entire coalition, [21–24] develops a game-theoretic concept of centrality based on the representation of a model of group interaction as a cooperative game in the form of a characteristic function.

At the same time, the Shapley vector, the Owen value, the Banzaf value, and others are used as one of the main game-theoretic metrics of centrality.

To estimate the communication properties of the elements of the NCS and their coalitions, a graph-theoretical concept of centrality is being developed [25–32] using a graph model of the NCS infrastructure.

The basic graph-theoretic metrics of centrality are degree centrality, eigenvector centrality, Katz centrality, betweeness centrality, proximity centrality, decay centrality, and others. All these metrics characterize the degree of control of the NCS elements over information flows.

Thus, the problem of forming a heterogeneous coalition structure of the NCS is a multi-criteria optimization problem. This class of problems has not been sufficiently studied in the modern literature, which makes it important to carry additional research.

2 The Problem Statement of NCS Multi-criteria Coalition-Structural Synthesis

Let's consider at the NCS infrastructure model in the form:

$$\Gamma_T = \langle \mathbf{N}, T, \mathbf{A}, v(K,T) \rangle. \tag{1}$$

The model (1) has the following features. From the one side Γ_T is an undirected weighted graph, where $\mathbf{N} = \{\overline{1,n}\}$ is the set of vertices of the graph; $\mathbf{A}$ is the adjacency matrix of the graph, characterizing the relationship between the vertices of the graph and their intensity. From the other side, it Γ_T is a cooperative game with a characteristic function $v(K,T)$, where $\mathbf{N} = \{\overline{1,n}\}$ is the set of players. It is assumed that cooperation between the players - elements of the NCS (vertices of the graph), is possible by forming coalitions $K \subset \mathbf{N}$ to achieve common goals in solving the task T. Thus, model (1) has a hybrid character and includes the communication and competence aspects of the functioning of the NCS.

It is advisable to evaluate the communication properties of the NCS elements on the basis of graph-theoretic centrality metrics. In particular, it is possible to use the normalized centrality indicator for betweeness [32]:

$$f_b(i) = \frac{2}{(|\mathbf{N}|-1)(|\mathbf{N}|-2)} \left(\sum_{(k,j)\in \mathbf{S}(i)} \frac{|\mathbf{N}_{kij}|}{|\mathbf{N}_{kj}|} \right), \tag{2}$$

where $|\mathbf{N}| = n$; $\mathbf{S}(i)$ is the set of all possible of graph nodes paired combinations (1) on the set $(\mathbf{N}\backslash i)$; $\mathbf{N}_{kji}$ is a shortest routes set for a paired combination of nodes (k,j) passing through the node i; $\mathbf{N}_{kj}$ is a shortest routes set for the nodes (k,j) paired combination; $\mathbf{N}_{kji} \subseteq \mathbf{N}_{kj}$.

The centrality indicator for betweenness characterizes the importance of the elements of graph (1) from point of view of control over the information flows of the infrastructure Γ_T.

It is proposed to evaluate the participant's competence properties in solving the task using an efficiency indicator calculated on the basis of the Shapley vector [19]:

$$f_{Sh}(i, T) = \sum_{\substack{K \subseteq N \\ i \in K}} \frac{(|K|-1)!(|N|-|K|)!}{|N|}, \qquad (3)$$

where $v(K, T)$ is the characteristic function of the cooperative game (1).

The type of characteristic function $v(K, T)$ reflects the specialization and relevant competencies of the player (the NCS element) and, as a result, his ability to form coalitions with other players (the NCS elements).

It is advisable to evaluate the significance of each element of the NCS $i \in N$ in terms of its impact on the entire infrastructure of the NCS when solving a task T using a vector criterion $\mathbf{F}(i, T) = [f_1(i), f_2(i, T)]^T$, where $f_1(i) = f_b(i)$, a $f_2(i, T) = f_{sh}(i, T)$.

$$\Phi(i, T) = \frac{1}{\left(1 + \frac{b_i(T)}{|N|-1}\right)^\psi}, \qquad (4)$$

where ψ is some positive integer; $b_i(T)$ - the number of participants $j \in N \setminus i$ for whom the condition

$$\mathbf{F}(i, T) \geq \mathbf{F}(j, T). \qquad (5)$$

The function $\Phi(i, T)$ has the following properties. For all elements $j \in N$ the inequality $\frac{1}{2^\psi} \leq \Phi(i, T) \leq 1$ is met. In this case, the maximum value $\Phi(i, T) = 1$ corresponds to the Pareto optimal elements $i \in N$. The minimum value $\Phi(i, T) = \frac{1}{2^\psi}$ corresponds to the element $i \in N$ with the minimum efficiency.

The efficiency index (4) allows for a multi-criteria ranking of elements of NCS $i \in N$.

It is proposed to formalize the statement problem of the coalition-structural synthesis of the NCS system in the form of a tuple:

$$\Gamma_S = \left\langle \mathbf{N}, \mathbf{T}, \mathbf{K}, \mathbf{F}^{\mathbf{K}(\mathbf{T})} \right\rangle. \qquad (6)$$

The components of task (6) have the following meaning:
$\mathbf{N} = \{\overline{1, n}\}$ - the set of elements (agents) of the NCS;
$\mathbf{T} = \{T_i, i \in \mathbf{L} = \{\overline{1, L}\}\}$ - the set of tasks assigned to the NCS, where each task $T_i, i \in \mathbf{L}$, is characterized by a set of competencies necessary for its solution

$\mathbf{K}$ - the coalition structure of the NCS, formed on a set $\mathbf{N}$, and defined as

$$\mathbf{k} = \left\{ K_1, ..., K_L \middle| K_i \bigcap_{i \neq j} K_j = \varnothing; \bigcup_{i \in \mathbf{L}} K_i = \mathbf{N_T} \subseteq \mathbf{N} \right\}, \qquad (7)$$

where K_i is an elements (coalition) group of the NCS combined to complete the task $T_i, i \in \mathbf{L}$;

$\mathbf{F^K(T)}$ - a vector criteria, that evaluates the effeciency of the coalition structure $\mathbf{K}$ in solving a set of tasks $\mathbf{T}$.

The vector criteria $\mathbf{F^K(T)}$ has the following structure:

$$\mathbf{F^K(T)} = \langle \mathbf{F}^{K_1 T}(T_1), ..., \mathbf{F}^{K_L T}(T_L) \rangle^T, \qquad (8)$$

$$\mathbf{F}^{K_i}(T_i) = \left[f_1^{K_i}(T_1), f_2^{K_i}(T_1) \right]^T, i \in \mathbf{L}. \qquad (9)$$

where $\mathbf{F}^{K_i}(T_i)$ - a vector criteria, that evaluates the effeciency of the coalition $K_i \mathbf{1 K}$ in solving the task (T - the symbol of the transposition operation).

The components of the vector $\mathbf{F}^{K_i}(T_i)$ are given, as

$$f_1^{T_i}(K_i) = v(K_i, T_i), \qquad (10)$$

$$f_2^{T_i}(K_i) = \sum_{j \in K_i} \Phi(j, T_i). \qquad (11)$$

The performance indicator (10) is interpreted as the readiness degree of the participants coalition K_i to perform the task T_i and characterizes the competence of the coalition K_i as a single player. The performance indicator (11) characterize the importance of a coalition K_i on a variety of possible coalitions $\mathbf{N^K}$.

Thus, task (6) is a discrete multi-criteria optimization problem in which it is required to form a coalition structure $\mathbf{K^*}$ of the form (7), providing the components of the vector efficiency criterion $\mathbf{F^K(T)}^T$ of the form (8)-(11) with the largest possible values.

3 The Method of the NCS Multi-criteria Lexicographic Coalition-Structural Synthesis

The main idea of this method is to introduce in problem (6) the tasks priority on the set $\mathbf{T} = \{T_i, i \in \mathbf{L}\}$. Let's say, for example,

$$T_1 \succ T_2 \succ ... \succ T_L \qquad (12)$$

Next, the initial task (6) is transformed into a multi-stage scheme, at the framework which the following sequence of tasks is solved.

Stage 1. The problem of an optimal coalition K_1^* forming on a set $\mathbf{N}_1 = \mathbf{N}$, designed to solve the problem is solving:

$$\Gamma_1 = \langle \mathbf{N}_1, T_1, K_1, \mathbf{F}^{K_1}(T_1) \rangle \Rightarrow K_1^*. \qquad (13)$$

Stage 2. The problem of an optimal coalition K_2^* forming on a set $\mathbf{N}_2 = \mathbf{N}_1 \setminus K_1^*$, designed to solve the problem is solving:

$$\Gamma_2 = \langle \mathbf{N}_2, T_2, K_2, \mathbf{F}^{K_2}(T_2) \rangle \Rightarrow K_2^*. \qquad (14)$$

Etc.

Stage L. The problem of an optimal coalition K_L^* forming on a set $\mathbf{N}_L = \mathbf{N}_{L-1}\setminus K_{L-1}^*$, designed to solve the problem is solving:

$$\mathbf{\Gamma}_L = \langle \mathbf{N}_L, \mathbf{T}_L, \mathbf{K}_L, \mathbf{F}^{K_L}(\mathbf{T}_L)\rangle \Rightarrow K_L^*. \tag{15}$$

The coalition structure

$$\mathbf{K}^* = \{K_1^*, ..., K_L^*\} \tag{16}$$

The key element of the multicriteria lexicographic coalition-structural synthesis method is the optimal coalition K_i^* forming task $\mathbf{\Gamma}_i$ designed to solve the problem $\mathrm{T}_i, i \in \mathbf{L}$.

The main problem $\mathbf{\Gamma}_i$ statements and algorithmic support are considered in [33, 34].

4 Conclusion

The article formulates the problem statement of an optimal forming of heterogeneous coalition structure in a network-centric system at the form of a discrete multi-criteria optimization problem.

A method of multicriteria lexicographic coalition-structural synthesis of a network-centric system has been developed.

A hypervector criteria for the effeciency of the coalition structure has been formed, which includes vector criteria for the effeciency of heterogeneous coalitions. Graph-theoretic metric of centrality by betweenness and game-theoretic metric of centrality by Shapley vector are used to calculate the components of the vector criteria for the effeciency of a heterogeneous coalition.

Solving the problem of forming an optimal heterogeneous coalition structure makes it possible to implement an intelligent mechanism of self-organization in network-centric systems under situational uncertainty.

References

1. Obnosov, B.V., Voronov, E.M., Mikrin, E.A., Serov, V.A., et al.: Stabilization, guidance, group control, and system modeling of unmanned aerial vehicles. Modern approaches and methods, vol. 2. Publishing House of Bauman Moscow state technical University, Moscow, p. 520 (2018). ISBN 978-5-7038-5058-9
2. Group control of moving objects in uncertain environments. In: Kh. Pshikhopov, V. M.: FIZMATLIT, p. 305. ISBN 978-5-9221-1674-9. (2015)
3. Kalyaev I.A. Models and Algorithms of Collective Control in Groups of Robots / Kalyaev I.A., Gayduk A.R., Kapustyan S.G. – Moscow, FIZMATLIT, 2009, p. 280. ISBN 978-5-9221-1141-6
4. Gorodetskiy, A., Tarasova, I. (eds.): Smart Electromechanical Systems. Situational Control/Studies in Systems, Decision and Control, vol. 261, p. 290. Springer, Cham (2020). 978-3-030-32710-1_9

5. Gorodetskiy A., Tarasova I. (eds.): Smart electromechanical systems. Behavioral Decision Making/Studies in Systems, Decision and Control, vol. 352, p. 240. Springer International Publishing (2021). 978-3-030-68172-2_13
6. Serov, V.A., Voronov, E.M., Kozlov, D.A.: Hierarchical population game models of machine learning in control problems under conflict and uncertainty. In: Gorodetskiy, A.E., Tarasova, I.L. (eds.) Smart Electromechanical Systems. SSDC, vol. 419, pp. 125–145. Springer, Cham (2022). https://doi.org/10.1007/978-3-030-97004-8_10
7. Gorodetskiy, A.E., Tarasova, I.L.: Smart Electromechanical Systems. SSDC, vol. 174. Springer, Cham (2019). https://doi.org/10.1007/978-3-319-99759-9
8. Smart Electromechanical Systems. Mathematical and Software Engineering / Studies in Systems, Decision and Control. Eds. Gorodetskiy A., Tarasova I. Vol. 544, Springer Nature, 2024, p 265. https://doi.org/10.1007/978-3-031-64277-7
9. Karpov I.E., Karpova I.P., Kulinich A.A. Social Communities of Robots. – Moscow, LENAND, 2019, p. 352. ISBN 978-5-9710-5757-4
10. Navarro I., Matha F.: Survey of Collective Movement of Mobile Robots. Int. J. Adv. Rob. Syst. 2013, vol. 10, No. 73 cdn.intechopen.com/pdfs/42383/InTech A survey of collective movement of mobile robots.pdf
11. Xu D., Zhang X., Zhu Z., Chen C., Yang P.: Behavior-Based Formation Control of Swarm Robots. Hindawi Publishing Corporation Mathematical Problems in Engineering. – 2014. – vol. 1. – P. 1–13
12. Ungureanu V. Pareto-Nash-Stackelberg Game and Control Theory: Intelligent Paradigms and Applications. - Springer, 2017. — 579, p. — (Smart Innovation, Systems and Technologies). — ISBN 10 3319751506, ISBN 13 978-3319751504
13. Smirnov A.V., Sheremetov L.B.: Models of forming coalitions of Cooperative Agents: state and Prospects of Research. Artif. Intell. Dec. Making. **2011**(1), pp. 36–48
14. Klusch M., Gerber A.: A dynamic coalition formation scheme for rational agents. IEEE Intell. Syst. pp. 42–47 (2002)
15. DeCristofaro, M., Lansdowne, C., Schlesinger, A.: Heterogeneous Wireless Mesh Network Technology Evaluation for Space Proximity and Surface Applications. SpaceOps 2014 Conference (2014). https://doi.org/10.2514/6.2014-1600
16. Diane S.A.K., Iskhakov A.Y., Iskhakova A.O.: Network-centric motion control algorithm for a group of mobile robots. Modeling, Optimization and Information Technology. 2022;10(1). Available from: https://moitvivt.ru/ru/journal/pdf?id=1146. https://doi.org/10.26102/2310-6018/2022.36.1.026. (In Russ.)
17. Mulyukha V.A., Guk M.Yu., ZaborovskyV.S.: Supervisory Network-Centric Control System for Robotic Objects // Robototehnika and Technicheskaya Kibernetika. 3(12)/2016, pp. 42–47
18. Sahingoz O.K.: Networking models in flying ad-hoc networks (FANETs): Concepts and challenges. J. Intell. Robo. Syst. – 2014. – T. 74. – (textnumero). 1-2. – pp. 513–527
19. Wang, J., Jiang, C., Han, Z., Ren, Y., Maunder, R., Hanzo, L.: Taking drones to the next level: cooperative distributed unmanned-aerial-vehicular networks for small and mini drones // IEEE Veh. Technol. Mag. **12**(3), 73–82 (2017)
20. Martynova L.A., Rozengauz M.B.: Efficient operation of a group of standalone unmanned submersibles in a network-centric system of underwater illumination. Informatsionno-upravliaiushchie sistemy [Information and Control Systems], 2017, no. 3, pp. 47–57 (In Russian). https://doi.org/10.15217/issnl684-8853.2017.3.47

21. Skibski O., Michalak T.P., Rahwan T.: Wooldridge M. Algorithms for the Shapley and Myerson values in graph-restricted games. Adaptive Agents and Multi-Agent Systems, 2014
22. Michalak T., Aadithya K., Szczepański P., Ravindran B., Jennings N.: Efficient Computation of the Shapley Value for Game-Theoretic Network Centrality. J. Artif. Intell. Res. — 2013. — vol. 46, pp. 607–650
23. Van Campen T., Hamers H., Husslage B., Lindelauf R.: A new approximation method for the Shapley value applied to the WTC 9/11 terrorist attack. Social Network Analysis and Mining, 2018, (textnumero)8(1)
24. Lindelauf R., Hamers H., Husslage B.: Game Theoretic Centrality Analysis of Terrorist Networks: The Cases of Jemaah Islamiyah and Al Qaeda. CentER Discussion Paper, 2011, (textnumero)107
25. Algaba E., Prieto A., Saavedra-Nieves A. Rankings in the Zerkani network by a game theoretical approach, p. 28 (2022). https://arxiv.org/abs/2202.07730v1
26. Newman, M.E.J.: Finding community structure in networks using the eigenvectors of matrices. Phys. Rev. E **74**(3), 036104 (2006). https://doi.org/10.1103/PhysRevE.74.036104
27. Van Mieghem P.: Graph spectra for complex networks. Camb.; N.Y.: Camb. Univ. Press, 2011. 346 p
28. Borgatti S.P., Everett M.G. A Graph-Theoretic Perspective on Centrality. Social Networks, 2005, (textnumero)28b, pp. 466–484
29. Yannick Rochat. Closeness centrality extended to unconnected graphs: the harmonic centrality index. Applications of Social Network Analysis, ASNA 2009. — 2009
30. Boldi P., Vigna S.. Axioms for Centrality. Internet Mathematics. — 2014. — T. 10, (textnumero). 3–4. — https://doi.org/10.1080/15427951.2013.865686
31. Mahendra Piraveenan. Percolation Centrality: Quantifying Graph-Theoretic Impact of Nodes during Percolation in Networks. PLOS ONE. — 2013. — T. 8, (textnumero). 1. — P. 53095. — https://doi.org/10.1371/journal.pone.0053095
32. Brandes U. A faster algorithm for betweenness centrality. J. Math. Soc. 2001, (textnumero)25, pp. 163–177
33. Serov V.A., Voronov E.M., Kurilenko Yu.Y.: Intellectual Model of Mobile Robots Coalitions Formation in Network-Centric Group Control Systems. In: Smart Electromechanical Systems. Mathematical and Software Engineering (Tarasova I.L., Kulik B.A., eds), vol. 544. Chapter 7, pp. 73–92. Springer Nature, (2024). https://doi.org/10.1007/978-3-031-64277-7
34. Serov V.A., Trubienko O.V., Skaev S.A., Tepsikoev S.A.: Evolutionary Computational Technology for a Heterogeneous Coalitions Multi-criteria Formation in Network-Centric Systems. (In this collection)

The Mathematical Model Identification Problem and Its Solving by Symbolic Regression

Askhat Diveev(✉), Sergey Kozlov, and Igor Prokopiev

Federal Research Center "Computer Science and Control" of the Russian Academy of Sciences, Moscow, Russia
aidiveev@mail.ru

Abstract. The control object mathematical identification problem is considered. It's supposed, that a mathematical model is an ordinary differential equations system in Cauchi form. Therefore, the problem consists of finding right parts of differential equations system. Machine learning by symbolic regression is used for solving this problem. An example of quad-rotor spatial motion mathematical model identification is presented. Synthetic data obtained from known mathematical model are used for identification of model in the example. To obtain the training data some optimal control programs are set, and then a right parts of differential equation system a found by symbolic regression. To check a quality of identification a new optimal control problem was solved for found mathematical model. After that the found optimal solution in the form of optimal control program is used for control of the known control object model. Results of simulation are compared with a solution of the same optimal control problem for the known mathematical model.

Keywords: identification problem · symbolic regression · optimal control · evolutionary computations

1 Introduction

Mathematical models of control objects are usually applied for construction of control systems and for simulation to study control object properties. Almost all mathematical models are derived from physical laws. To recently time it was considered, that mathematical model of any control object should be how can be more simpler. Then this model easier to study and to use. Despite the development of computer technology in our time these requirements to mathematical models have not changed much. Everybody knows that any mathematical model always describes a real object with some error. Then why is need complex high accuracy mathematical model, if it isn't in any case accuracy. It means, that calculated on a model a control system at implementing on a real object will always be refined. Last studies showed, that an accuracy mathematical model of a control object has the following advantages. If control problem was solved for a more accurate model, then fewer changes will be required when implementing this solution in a real object. The results of modeling the solution of the control problem for the exact model of the control object show more realizable behavior of the object than solutions for the simplified model.At last, a high accuracy control object mathematical model of sometimes can be used in open loop control system for defining coordinates of control object instead of the navigation system during some small time. Today to obtain the accuracy mathematical control object model artificial neural network is often used [1–3]. In this work symbolic regression is used [4].

The main requirements for building a simplified model of a control object are related to the stage of implementing control in a real object. Control synthesis problem is solved for control object model, i.e. control is defined as a function of control object state vector component. In most cases, the control synthesis task is solved manually. The control system designer analyzes the model of the control object and establishes control channels, i.e. determines which controls affect a certain movement of the object. Further, the developer inserts certain regulators into these control channels, which must generate control signals based on the magnitude of the object movement error to compensate for this error. In some cases, analytical methods of synthesis [5,6] are used to synthesize the control. In both cases, it is better to use a simple model of the control object for both manual analysis of the model and for analytical methods of synthesis.

Today, the control synthesis problem can be solved automatically based on machine learning of symbolic regression methods. In this case, for the computational method of synthesis, the mathematical model of the control object can be quite complex. If the model of the control object evaluates the behavior of the object with high accuracy, then the automatic synthesis of the control system using such a model allows you to obtain control functions, the installation of which in the control system of a real object does not require or almost does not require their clarification.

2 The Identification of Control Object Model Problem Statement

It is known, that a mathematical model of control object has the form of ordinary differential equation system. There is a control vector In the right side of the system. The control object movements ibn 3D geometric space. The control vector acts to the control object and changes second derivation of its motion path. In general case a mathematical model of control object has the following form

$$\ddot{\mathbf{x}} = \mathbf{f}(\mathbf{u}), \tag{1}$$

where $\mathbf{x}$ is a state vector of control object, $\mathbf{x} = [x_1 \ x_2 \ x_3]^T$, $\mathbf{u}$ is a control vector, in considered problem dimension of control vector is known, it is $m = 4$, $\mathbf{u} = [u_1 \ldots u_4]^T$. Constant gravity acceleration is contained on the right side of the mathematical model. Therefore, a mathematical model of control object spatial motion has the following preliminary form

$$
\begin{aligned}
\dot{x}_1 &= x_4, \\
\dot{x}_2 &= x_5, \\
\dot{x}_3 &= x_6, \\
\dot{x}_4 &= f_1(\mathbf{u}, g), \\
\dot{x}_5 &= f_2(\mathbf{u}, g), \\
\dot{x}_6 &= f_3(\mathbf{u}, g),
\end{aligned}
\tag{2}
$$

where g is constant gravity acceleration, $g = 9.80665$ m/s^2.

The restrictions on the control vector are known.

$$\mathbf{u}^- \leq \mathbf{u} \leq \mathbf{u}^+, \tag{3}$$

where $\mathbf{u}^-$, $\mathbf{u}^+$ are the given low and up restrictions on the control vector values.

The set of control programs is given

$$U^* = \{\mathbf{u}^{*,1}(t), \ldots, \mathbf{u}^{*,K}(t)\}, \tag{4}$$

where $0 \leq t \leq t_{f,i}$, $t_{f,i}$ is a given limit time for the program control i, $i = 1, \ldots, K$.

The set of initial states of the control object for each the control program is given

$$X^* = \{\mathbf{x}^{*,1}(t), \ldots, \mathbf{x}^{*,K}(t)\} \tag{5}$$

Two quality criteria are given

$$J_1 = \sum_{i=1}^{K} \left(\int_0^{t_{f,i}} \|\mathbf{x}^{*,i}(t) - \mathbf{x}(t, \mathbf{x}^{0,i})\| dt \right) \to \min, \tag{6}$$

$$J_2 = \sum_{i=1}^{K} \left(\max_{0 \leq t \leq t_{f,i}} \|\mathbf{x}^{*,i}(t) - \mathbf{x}(t, \mathbf{x}^{0,i})\| \right) \to \min, \tag{7}$$

where $\mathbf{x}(t, \mathbf{x}^{0,i})$ is a particular solution of the system (1) from the initial state $\mathbf{x}^{0,i}$, $i \in \{1, \ldots, K\}$.

It is necessary to find functions $f_j(\mathbf{u}, g)$, $j - 1, 2, 3$ to minimize both of the criteria (6), (7)

To solve this problem the network operator method is used

3 The Network Operator Method

The network operator method is one of symbolic regression methods, that uses the principal small variation of the basic solution [4]. This allows to save a inherit property for new possible solutions after evolutionary transformation of selected possible solutions as parents. Saving inherit property provides convergence of the evolutionary algorithm better random search.

Symbolic regression methods are used for coding the mathematical expression in a special form and search for the optimal solution according to the given criterion on the space of codes. The special genetic algorithm is used for finding an optimal solution. The main crossover operation of the variation genetic algorithm is developed for code of each The special genetic algorithm performs the crossover operation for two selected parents codes of symbolic regression such, that after this operation both offsprings had correct codes of mathematical expressions. The network operator method codes the mathematical expression in the form of an oriented graph. For this purpose, the network operator uses special alphabet of elementary functions with one and two arguments, The alphabet also includes a set of arguments of searched mathematical expression. The set of arguments contains also vector of parameters for extend of search space. values of the parameters are searching at the same time as structure of mathematical expression by classical genetic algorithm.

Further an example of coding the mathematical expression by the network operator is presented. Let the following mathematical expression is given

$$y = \sin(q_1 x_1 + q_2) + \exp(-q_1 \cos(x_2)), \tag{8}$$

where x_1, x_2 are variables, q_1, q_2 are parameters.

To code this mathematical expression, it is enough the following alphabet The set of arguments

$$\mathbf{F}_0 = \{f_{0,1} = x_1,\ f_{0,2} = x_2,\ f_{0,3} = q_1\ f_{0,4} = q_2\}. \tag{9}$$

The set of functions with one argument

$$\mathbf{F}_1 = \{f_{1,1}(z) = z,\ f_{1,2}(z) = -z, \\ f_{1,3}(z) = \sin(z), f_{1,4}(z) = \exp(z),\ f_{1,5}(z) = \cos(z)\}. \tag{10}$$

The set of functions with two arguments

$$\mathbf{F}_2 = \{f_{2,1}(z_1, z_2) = z_1 + z_2,\ f_{2,2}(z_1, z_2) = z_1 z_2\}. \tag{11}$$

All functions with two arguments are commutative, associative, and have unit elements. The function of summarization has unit element 0. The function of multiplication has unit element 1.

Let's write the mathematical expression (8) in codes of the alphabet

$$y = f_{2,1}(f_{1,3}(f_{2,1}(f_{2,2}(q_1, x_1), q_2)), f_{1,4}(f_{2,2}(f_{1,2}(q_1), f_{1,4}(x_2)))) \tag{12}$$

The graph of the network operator for the mathematical expression (8) is presented in the Fig. 1.

In the figure arguments if the mathematical expression are placed in the nodes-sources, the numbers of functions with two arguments are plced in other nodes, the numbers of functions with one argument are placed near arcs.

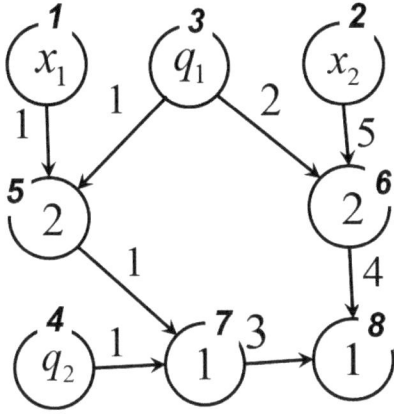

Fig. 1. The Network operator graph of the mathematical expession (8)

The numbers of nodes are placed in upper parts of the nodes. The network operator graph is stored in a memory of computer in the form of integer upper triangular matrix

$$\Psi = [\psi_{i,j}] = \begin{bmatrix} 0 & 0 & 0 & 0 & 1 & 0 & 0 & 0 \\ 0 & 0 & 0 & 0 & 0 & 5 & 0 & 0 \\ 0 & 0 & 0 & 0 & 1 & 2 & 0 & 0 \\ 0 & 0 & 0 & 0 & 0 & 0 & 1 & 0 \\ 0 & 0 & 0 & 0 & 2 & 0 & 1 & 0 \\ 0 & 0 & 0 & 0 & 0 & 2 & 0 & 4 \\ 0 & 0 & 0 & 0 & 0 & 0 & 1 & 3 \\ 0 & 0 & 0 & 0 & 0 & 0 & 0 & 1 \end{bmatrix}, \; i,j = 1,\ldots,L = 8. \qquad (13)$$

The main diagonal of the matrix contains zeros or the function numbers with two arguments. Zero corresponds node-sourc of the graph. Remain non-zero elements in the matrix are the function numbers of with one argument. The network operator method uses commutative and associative functions with two arguments, therefore these functions can be used as functions with any number of arguments. All non-zero elements in one column with non-zero element on the main diagonal are arguments of the function specifies in the main diagonal. If in the network operator matrix an element $\psi_{i,j} \neq 0$, then between the node i and node j is the arc from node i to node j related to the function with one argument under number $\psi_{i,j}$.

In order to calculate the mathematical expression by the network operator matrix, a nodes vector is used. Each component of nodes vector is associated with a network operator node. The number of elements in the nodes vector is equal to the number of nodes. The initial value of the nodes vector consists of arguments of a mathematical expression and unit elements of functions with two arguments.

The initial value of the nodes vector of the network operator matrix (13) is

$$\mathbf{z}^{(0)} = [z_1^{(0)} \ldots z_8^{(0)}]^T = [x_1\ x_2\ q_1\ q_2\ 2\ 2\ 1\ 1]^T. \tag{14}$$

The component values of the nodes vector are changed according to the following formula

$$z_j^{(i)} \leftarrow \begin{cases} f_{2,\psi_{j,j}}(z_j^{(i-1)}, f_{1,\psi_{i,j}}(z_i^{(i-1)})),, & \text{if } \psi_{i,j} \neq 0 \\ z_j^{(i-1)}, & \text{otherwise} \end{cases} \tag{15}$$

At the search of the optimal mathematical expression the variation genetic algorithm is used. This algorithm uses the principle of small variations of basic solution. According this principle initially one basic solution is given by researcher as the best possible solution. This basic solution is codded by the network operator method. All other possible solutions are codded in other fotm. Codes of remain possible solutions are ordered sets of small variations codes.

Small variations are such changes of the network operator matrix, that small changes the corresponding mathematical expression. The small variation code is an integer vector with four components

$$\mathbf{w} = [w_1\ w_2\ w_3\ w_4]^T, \tag{16}$$

where w_1 is a type of small variation, w_2 is the line number of the network operator matrix, w_3 is the column number of the network operator matrix, w_4 is a new element value of the network operator matrix.

The network operator method uses four types of small variation: $w_1 = 0$ is an exchange of the function with one argument: if $\psi_{w_2,w_3} \neq 0$, then $\psi_{w_2,w_3} \leftarrow w_4$; $w_1 = 1$ is an exchange of the function with two arguments: if $\psi_{w_2,w_2} \neq 0$, then $\psi_{w_2,w_2} \leftarrow w_4$; $w_1 = 2$ is an insertion of the additional function with one argument: if $\psi_{w_2,w_3} = 0$, then $\psi_{w_2,w_3} \leftarrow w_4$; $w_1 = 3$ is an elimination of the function with one argument: if $\psi_{w_2,w_3} \neq 0$ and $\exists \psi_{w_2,j} \neq 0$, $j > w_2$, $j \neq w_3$ and $\exists \psi_{i,w_3} \neq 0$, $i \neq w_2$, then $\psi_{w_2,w_3} \leftarrow 0$.

In the population any possible solution is coded as ordered set of small variation vectors

$$W_i = (\mathbf{w}^{i,1}, \ldots, \mathbf{w}^{i,d}),\ i = 1, \ldots, H, \tag{17}$$

where d is a number of small variations applied to the basic solution at one time, it is named depth of variations, H is an population power of possible solutions without the basic solution.

Any possible solution $\mathbf{\Psi}_i$ is a result of application the set of small variations to the basic solution

$$\mathbf{\Psi}_i = W_i \circ \mathbf{\Psi}_0 = \mathbf{w}^{i,d} \circ \mathbf{w}^{i,d-1} \circ \ldots \circ \mathbf{w}^{i,1} \circ \mathbf{\Psi}_0, \tag{18}$$

where $\mathbf{\Psi}_0$ is a network operator matrix of the basic solution.

At the performing crossover operation two possible solutions in the form of ordered sets of small variation vectors are selected randomly

$$\begin{aligned} W_\alpha &= (\mathbf{w}^{\alpha,1}, \ldots, \mathbf{w}^{\alpha,d}), \\ W_\beta &= (\mathbf{w}^{\beta,1}, \ldots, \mathbf{w}^{\beta,d}). \end{aligned} \tag{19}$$

A crossover point is determined randomly, $c \in \{1, \ldots, d\}$. Two new possible solutions are obtained by the exchange of tails after the crossover point of selected possible solutions

$$\begin{aligned} \mathbf{W}_{H+1} &= (\mathbf{w}^{\alpha,1}, \ldots, \mathbf{w}^{\alpha,c}, \mathbf{w}^{\beta,c+1}, \ldots, \mathbf{w}^{\beta,d}), \\ \mathbf{W}_{H+2} &= (\mathbf{w}^{\beta,1}, \ldots, \mathbf{w}^{\beta,c}, \mathbf{w}^{\alpha,c+1}, \ldots, \mathbf{w}^{\alpha,d}). \end{aligned} \quad (20)$$

4 Computational Experiment

To solve identification problem by the network operator method a training sample is needed. For this purpose the known mathematical model of quadcopter spatial motion instead of a real physical control object is used. This mathematical is well known. In the computational experiment this model is set as unknown, but it is used for calculating training sample. The real mathematical model of quadcopter spatial motion has the following form

$$\begin{aligned} \dot{x}_1 &= x_4, \\ \dot{x}_2 &= x_5, \\ \dot{x}_3 &= x_6, \\ \dot{x}_4 &= u_4(\sin(u_3)\cos(u_2)\cos(u_1) + \sin(u_1)\sin(u_2)), \\ \dot{x}_5 &= u_4\cos(u_3)\cos(u_1) - g, \\ \dot{x}_6 &= u_4(\cos(u_2)\sin(u_1) - \cos(u_1)\sin(u_2)\sin(u_3)), \end{aligned} \quad (21)$$

The restriction on control is

$$\begin{aligned} -\pi/12 = u_i^- &\leq u_i \leq \pi/12 = u_i^+, \ i = 1, 3, \\ -\pi = u_2^- &\leq u_2 \leq \pi = u_2^+, \\ 0 = u_4^- &\leq u_4 \leq 12 = u_4^+. \end{aligned} \quad (22)$$

In the experiment the twelve programs of control were given. Each program lasted $t_f = 3$ s,

The network operator found the following mathematical expression of the control object model.

$$f_1(\mathbf{u}, g) = H + u_3^2, \quad (23)$$

$$f_2(\mathbf{u}, g) = f_1^{-1}(\mathbf{u}, g) + \ln(|G|) +$$

$$\ln(|D|) + \sin(q_4 u_4) + \sin(q_3 u_3) + (q_4)^{-1} + \arctan(u_1) \quad (24)$$

$$f_3(\mathbf{u}, g) = \arctan(f_2(\mathbf{u}, g)) + \cos(f_1(\mathbf{u}, g)) +$$

$$\sin(G) + \sin(F) + \cos(q_4 u_4 + B) + (q_3)^{-1}, \quad (25)$$

where

$$A = \tanh(q_2 + u_2 + u_1 - u_1^3) + q_1 u_1 - (q_1 u_1)^3 + g,$$

$$B = A + q_3 u_3 + u_3^3,$$

$$C = \cos(A) + u_1^2,$$

$$D = C + \sin(q_4 u_4 + B),$$

$$E = D + (q_4 u_4 + B + \cos(A))^{-1},$$
$$F = E - q_3 u_3,$$
$$G = F \tanh(u_4),$$
$$H = G + \cos(q_4 u_4 + B) + \tanh(u_3),$$

$q_1 = 8.98462$, $q_2 = 10.54468$, $q_3 = 11.74390$, $q_4 = 15.72803$.

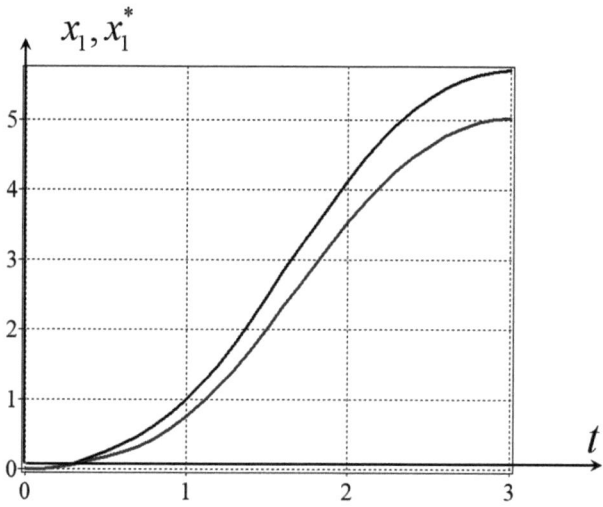

Fig. 2. Plots of x_1 (black) and the program variable x_1^* (blue) (Color figure online)

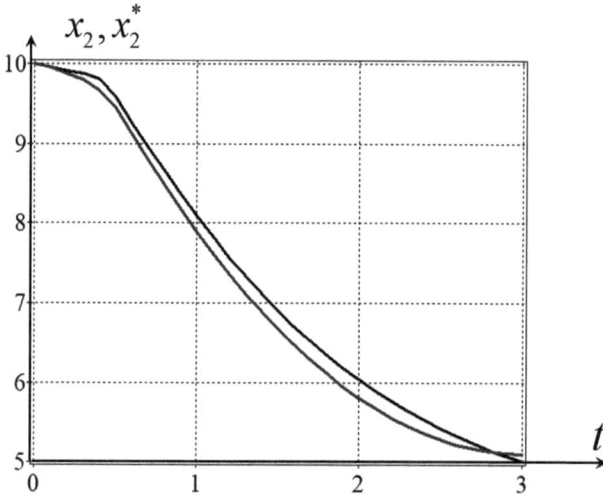

Fig. 3. Plots of x_2 (black) and the program variable x_2^* (blue) (Color figure online)

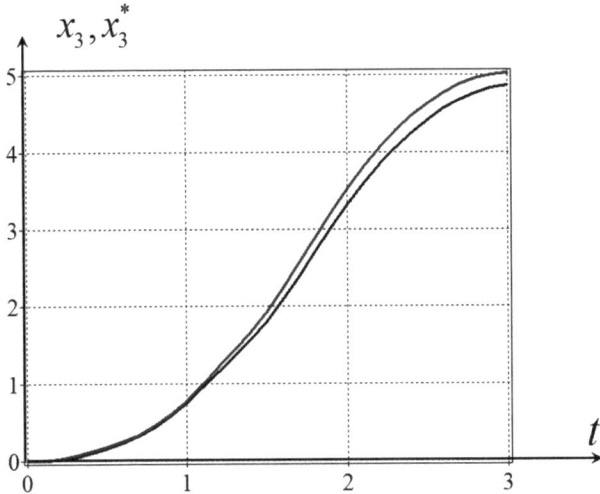

Fig. 4. Plots of x_3 (black) and the program variable x_3^* (blue) (Color figure online)

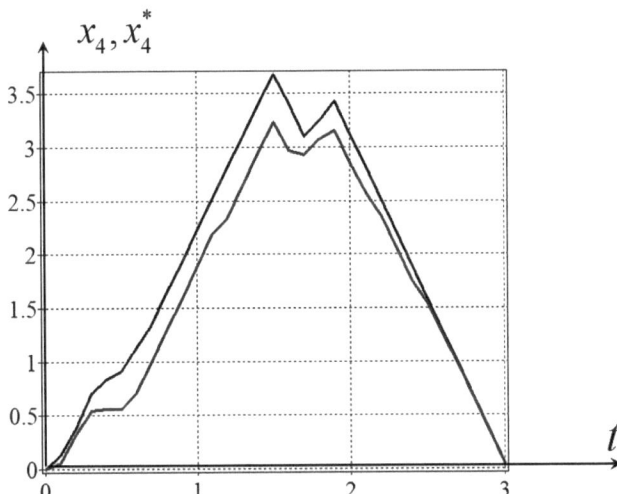

Fig. 5. Plots of x_4 (black) and the program variable x_4^* (blue) (Color figure online)

In the Figs. 2, 3, 4, 5, 6 and 7 the plots of components of state space vector of the model (2) (black) and one of the program trajectory (blue) are presented

The graphs of the computational experimental results show that the mathematical expressions of the functions of the right parts of the system of ordinary differential equations of the quadcopter model obtained by the network operator method give close solutions to the solutions obtained by integrating the model (21) used in work as a physical model of the control object.

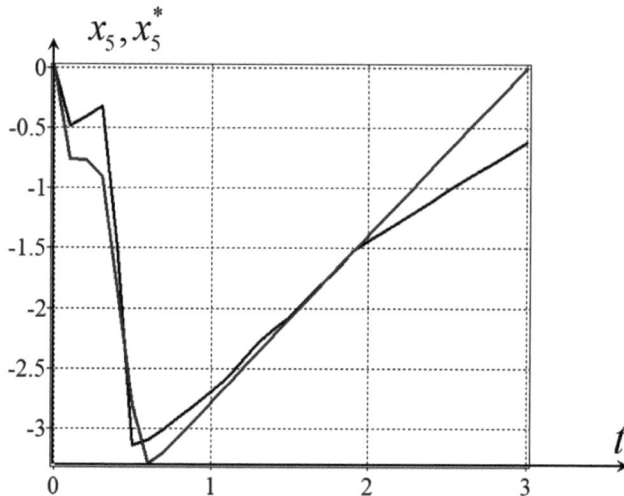

Fig. 6. Plots of x_5 (black) and the program variable x_5^* (blue) (Color figure online)

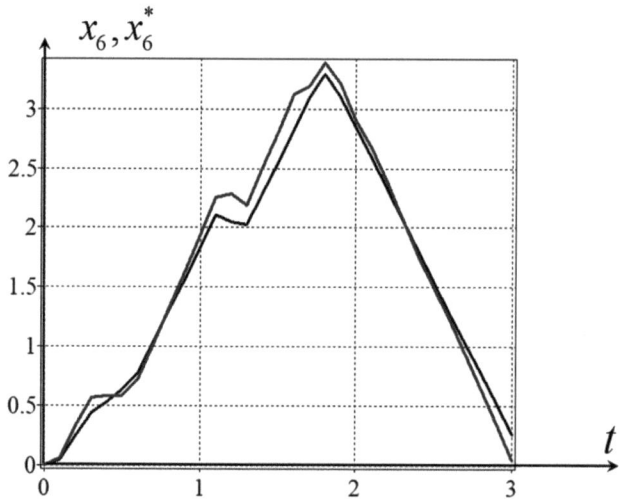

Fig. 7. Plots of x_6 (black) and the program variable x_6^* (blue) (Color figure online)

5 Conclusions

Control object mathematical model identification problem was considered in this paper. In the problem it was known that the control object moves in three-dimensional geometric space under the influence of external mechanical forces. Therefore, the dimension of the model was known and equal to twice the dimension of space, in wich the control object movement. In the first half of the differential equations of the model, the right parts were equal to the speeds of the components of the control object state space vector. To solve the identification problem, the network operator method was used. To obtain a training sample for the computational experiment instead of real physical control object the quadcopter spatial motion mathematical model was used. The results of the computational experiment showed that the mathematical model receivwd by the network operator mathod, has solutions aproximately coinciding to those of the model used as a physical control object.

References

1. Liu, G.P.: Nonlinear Identification and Control: A Neural Network Approach, p. 2016. Springer VerLag, London (2001)
2. Xu, J., Zhang, M., Zhang, J.: Kinematic model identification of autonomous mobile robot using dynamical recurrent neural networks. In: Proceedings of the IEEE International Conference on Mechatronics & Automation Niagara Falls, Canada, pp. 1447–1450 (2005)
3. Roy, R., Barai, R.K., Dey, R.: Identification of differentially driven wheeled mobile robot using neural networks. Int. J. Electr. Electron. Comput. Eng. **2**(2), 38–45 (2013)
4. Machine Learning Control by Symbolic Regression. Springer, Cham (2021). https://doi.org/10.1007/978-3-030-83213-1_5
5. Tran, T., Newman, B.: Integrator-backstepping control design for nonlinear flight system dynamics. In: American Institute of Aeronautics and Astronautics, Inc. AIAA SciTech, 5–9 January 2015, Kissimmee, Florida (2015)
6. Shchegoleva, A., Polyak, M.: Department of computer technologies and application of analytical design of aggregated regulators method to toxin-producing- phytoplankton models. In: Conference: 2023 XXVI International Conference on Soft Computing and Measurements (SCM), pp. 301–304

Hierarchal Compromise in Decision Making

Felix Ereshko[✉]

Federal Research Center "Computer Science and Control" of the Russian Academy of Sciences,
Moscow, Russia
fereshko@yandex.ru

Abstract. We consider decision making problem in player interaction model with two-level organization system. We describe global and Russian experience in the research of intergovernmental economic and ecologic interactions and balance system forming.

Keywords: coalition · organization · Center system · games · non-contradicting interests · Pareto system analysis · models · balance · procedure · algorithm

1 Introduction

Player unification is one of the main problems in decision making. This problem is considered in many publications. One can consider player unification during the research even for two-player game. Moreover, in three-player game the process of player unification is inevitable.

The goal of this paper is to demonstrate the features of hierarchal games theory in constructing and using adequate mathematical models in decision making for organization systems.

We use analytical constructions considering decision making asymmetry [1].

In the scope of the theory of hierarchal games one uses the following characteristics as the main ones: (a) the existence of specific player (Center) who has the right to first choose strategy depending on known or assumed information about the actions of its subordinate players; (b) Center chooses strategy based on the principle of the best guaranteed result. We consider the problem of choosing the best Center strategy with active behavior of subordinate systems. These systems tend to achieve their own goals within the rules of Center.

Let us present the main mathematical constructions in the scope of hierarchal games theory [1].

Let Π_0 (Center) maximize its efficiency criterion $f_0(x, u)$ where $u = (u_1, u_2, \ldots, u_n)$, $u_i \in U_i$, $u \in U$, $U = U_1 \times \cdots \times U_n$ is Center choice and $x_i \in X_i$, $x = (x_1, \ldots, x_n)$ is ith subsystem choice. Lower-level subsystem Π_i maximizes efficiency criterion $f_i(x_i, u_i)$, $i = 1, 2, \ldots, n$, $x_i \in X_i$ which is considered to be continuous on $U \times X_1 \times \cdots \times X_n$. We consider several Center awareness cases and show how regulation mechanisms are formed in subsystem control.

First type mechanisms (direct mechanisms, Game Γ_1). Center does not know the actions of its subsystems. Then its control mechanism is aimed to setting specific values $u \in U$ and sending this information to the subsystems. The best control values can be found from the problem

$$G_1 = \sup_{u \in U} \min_{x_i \in B_i^1(u_i)} f_0(x, u),$$

where B_i^1 is the set of optimal subsystem responses

$$B_i^1 = \left\{ x_i \in X_i | f_i(x_i, u_i) = \max_{y_i \in X_i} f_i(y_i, u_i) \right\}.$$

Such direct type control mechanisms are: setting the planning target, resource management, price formation, quota formation and other production constraints.

Second type mechanisms (inverse mechanisms, Game Γ_2). Center considers subsystem choice information during choosing its own strategy using functions $\tilde{u}_i = u_i(x_i)$. Then

$$B_i^2(\tilde{u}) = \left\{ x_i \in X_i | f_i(x_i, u_i(x_i)) = \sup_{y_i \in X_i} f_i(y_i, \tilde{u}_i) - \delta_i(\tilde{u}_i) \right\},$$

$$\delta_i(\tilde{u}_i) \geq 0,$$

and

$$G_2 = \sup_{\tilde{u} \in \tilde{U}} \inf_{x_i \in B_i^2(\tilde{u}_i)} f_0(x, \tilde{u}),$$

with $\delta_i\left(\widetilde{u_i}\right) = 0$ $\delta_i(\tilde{u}_i) = 0$ iff $\max_{y_i \in X_i} f_i(y_i, \widetilde{u_i})$ is reachable.

Such mechanisms are: tariff systems, reward systems, penalty systems, tax policy, stimulations.

Next, in game state research we must consider the question on correlation of target functions of players and their comparison.

If criteria are in different scales, the research is based on the principle of guaranteed result. Such questions are considered in [1].

On of the first publications on this matter using criteria convolution if the work of Germeyer and Vatel [2]. Later this state was included in the book under the name "the journey in one boat" [3]. Players' criteria were transformed to the form $f_k(x) = \min[F_0(x), F_k(x)]$ where $F_k(x)$ is the kth player initial function and $F_0(x)$ is the common target function for all players.

Its feature is that all players, each with its own interest, are connected with global target ("reaching the shore"). To achieve this target, each player must spend the part of their own resources – food, drinking water, physical power, clothes – to the common pool. Otherwise, they cannot "reach the shore".

The mathematical feature of "the journey in one boat" situation is the existence of some monotonical correlation between achieving global target and players common resource pool. In other words, the more resources will be spent to the common pool, the faster and easier the global target will be achieved [3].

Therefore, we consider a compromise in economic systems in the case when each player has two-level criteria: the upper-level global criterion for all systems and the lower-level criterion for their own system. Moreover, the global criterion is the same for all players.

The hierarchal compromise between these criteria can be considered in the form of the following two-step procedure (game Γ_1): first we construct an effective set for local criterion (lower level) and the we optimize global criterion in interest of meta-player on this effective set. This procedure is called meta-target compromise in [1].

We illustrate it on the problem of planning target balancing in economical scope.

Consider the interaction between the central planning department Π_0 and the complex of n interconnected natural economy sectors Π_i. Let Center control be the choice of vectors $\{u_i\}$, $i = 1,2,\ldots,n$ where their components are production planning targets for each sector. The set Z_i of possible states is defined as the set of all admissible intensities for applied technology. As in the resource problems, we assume that the union of production order vectors for other sectors are formed by each sector solving the following linear optimization problem

$$\min(c_i, z_i) = \min \sum_{p=1}^{m_i} \sum_{r=1}^{R_{ip}} c_{ip}^r z_{ip}^r$$

with

$$\sum_{p=1}^{m_i} \sum_{r=1}^{R_{ip}} d_{ipt}^r z_{ip}^r \geq u_{it}, \quad t = 1, \cdots, T_i,$$

$$\sum_{r=1}^{R_{ip}} l_{ip}^r z_{ip}^r \leq L_{ip}, \quad p = 1, \cdots, m_i,$$

$$z_{ip}^r \geq 0, \quad p = 1, \cdots, m_i, \quad r = 1, \cdots, R_{ip}$$

where z_{ip}^r is the intensity of the application of rth technology on pth company of ith sector, d_{ipt}^r is the production standard of tth type good on pth company using rth technology with unit intensity, c_{ip}^r is the preceding expense, l_{ip}^r is the labor intensity of using rth technology with unit intensity, L_{ip} is the cash labor resource on pth company, u_{it} is the planning target of the production of tth type good for the whole sector.

We consider the case when intermediate good operator has the linear form: $v_{ik} = \sum_{p=1}^{m_i} \sum_{r=1}^{R_{ip}} a_{ip}^{kr} z_{ip}^r$. Hence, the corresponding map $u_i \to V_i(\hat{z}_i, \hat{z}_i \in \hat{Z}_i)$ where $\hat{z}_i = \text{Argmin} c_i(z_i) \hat{z}_i = \text{Arg min } c_i(z_i)$ may not be single-valued. Therefore, intermediate product operator of the whole sector.

$$P(u) = \sum_{i=1}^{n} V_i(z_i),$$

where $u = (u_1, \ldots, u_i, \ldots, u_n)$ is the input vector for multi-sector system consisting of input for each sector, is also may not be single-valued.

In such conditions Center must consider additional hypothesis for planning target balance estimation and product order. This hypothesis allows to formally write such planning target union that may provide the production covering the needs of the whole natural economy sector.

If we fix the final product value $y \geq 0$, then in the scope of guaranteed result approach we obtain that the balance equation has the form.

$$\min_{v \in P(u)} (u - P(u)) \geq y$$

i.e.,

$$\sum_{i=1}^{n} u_{ik} \geq \sum_{i=1}^{n} v_{ik}(u_i) + y_k, \quad k = 1, \ldots, N,$$

for all.

$$v_i(u_i) \in V_i(\hat{z}_i), \quad i = 1, \ldots, n.$$

Next, if such balance is non-unique, Center can consider the following problem, e.g., pollution minimization problem.

$$\min \sum_{i=1}^{n} \mu_i \sum_{p=1}^{m_i} \sum_{r} b_{ip}^{lr} z_{ip}^{r}.$$

However, such formulation of plan balancing may be too constraining. We can consider another hypothesis which is called goodwill hypothesis. This hypothesis states that the balance holds when.

$$\sum_{i=1}^{n} u_{ik} \geq \sum_{i=1}^{n} v_{ik}(u_i) + y_k, \quad k = 1, \ldots, N.$$

holds for only one $v_i \in V_i$. Then as in the previous case, we can minimize some functional, (e.g., total pollution) or increase Center awareness in order to apply wider strategy class.

In the considered cases Center actually achieves two targets. The first one is qualitative: to achieve balance. The second one is quantitative: to reduce the pollution.

Therefore, we consider game Γ_1 with explicit states at which the players have different common admissible domains. Player 1 chooses control $u \in D \subset E^k$. Knowing that, Player 2 makes a choice $x \in X(u) = \{x \in E^n | Ax = b + Bu, x \geq 0\}$ maximizing their payoff function (c, x). Player 1 needs the choice of Player 2 to belong to the set $G = \{x \in E^n | (g_i, x) \geq q_i, i = 1, \ldots, l\}$. If such Player 2 choice set is non-empty, then Player 1 maximizes their payoff function (g_0, x) on this set. Here $c, g_0, g_1, \ldots, g_r \in E^n$; $b \in E^m$; A is $(m \times n)$ matrix of the rang m; B B is $(m \times k)$ matrix; D is multifaceted set defined by inequality system.

By the hypotheses mentioned above we consider two problem statements.

1. Non-goodwill lower level: find

$$\max_{u \in \mathbf{D} \cap T'} \min_{x \in X_c(u)} (g_0, x)$$

where

$$X_c(u) = \{x \in X(u) | (c, x) = \max_{y \in X(u)} (c, y)\}$$

$$X(u) = \{x \in E^n | Ax = b + Bu, x \geq 0\}$$

$$T' = \{u \in T | X_c(u) \subset G\}, T = \{u \in E^k | X(u) \neq \emptyset\};$$

2. Goodwill lower level: find

$$\max_{u \in \mathbf{D} \cap T''} \max_{x \in X_c(u) \cap G} (g_0, x)$$

where

$$T' = \{u \in T | X_c(u) \cap G \neq \emptyset\}$$

The idea of finding the best guaranteed result for the upper level in this case is similar to the idea used in constructing the series of algorithms by Center's rational resource strategies. First, we construct the set of admissible upper-level choices in the sense that lower-level optimal solutions satisfy additional conditions. Then on this set we maximize upper-level criterion.

2 Hierarchal Structure Models Examples

2.1 Example 1. Interstate Conflict

This game was designed by Pavlovskiy [4, 5]. The author of this paper took part in discussing and designing the rules of the game, and also applying this game into one state.

This game was based on unified imitation approach, in the scope of which mathematical models are used to describe economic and military processes. The decision making was conducted by real imitation game players.

Mediator. This game had hierarchal structure. The role of center in this game was taken by Mediator. Their role was to organize the game, to calculate economical and military state of players-countries based on their decisions, and game rules formulation. The strategies of mediator were the rules of the game.

Three counties imitation game. Three counties game rules. We Consider virtual one-dimensional closed world on a circumference (Fig. 1).

The states may provide each other with humanitarian and military products by any agreed price, launch production and deploy neighbor's army on their territory and vise-versa, form alliances, start wars and sign peace treaty.

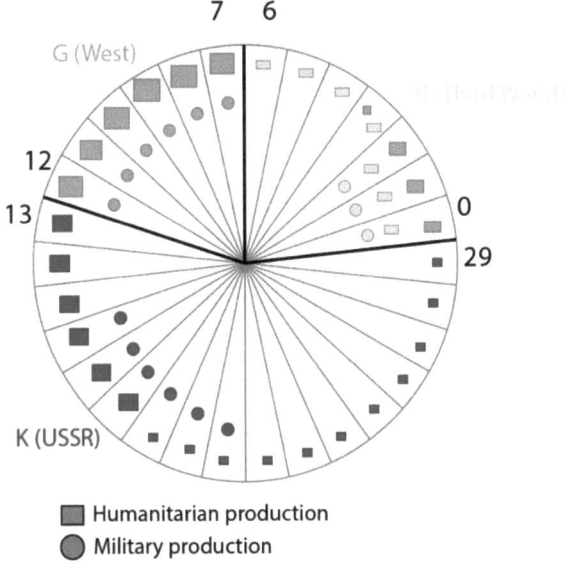

Fig. 1. Formalized description of the countries in the game.

Dynamical economic and military state of the countries is calculated by Mediator based on its model description below.

Each country has its own economy. The economy has two sectors: humanitarian sector and military sector. Military sector produces arms. Humanitarian sector produces high-tech goods which can be invested either into military or humanitarian sectors.

We assume that the production unit can be converted from one sector to another. During this process the unit loses 10% of its power and 25 days it is not able for production.

The mentioned parameters belong to Mediator choice.

The dynamics equation of economic variables:

$$P_{\alpha,\beta,i,t+1} = P_{\alpha,\beta,i,t} - \mu_{\alpha,\beta,i}P_{\alpha,\beta,i,t} + \rho_{\alpha,\beta,i}I_{\alpha,\beta,i,t}, \quad T_{\alpha,\beta,i,t+1} = \eta_{\alpha,\beta,i}P_{\alpha,\beta,i,t}.$$

Here α is country index (G, W and K), β is economy sector index ("humanitarian" and "military"), i is "geographical" sector index ($0 \leq i \leq 29$); $P_{\alpha,\beta,i,t}$ is production unit power; $T_{\alpha,\beta,i,t+1}$ is production output; $I_{\alpha,\beta,i,t}$ is production investment, $\mu_{\alpha,\beta,i}$ is amortization, $\rho_{\alpha,\beta,i}$ is fund intensity, $\nu_{\alpha,\beta,i}$ is capital productivity.

Mediator's Purpose. The Mediator's functional is total humanitarian production at final time instant divided by total humanitarian production at initial time instant. This is similar to people good during the game.

$$\Phi_\alpha = \frac{\sum_i P_{\alpha,\text{humanitarian},i,T} - \sum_i P_{\alpha,\text{humanitarian},i,0}}{\sum_i P_{\alpha,\text{humanitarian},i,0}} \to \max.$$

Mediator's Strategies. Mediator's strategies include the thoughts of Lev Tolstoy on War and Peace. "The power (motion quantity) is a product of mass on something unknown called X. War science vaguely admits the existence of this term by numerous historical examples of arms mass not coinciding with enemy power, of small squads winning over large armies. It tries to find this term in geometric conclusions, in arms design or, which is more common, in colonel's genius...This unknown X is army's spirit, it is a larger or smaller wish to bring one's life to danger of all men in the army, completely independent on who leads them – a genius or a non-genius; in three or two lines, with clubs or rifles firing 30 times per minute. men having a wish to fight always put themselves in the best position to do it. The spirit of army is a term which multiplied by the mass gives us the power."

Regular Arms War. The countries may fight with each other which means that the arms fighting on the boundary between them not until total win or lose but until the loss of combat capability. When one country loses combat capability, the number of operations decreases, and it changes the combat state from offense to defense, from defense to retreat. Hence, the frontline moves. We introduce main formulas to model military actions. We highlight three main ones: offense – we consider it as rang 3 operation; defense – rang 2 operation and retreat – rang 1 operation. On frontline countries suffer losses according to 1st type Osipov–Lanchester equation considering reinforcements. Efficiency coefficients significantly differ in offense and defense operations:

$$\frac{dN}{dt} = -\beta M + X_N, \quad \frac{dM}{dt} = -\alpha N + X_M.$$

Combat operations are conducted until one side loses psychological resistance. The decrease of this resistance correlates with relative losses of human resource. The loss of resistance decreases the operation rang by 1 – a new operation combination occurs, some of which cause the motion of the frontline.

Stability estimation is formed by Mediator strategy according to Tolstoy.

The war is the order to attack, defend or retreat on one or both sides of the conflict. The motion of arms from other sectors to the frontline may occur.

Nuclear Arms War. Countries G and K has so-called increased power arms. This is not completely equivalent to nuclear arms, not so powerful. mediator formulated its own rules on this type of arms according to the time trends connected to analogue rocket control systems when the time of re-aiming was longer than the time of target reach. One could know that they were attacked and either initiate counter-attack using known target distribution while the rocket had not reached the target or wait until they had been hit and then attack using the remaining arms and new target distribution.

In 179 steps by Game Rules countries conducted diplomatic negotiations, formed alliances, lost, developed and restored economical potentials; exchanged goods, moved their arms, conducted regular arms and increased power arms wars, conquered and lost territories.

Mediator. Finally, in conditions of loss of conflict spirit imitation game players began to diplomacy, and after 179 days mediator chose to stop the game.

In this description Mediator on each step may check how players deviate from rational (in the sense of meta-target) behavior using process development equations system.

In this real-time experiment Mediator took role of passive spectator and was on the other side of the players.

Next, it was assumed that Mediator may imply their own strategy in their own interest, solving two-layer system control problem with meta-target described below as hierarchal compromise procedure.

Formally Country W won.

Conclusion. The Game showed that the introduced procedure in digital implementation may be successfully applied not only as training complex but also real-time decision making.

The analysis of chosen articles by players allows to estimate the competence quality of decision makers in real-time strategic conditions.

2.2 Example 2. Ecological Compromise

The described above player interaction model naturally occurs in ecology in analyzing and choosing of various global development strategies: e.g., choose a common minimum pollution strategy on the set of various effective alternatives of separate regions economic development. The corresponding formal description below is based on the global inter-sector model developed by a team of American economists guided by Leontief [8] in the scope of UN global development and international cooperation research to determine development parameters within 1970–2000. By its structure the model is a union of regional blocks connected by production and financial balances. The regional block consists of two parts: inter-sector input–output balance and macro-economic equations. In the works of Institute of Economics and Industrial Engineering of the Siberian Branch of the RAS and Computing Centre of the Academy of Sciences multi-criteria modifications of Leontief model are considered [7, 9–11].

The following consideration is based on the simplest 4×6 model in which the world in split into two developed regions (I is North America and II is the rest of the developed countries) and two developing regions (III is Latin America and IV is the rest of the developing countries). Macro-economic variables of the model include investments I, capital K, employment L and consumption λ. Output vector x consists of four goods involved in international trade (natural economy, extractive industry and heavy industry), service and pollution refinement. Transport sector is included into the service sector which, thus, require money for international transportation. Export and import are denoted by $E = (E_1, \ldots, E_4)$ and $M = (M_1, \ldots, M_4)$ respectively.

Input–output equations for region s have the form:

$$x_i^s = \sum_{i=1}^{6} a_{ij}^s x_j + \gamma_i^s I^s + c_i^s \lambda^s + \sigma_i^s p^s + E_i^s - M_i^s, \, i = 1, \ldots, 4;$$

$$x_5^s = \sum_{i=1}^{6} a_{5j}^s x_j + \gamma_5^s I^s + c_5^s \lambda^s + \sigma_5^s p^s + \sum_{i=1}^{4} a_{tj}^s (E_j^s + M_j^s),$$

$$x_6^s = \sum_{i=1}^{6} a_{6j}^s x_j + \gamma_6^s I^s + c_6^s \lambda^s + \sigma_6^s p^s.$$

Here $A^s = \|a_{ij}^s\|$ is the technological expense matrix, a_{tj}^s is the row of transport expenses, γ^s is the vector of investment structure. By P^s we denote the number of people in sth region – this is the parameter which varies from one variant to another. c^s and σ^s are consumption structure coefficients vectors depending on consumption level and number or people. Besides, there may be upper and lower constraints on the output:

$$x_j \leq \bar{x}_j, \ j \in \bar{J}, \ x_j \geq \underline{x}_j, \ j \in \underline{J}$$

Here $\bar{J}$ contains extractive industry and J contains production industry.

Macro-economic constraints consist of labor constraints, capital–output connections and investment–capital connections:

$$\sum_{i=1}^{6} l_j^s x_j^s + c_l^s \lambda^s + \sigma_l^s p^s \leq L^s,$$

$$\sum_{i=1}^{6} k_j^s x_j^s + c_k^s \lambda^s + \sigma_k^s p^s \leq K^s,$$

$$c_l^s \lambda^s + r^s k^s + \sigma_l^s p^s = I^s.$$

Here (c_l^s, c_k^s, c_l^s) and $(\sigma_l^s, \sigma_k^s, \sigma_l^s)$ are final consumption structure coefficients depending on consumption level and number of people by labor resource, main funds and investments, respectively; l^s and k^s are labor and capital expense coefficients respectively. L^s is the total labor which is fixed in the model.

Export and import satisfy balance constraint:

$$\sum_{i=1}^{4} p_i(E_i^s - M_i^s) \geq 0, \ \sum_{s=1}^{4} M_i^s = \sum_{s=1}^{4} E_i^s, \ i = 1, ..., 4.$$

where p is exchange prices vector.

As regions target functions (lower-level local criteria) we take consumption values λ^s.

We complete model description by formulating global criterion – total pollution:

$$F_0 = \sum_{s=1}^{4} c^s x_6^s$$

3 Compromise Procedure

3.1 Main Definitions

Let bounded set X be defined by a system of linear constraints: $Ax = b, x \geq 0$ where A is $(m \times n)$ matrix, b is a vector, $x \in E^n$ and there exist k linear functionals $F_1(x) = (c_1, x), \ldots, F_k(x) = (c_k, x)$. The problem is to find the maximum of $F_0(x)$ on Pareto set for functionals $F_1(x), \ldots, F_k(x)$ defined on X.

The following is based on the Theorems from [1]:

Definition. If x* is an effective point, then there exists a vector $\lambda \in E^k$, $\lambda > 0$, $\sum_{i=1}^{k} \lambda_k \geq 1$ such that x* is a solution to linear programming problem

$$\sum_{i=1}^{k} (\lambda_i c_i, x) \to \max, x \in X$$

and for any $\lambda \in E^k$, $\lambda > 0$, $\sum_{i=1}^{k} \lambda_k \geq 1$ the solution x^* is an effective point.

Definition. Optimality domain T_J of a basis J is a set.

$$T_J = \{\lambda \in E^k | \Delta_l^J(\lambda c) \geq 0, l \in \tilde{J}\}$$

Definition. $\tilde{J}$ is an addition to J, i.e., $J \cap \tilde{J} = \emptyset$ and $J \cup \tilde{J} = \{1, ..., k\}$. If $x_J = A_J^{-1} b \geq 0$ and $T_J \cap \mathbf{D} \neq \emptyset$, then basis J is called optimal. Extreme point x corresponding to this basis is effective since it is a solution to the linear programming problem for any $\lambda \in T_J \cap D$.

Definition. For any admissible basis a neighboring basis is any admissible basis differing from initial one by one coefficient.

The initial problem has the formal form.

$$\max_{\lambda \in D} \max_{x \in X_c(\lambda)} (g, x)$$

$X = \{x \in E^n | Ax = b, x \geq 0\}$, $\mathbf{D} = \{\lambda \in E^k | \lambda_i \geq 1, i = 1, ..., k\}$.

3.2 Principal Algorithm Scheme

Step 1. For $\lambda = \lambda_0 \in D$ find optimal basis J of the problem $\max_{x \in X} \sum_{i=1}^{k} (\lambda_i c_i, x)$, write J into the queue P.

Step 2. For each basis $J \in P$ execute Steps 3–4.

Step 3. Calculate $F_J = \sum_j = 1_i^{ng} x_i|_J$.

Step 4. For basis J by direct simplex method construct admissible bases including indexes $\{1, ..., n\} \setminus J$. Write to P such bases J for which $\mathbf{D} \cap T_I \neq \emptyset$. Put $F^* = \max_{J \in P} F_J$..

Proposition. The maximum value of linear criterion equals F* on an effective set.

The proof follows from the fact that Pareto set is a union of finite number of multifaceted sets [12], hence the maximum of linear criterion is reached in one of extreme points. Let us show that the introduced algorithm gives all Pareto extreme points. Let λ^* be an arbitrary point of D. Consider a linear programming problem depending on one parameter $\beta \in [0,1]$:

$$\sum_{i=1}^{k} \left(\left[\lambda_i^0 c_i + (-\lambda_i c_i^0 + \lambda_i^* c_i)\beta \right], x \right) \to \max, \quad x \in X.$$

Using parametrical research algorithm for this problem [12] we obtain a sequence of neighbor admissible bases. In this sequence the first basis is optimal in the problem on Step 1 for $\lambda = \lambda^0$, and the last one is optimal for $\lambda = \lambda^*$. In the introduces algorithm each found basis of the sequence is written into queue, hence for it a neighboring basis can be found. Finally, in the queue P we obtain a basis which is optimal for $\lambda = \lambda^*$. If for $\lambda = \lambda^*$ several bases are optimal then the procedure described on Step 4 allows us to consider all these bases.

3.3 The Realization of the Principal Algorithm Scheme to Numerical Experiments

```
Function findOptimal(){
    Input: D
```
$$S(\lambda) = \max_{x \in X} \sum_{i=1}^{k}(\lambda_i c_i, x)$$
```
    Output:
```
F^*
$$\lambda = \lambda^0 \in D$$
$$J = findOptimalBasic(S(\lambda))$$
$$P = \{J\}$$
```
    while
```
$P \neq \emptyset$
$$J = P.pop()$$
$$F_J = \sum_{j=1}^{n} g_i x_i \mid_J$$
$$I = findNeighborBasises(J)$$
```
    for
```
$\forall i \in I$
```
        if
```
$D \cap T_i \neq \emptyset$
$$P = P \cup i$$
$$F^* = \max_{J \in P} F_J$$
```
    return
```
F^*
```
}.
```

4 Conclusion

In this paper we share the experience of hierarchal compromise development and show several examples of statement of meaningful problems, purposes, information space and mathematical modelling complexes which allows us to make research and obtain planning alternatives for decision making.

It follows from the above that imitation games can be used to construct rational strategies for Center.

On the other hand, if real players choose strategies in the game, they may use hierarchal process representation in analytical forecast.

References

1. Germeyer, Y.: *Igri s neprotivopolozhnimy interesami* [Games with non-opposite interest]. Nauka, Moscow, Russia (1976)
2. Germeyer, Y., Vatel, I.A.: Igri s ierarhicheskim vektorom interesov [Games with hierarchal interest vector]. Technicheskaya kibernetika **3**, 54–69 (1974)
3. Moiseev, N.N.: *Matematicheskie zadachi sistemnogo analiza* [Mathematical problems of system analysis]. Nauka, Moscow, Russia (1981)
4. Pavlovskiy, Yu.N.: *Imitacionnie sistemy i modeli* [Imitation systems and models]. Moscow, Russia: Znanie (1990)
5. Belotelov, N.W., Bdrodskiy, Y., Olenev, N.N., Pavlovskiy, Y.: *Opit imitacionnogo modelirovaniya pri analize socialno-ekonomicheskih yavleniy* [Imitation modelling experience in social-economic events analysis]. MZ Press, Moscow, Russia (2005)
6. Ereshko, F.I., Chemezov, S.W., Turko, N.I.: Modelirovanie vzaimodeistviya stran kak instrument virabotki kompromissnih politicheskih resheniy: praktika teoretiko-igrovogo podhoda [Modelling of interstate interaction as politic compromise construction instrument: the application of theoretical-gaming approach]. *Proc. of Mejgosudarsvtennoe protivoborstvo v usloviyah globalizacii i ego vliyanie na upravlenie nacionalnoi oboronoi Rossiyskoi Federacii* (Moscow, Russia). Izdatelskiy dom "UMTs" (2023)
7. Ereshko, F.I.: Matematicheskie modeli i metodi prinyaniya soglasovannih resheniy v aktivnih ierarhicheskih sistemah [Mathematical models and agreed decision-making methods in active hierarchal systems]. Ph.D. thesis. Moscow, Russia, ISC RAS (1998)
8. Leontief, W.W., et al.: *Budushee mirovoi ekonomiki* [The future of modern economics]. Moscow, Russia: Izdatelstvo "Mejdunarodnie otnosheniya" (1979)
9. Granberg, A.G., Rubinshtein, A.G.: Eskperimenti s agregirovannoi mejregionalnoi modeliy mirovoi ekonomiki [Experiments with aggregated interregional global economics model]. Izvestiya Sibirskogo otdeleniya Akademii nauk SSSR, seriya obschestvennih nauk **6**(2), 25–36 (1978)
10. Zlobin, A.S., Menshikov, I.S.: Issledovanie effektivnih variantov razvitiya mirovoi ekonomiki s pomoschiy dialogovoi sistemy [Effective global economics development research using dialogue system]. Proc. of Konferenciya molodih ychenih I sociologov (Novosibirsk, Russia), pp. 111–115 (1979)
11. Ereshko, F.I., Zlobin, A.S.: Optimizaciya lineinoi formi na effektivnom mnojestve [Linear form optimization on an effective set]. *Proc. of Vsesoyuzniy seminar "Chislennie metodi nelineinogo programmirovaniya"* (Harkov, USSR), pp. 167–171 (1976)
12. Goldshtein, E.G., Yudin, D.B.: *Novie napravleniya v lineinom programmirovanii* [New branches of linear programming]. Sovetskoe Radio, Moscow (1966)
13. Ereshko, F.I., Mushkov, A.Y., Turko, N.I., Tsvirkun A.D.: Managing large-scale projects in a mixed economy. Autom. Remote Control **83**, 755–779 (2022). https://doi.org/10.1134/S0005117922050083
14. Ereshko, Gorelov, M.: Prices and Decentralization of Control. In: IEEE Xplore Digital Library. 15th International Conference on «Management of Large-Scale System Development» (MLSD'2022), Moscow, Russia, pp. 1–4 (2022). https://doi.org/10.1109/MLSD55143.2022.9934208. Scopus. https://ieeexplore.ieee.org/xpl/conhome/9933724/proceeding?isnumber=9933990&pageNumber=2
15. Ereshko, F.I., Gorelov, M.A., Budzko, V.I.: An intelligent approach to decentralized control in the agro-industrial complex. Comm. Comput. Inform. Sci. **1539**, 254–264 (2022). https://doi.org/10.1007/978-3-030-95494-9_21 (Ядро РИНЦ, Scopus)
16. Ereshko, F., Kamenev, I., Shimansky, A.: Economic modeling of the impact of human capital on economic growth and competitiveness. In: Gibadullin, A. (eds.) Digital and Information

Technologies in Economics and Management. DITEM 2021. Lecture Notes in Networks and Systems, vol. 432. Springer, Cham (2022). https://doi.org/10.1007/978-3-030-97730-6_1
17. Ereshko, F., Karanina, E.: A systematic approach to the diagnosis of economic security of a transport enterprises. Transp. Res. Proc. **63**, 322–328 (2022). ISSN 2352-1465. https://doi.org/10.1016/j.trpro.2022.06.019. https://www.sciencedirect.com/science/article/pii/S2352146522002745

Author Index

A

Abgaryan, K.K. III-224, III-216
Abramenko, Georgii I-257
Adamatzky, Andrew I-1
Albychev, Alexander II-250
Aleksandrov, A. III-41
Andrikov, Denis A. III-118
Arun, Vanishri III-53
Avdeenko, Tatiana I-168

B

Barinov, Arseniy II-238
Bazylev, Dmitry II-224, II-275, III-167
Belotelov, Vadim I-10, II-314
Borisov, A.V. I-115
Borisov, Andrey V. I-217
Bosov, Alexey V. I-130
Boyko, S. V. I-155
Buinevich, Mikhail I-231
Bureneva, Olga I. I-182
Bzhikhatlov, Islam II-113

C

Cao, Yu III-15
Checkin, A. U. II-139
Cherkasov, V.V. II-99
Cherkasov, Vladislav III-69
Chernenko, P. G. I-155
Chebotarev, Petr Aleksandrovich I-300
Chervyakov, Alexander II-250

D

Daryina, A. N. III-159
Denisov, Sergeevich Maksim I-300
Devyatkin, Daniil D. I-246
Dhar, Narendra Kumar II-99
Diveev, Askhat I-10, III-255
Dorofeev, Nikolay V. I-205

Dostovalova, Anastasia M. II-31
Dotsenko, Anton I-10
Drogovoz, Viktor A. III-235

E

Ereshko, Felix III-266

F

Frolov, Oleg I-311

G

Gabdrakhmanova, Nailia III-143
Galkina, D. III-131
Gavriliev, Erchimen I-168
Gavrilov, E.S. III-216
Gazanova, Nurziya II-250
Geyda, Alexander S. II-124
Gorkavyy, M. A. I-285
Gorkavyy, M.A. II-259
Gorshenin, Andrey K. II-31
Goryachev, Maxim S. I-205
Grabar, D.M. II-259
Gromov, Ivan III-108
Grusho, A. I-102
Grusho, N. I-102
Gruzlikov, Alexander II-19

H

Ha, Muon I-37
Hung, N. N. III-29

I

Ilchenko, E.S. II-259
Ilin, Dmitry II-250
Iureva, Radda II-71, II-275, III-167
Ivanov, Alexey V. I-130
Ivanov, Sergey I-68
Ivanov, Y. S. I-285, II-259

© The Editor(s) (if applicable) and The Author(s), under exclusive license to Springer Nature Switzerland AG 2026
A. Diveev et al. (Eds.): INTELS 2024, CCIS 2605, pp. 279–281, 2026.
https://doi.org/10.1007/978-3-032-04764-9

Izrailov, Konstantin I-231

J
Jose, K. P. II-85, II-195
Joseph, Binumon II-195

K
Karaulov, Vladislav II-19
Khanh, N. D. III-29
Khramov, Aleksandr E. III-118
Kitaeva, Anna III-15
Klimenko, Anna II-238
Knyazeva, V. III-41
Kokoulin, Andrey N. III-83
Kokoulin, Rostislav A. III-83
Kolmogorova, Svetlana I-68
Konashenkova, Tatyana I-142
Konovalov, D. III-131
Konstantinov, Sergey I-10
Korchagin, V. I-87
Korepanov, Eduard I-142
Kotenko, Igor I-231, I-257
Kovshov, E. I-87
Kozlov, D. A. III-247
Kozlov, Sergey III-255
Kremlev, A. III-131
Kropotov, J. III-41
Ksemidov, B. S. III-224
Kudryavtsev, S. N. III-95
Kulchenkov, V. I-102
Kurbanov, Sinan V. III-118
Kurinov, Yuri N. I-217
Kuvshinnikov, V. I-87

L
Latypova, Viktoriya A. II-182, II-289
Lavanya, M. S. III-53
Levina, A. B. I-155
Levina, Alla II-59
Lipkovich, M. III-41
Litvinenko, Yulia II-19

M
Ma, Danting I-37
Magid, Evgeni I-311
Malinetsky, Georgy III-182
Malyshev, Dmitry III-69
Margun, Alexey II-71, II-113, II-275, III-167

Martínez, Genaro J. I-1
Martinez-Garcia, Edgar A. I-311
Mattam, Bobina J. II-85
Melnik, Maxim I-257
Mikhailov, Ilia III-1
Morita, Kenichi I-1

N
Nair, Salini S. II-85
Nikulchev, Evgeny II-250
Novikov, V. M. II-139
Nozdrachev, Alexey III-69
Nozdracheva, Anna II-210, III-69

O
Otradnov, Konstantin K. III-196

P
Pavlov, Vadim I-68
Petrov, Vladislav III-1
Pilgun, Maria III-143
Podoprosvetov, Alexey II-45, III-182
Poleshchenko, Dmitriy III-1
Polyakov, V.M. II-99
Ponomarev, V. III-41
Prokopiev, Igor III-255
Prokopyev, Igor I-10
Pronina, M. III-41
Proshunin, A. I. II-139

R
Romanov, Nikita I-68
Rybak, L.A. II-99
Rybak, Larisa II-210, III-69

S
Sabbry, Nawras H. II-170
Safin, Ramil I-311
Sagatdinov, A. III-41
Selvesyuk, N. I. II-139
Semenikhin, K.V. I-115
Semenov, M. E. II-139
Serebrenny, Vladimir I-193
Serov, V. A. III-247
Shabanov, A. P. II-1
Shanarova, N. III-41
Sharapov, Ruslan V. I-205
Shereuzhev, Madin I-193
Shichkina, Yulia I-37, III-53

Shit, Santu II-99
Sinitsyn, Igor I-142
Sinitsyn, Vladimir I-142
Skaev, S. A. III-247
Smelyansky, R.L. I-115
Smolin, Vladimir II-45, III-182
Sobolevskii, V. A. II-157
Sofronova, Elena I-10
Sokolov, Sergey II-45
Solovoyv, A. M. II-139
Stepanov, E.P. I-115
Stepanov, Oleg II-19
Susoev, Nikolay I-54

T

Tapkire, Mayura III-53
Tepsikoev, S. A. III-247
Tianci, Gao I-25
Timonina, E. I-102
Tolok, A. V. II-301
Tolok, N. B. II-301
Trubacheev, Boris II-59
Trubienko, O. V. III-247
Tsoy, Tatyana I-311
Tuan, Nguyen Tat I-273
Tuan, Pham Van III-29

V

Van Hoa, Nguyen I-273
Van Tuan, Pham I-273
Vaschenko, Anna I-311
Vinh, Nguyen Quang I-273
Voroshchenko, V. D. I-285

W

Wu, Guo I-193
Yashina, Marina I-54

Y

Yashina, Marina I-54

Z

Zabezhailo, M. I-102
Zatsarinny, A. I-102
Zatsarinnyy, A. A. II-1
Zelinka, Ivan I-1
Zhdanov, V. III-131
Zhdanov, Viktor II-113
Zhiganov, S.V. II-259
Zhukov, Dmitry O. III-196
Zhukovskaya, Alexandra III-15
Zimenko, K. III-131
Zimenko, Konstantin II-71
Zuev, S.V. II-99
Zuev, Sergei II-210

If you have any concerns about our products,
you can contact us on
ProductSafety@springernature.com

In case Publisher is established outside the EU,
the EU authorized representative is:
**Springer Nature Customer Service Center GmbH
Europaplatz 3, 69115 Heidelberg, Germany**

Printed by Libri Plureos GmbH
in Hamburg, Germany